Breeding Approaches in Plantation, Spice, Medicinal and Aromatic Crops

Breeding Approaches in Plantation, Spice, Medicinal and Aromatic Crops

Richa Sao
Department of Genetics and Plant Breeding,
Mahatma Gandhi University of Horticulture & Forestry
Sankara, Patan, Durg, Chhattisgarh, India

Anita Kerketta
Department of Vegetable Science,
Mahatma Gandhi University of Horticulture & Forestry
Sankara, Patan, Durg, Chhattisgarh, India

Bharti Sao
Department of Floriculture and Landscaping
Mahatma Gandhi University of Horticulture & Forestry
Sankara, Patan, Durg, Chhattisgarh, India

CRC Press
Taylor & Francis Group
Boca Raton London New York

CRC Press is an imprint of the
Taylor & Francis Group, an **informa** business

-EPH-

Elite Publishing House

First published 2027
by CRC Press
4 Park Square, Milton Park, Abingdon, Oxon, OX14 4RN

and by CRC Press
2385 NW Executive Center Drive, Suite 320, Boca Raton FL 33431

CRC Press is an imprint of Informa UK Limited

British Library Cataloguing-in-Publication Data
A catalogue record for this book is available from the British Library

Print edition not for sale in India.

ISBN: 9781041371427 (hbk)
ISBN: 9781041378761 (pbk)
ISBN: 9781042010332 (ebk)

DOI: 10.1201/9781042010332

Typeset in A Caslon Pro
by Elite Publishing House

-EPH-

Contents

Preface

The agricultural landscape of the twenty-first century demands an ever-expanding knowledge base that merges traditional wisdom with modern scientific precision. Plantation, spice, medicinal, and aromatic crops together represent the cultural and economic backbone of tropical agriculture, providing the oils, fibres, beverages, medicines, and fragrances that sustain industries and livelihoods across the world. Yet these crops share a distinctive challenge, perennial growth, genetic heterogeneity, and complex reproductive systems that make genetic improvement both intellectually demanding and strategically long-term.

The book ***Breeding Approaches in Plantation, Spice, Medicinal and Aromatic Crops*** was conceived to bridge the widening gap between the fundamentals of plant-breeding science and its specialized application to long-lived, high-value perennials. While annual crops have benefited from rapid-cycle selection, genomic prediction, and transgenic advances, perennial crop breeding continues to rely on patient observation, multi-year evaluation, and the integration of conventional and molecular approaches. The purpose of this volume is therefore to provide a comprehensive, methodical, and updated treatment of breeding theory and practice for these diverse crop groups, ranging from coconuts and coffee to cardamom, ashwagandha, and lavender.

This book grew from years of teaching, research, and field experience within India's plantation and spice research networks under the Indian Council of Agricultural Research and partner universities. Each chapter blends empirical results from institutional programs with recent innovations in molecular breeding, biotechnology, and climate-smart selection. The intent is not only to describe existing cultivars and breeding successes but also to illuminate the evolutionary logic behind crop diversity, adaptation, and domestication. Emphasis is placed on how the understanding of genetic resources, cytogenetics, pollination biology, and incompatibility systems can be translated into workable breeding strategies for sustained genetic gain.

Every section of the book has been structured with clarity and progression. The initial chapters establish the conceptual foundation, defining breeding objectives, hybridization principles, and crop-specific methodologies for major plantation species such as coconut, arecanut, oil palm, cashew, coffee, tea, cocoa, and rubber. Subsequent chapters expand into the intricate world of spices, medicinal plants, and aromatic crops, domains where biochemical diversity and secondary-metabolite genetics dominate breeding goals. Later sections focus on genetic resources, evolutionary mechanisms, and biotechnological interventions, highlighting cytological behaviour, sterility, and incompatibility systems that govern hybrid development. The final chapters synthesize global breeding achievements, ideotype concepts, and molecular innovations that redefine the future of perennial crop improvement.

This book is intended equally for postgraduate students, young researchers, and professional breeders who aspire to work in these sectors. It encourages readers to appreciate the continuum between biodiversity conservation and genetic enhancement, showing that responsible breeding is both a scientific and ethical enterprise. In an era of climate instability and resource constraints, the long-term sustainability of tropical agriculture depends on breeding programs that deliver resilient, high-yielding, and quality-oriented cultivars without eroding ecological balance.

I acknowledge with gratitude the collective contributions of scientists, field technicians, farmers, and institutional mentors whose decades of work in germplasm maintenance, crossing programs, and molecular analysis have made such a synthesis possible. Their dedication to maintaining field gene banks, recording flowering synchrony, and conserving rare alleles forms the silent infrastructure of every breeding success documented here.

It is my hope that this book will inspire a new generation of scientists to pursue innovative, multidisciplinary, and eco-centric breeding approaches, ensuring that the genetic wealth of tropical crops continues to serve humanity with productivity, resilience, and harmony with nature.

Authors

About the Authors

Dr. Richa Sao is currently working as an Assistant Professor at Mahatma Gandhi University of Horticulture & Forestry, Sankara, Patan, Durg, Chhattisgarh. She has completed her M.Sc.(Ag.) and Ph.D. (Ag.) in Genetics and Plant Breeding from *Indira Gandhi Krishi Vishwavidyalaya, Raipur* and Ph.D. research work was done in collaboration with *Bhabha Atomic Research Centre, Mumbai, India.* She has received 18th Chhattisgarh Young Scientist Award 2023 and the Best Ph.D. thesis award 2023 by ICAR NRRI in 2nd Indian Rice Congress. Besides, she has published many research articles in reputed journals (Three in NAAS ratings above 10), including book chapters in Springer Publications and popular articles. She has qualified the ASRB NET and her interest of area lies on research involving on crop germplasm characterization, mutation breeding, molecular marker based profiling, NGS based genomics & transcriptomic profiling, in addition to the conventional

Dr. Anita Kerketta completed her B.Sc. (Ag.) I.G.K.V., Raipur, Chhattisgarh in 2013 and M.Sc. (Ag.) Horticulture with specialization in Vegetable Science from Sam Higginbottom Institute of Agriculture, Technology & Sciences, Allahabad, U.P. in 2015 and Ph.D. 2020. She has been awarded the University Gold Medal in M.Sc. (Ag.) Horticulture with specialization in Vegetable Science during 2015. She also cleared ICAR-ASRB-NET in 2017. Dr. Anita has contributed to Horticulture research with more 25 published research paper and Article and radio talks, attended many conference at the National or International Level. She has worked as an Assistant Professor in the Department of Horticulture, Naini Agricultural Institute, Sam Higginbottom University of Agriculture, Technology & Sciences, Prayagraj, U.P. with a teaching experience of 5 years. Presently joined as Assistant Professor in Mahatma Gandhi University of Horticulture and Science, Durg, Chhattisgarh. She has a good experience in field of Horticulture with academic research and practical application making valuable contributions to Horticulture development.

Dr. Bharti Sao is currently serving as an Assistant Professor at Mahatma Gandhi University of Horticulture and Forestry, Sankara, Patan, Durg, Chhattisgarh. She obtained her Ph.D. in Horticulture (Floriculture and Landscape Architecture) from *Indira Gandhi Krishi Vishwavidyalaya (IGKV), Raipur* in 2021, and her M.Sc. (Ag.) in Horticulture (Floriculture and Landscape Architecture) from *Sam Higginbottom Institute of Agriculture, Technology & Sciences (SHIATS), Allahabad (U.P.)* in 2014. She holds a Diploma in AutoCAD and qualified the Agricultural Scientists Recruitment Board–National Eligibility Test (ASRB-NET) in *Floriculture and Landscaping* in 2016. Dr. Sao has over five years of academic experience as a faculty member at reputed agricultural and horticultural institutions. Her research interests lie in floriculture and landscape architecture, with a special emphasis on plant mutation breeding, ornamental crop improvement, post-harvest management, and sustainable landscaping practices. Dr. Sao has authored over 12 scientific works, encompassing research articles, review papers, book chapters (including one published by Springer), as well as popular and extension articles in esteemed national and international journals and magazines. She has been honored with several distinguished awards, notably the National Fellowship for OBC Candidates (2018–19) conferred by *the University Grants Commission (UGC), New Delhi*, in recognition of her academic excellence and research potential. Beyond her research and teaching responsibilities, Dr. Sao actively participates in academic training, conferences, and extension activities.

Introduction

Introduction

Plantation, spice, medicinal, and aromatic crops together constitute one of the most culturally, economically, and ecologically important segments of tropical and subtropical agriculture. Their significance transcends commercial value; they are enmeshed in the social histories, culinary traditions, healing systems, and industrial development of many civilizations. For centuries, these crops have shaped human–plant interactions across continents, triggering global explorations, trade expansions, and technological innovations. In the present era, they remain indispensable for generating rural employment, supporting agro-industries, expanding exports, and maintaining ecological balance. Unlike annual crops, which complete life cycles within a short span, these perennial and semi-perennial species persist for many years in the landscape, interacting with soils, climate, and communities over extended periods. Their improvement through scientific breeding, therefore, demands a comprehensive and long-term understanding of biological intricacy, genetic diversity, evolutionary processes, and agronomic responses.

The fundamental distinctiveness of these crops lies in their perennial nature, long juvenile phases, high heterozygosity, and frequently complex reproductive biology. These characteristics impose unique challenges on crop improvement but also offer opportunities to exploit naturally occurring variation, heterosis, and clonal potential. Plantation crops such as coconut, oil palm, rubber, coffee, tea, cocoa, and cashew form the economic backbone of vast tropical regions. They yield products like edible oils, beverages, natural rubber, and high-value nuts that sustain global industries. Each crop has a long history of domestication and commercial cultivation, shaped by centuries of farmer selection, exploration, colonial introductions, and scientific breeding. Breeding programmes must therefore operate with an understanding that planting material introduced today will remain productive for decades. A single genetic decision can influence national productivity, rural incomes, industrial supply chains, and ecological sustainability for generations.

Spice crops such as black pepper, cardamom, nutmeg, cinnamon, ginger, and turmeric have historically fuelled global trade networks. While their value originally lay in flavour, preservation, and ritual usage, today they are equally prized for their nutraceutical, antioxidant, and antimicrobial properties. These crops often grow in complex agroforestry systems, under the shade of tall plantation trees or along humid forest belts. Their suitability to intercrop systems and their perennial nature make them unique candidates for integrated land use strategies. Breeding for spices requires a deep

DOI: 10.1201/9781042010332-1

appreciation of phytochemical variability, as quality depends heavily on volatile oil composition, alkaloid content, colour, aroma, and flavour. In crops like black pepper, the interplay of spike morphology, berry size, piperine content, and disease resistance determines economic value. Cardamom breeding demands the integration of capsule size, aroma compounds, drought tolerance, and resistance to pests and diseases that thrive in humid environments. These complexities make spice-crop breeding an intricate blend of classical genetic manipulation and biochemical improvement.

Medicinal plants form a cornerstone of traditional healing systems and are increasingly central to pharmaceutical research and wellness industries. Species like senna, periwinkle, ashwagandha, coleus, safed musli, and isabgol supply active ingredients that feed into the formulation of herbal drugs, phytopharmaceuticals, and nutraceuticals. However, many medicinal-crop species are only partially domesticated, and their production still relies on heterogeneous landraces or wild populations. This creates variability in bioactive compound levels, yield stability, and disease responses. Breeding is challenged by inconsistent flowering patterns, partial self-incompatibility in certain species, chemotype diversity, and the environment's strong influence on metabolite expression. Enhancing medicinal crops involves not only improving biomass or root yield but stabilizing and increasing concentrations of key metabolites such as alkaloids, glycosides, withanolides, terpenoids, or saponins. Breeding in these crops must combine classical genetic improvement, metabolite profiling, in vitro culture techniques, and selection of elite chemotypes that respond consistently across environments.

Aromatic crops, particularly essential oil-bearing grasses such as lemongrass, palmarosa, citronella, vetiver, and Japanese mint, contribute substantially to global fragrance, cosmetic, soap, detergent, insect-repellent, and aromatherapy industries. These crops produce oils rich in compounds like citral, geraniol, citronellal, menthol, and other terpenoids of industrial interest. Their hardiness, suitability for marginal lands, ability to thrive with minimal inputs, and capacity for multiple ratoons make them highly profitable for smallholder cultivation. Yet, aromatic-crop improvement requires careful attention to essential oil yield, herbage biomass, chemotype stability, tolerance to diseases, and resilience under rainfed or stress-prone conditions. Many aromatic grasses are polyploid, highly cross-pollinated, and poorly seed-producing, which limits the use of sexual reproduction and necessitates clonal selection, mutation breeding, and tissue-culture-based propagation. Maintaining chemotype purity and ensuring high industrial-grade oil composition are central breeding priorities.

Although the biological features of these crop groups differ widely, they share several challenges that unify their breeding approaches. Long juvenile phases delay the assessment of hybrid progenies and slow the overall pace of genetic gain. High levels of heterozygosity complicate trait fixation but simultaneously offer immense variation for selection. Large plant sizes restrict population numbers and demand optimization of space in breeding nurseries. Many crops have complex flowering behaviours, ranging from asynchronous flowering to protogyny, protandry, or dioecy, which complicates controlled pollination. Seed viability issues, including recalcitrant seeds or poor germination, limit breeding options in several species. Environmental variability exerts significant influence on yield, quality, and phytochemical profiles, making multi-year and multi-location evaluation essential. These challenges have encouraged breeders to integrate classical selection, hybridization, clonal propagation, mutation breeding, polyploidy manipulation, biochemical assays, and molecular techniques into coherent long-term improvement strategies.

The economic importance of plantation, spice, medicinal, and aromatic crops continues to grow with global population expansion, rising wellness consciousness, increasing urban markets, and diversified industrial applications. Plantation crops remain major foreign exchange earners and provide raw materials for large-scale industries. Beverage crops such as tea and coffee enjoy persistent global demand. Natural rubber remains indispensable for tyres, medical equipment, and engineering products. Coconut and oil palm produce oils central to food, cosmetic, biofuel, and oleochemical industries. Spices remain staples in global cuisines, with rising demand for high-quality, residue-free, organically grown products. Medicinal plants fuel booming nutraceutical, herbal drug, and phytopharmaceutical sectors. Aromatic crops support rapidly expanding aromatherapy, natural-perfume, and wellness industries. The rising preference for natural products over synthetic alternatives also elevates the value of phytochemically rich varieties. As demand increases, breeding becomes crucial for ensuring a consistent, high-quality, sustainable supply.

Genetic resources are the foundation of breeding programmes for all these crop groups. Understanding the evolutionary origin, geographic distribution, domestication pathways, and natural variation of each species allows breeders to tap into useful alleles for yield, quality, resistance, and adaptability. Many plantation and spice crops have dispersed widely from centres of origin, creating secondary centres of diversity rich in adaptive traits. Medicinal plants often preserve valuable chemotypes within wild or semi-cultivated populations. Aromatic grasses display enormous variation in essential oil composition across ecotypes. Conservation and characterization of these genetic resources, through field gene banks, in situ reserves, tissue culture, and molecular techniques, are vital to breeding progress. Detailed morphological, physiological, biochemical, and molecular characterisation creates a map of diversity that breeders can use to design crosses, identify elite parents, and broaden genetic bases that may have eroded through monoculture or narrow selection in commercial plantations.

Breeding methods in these crops have evolved from simple visual selection to modern genomic approaches. Traditional methods such as mass selection, pure-line selection, and progeny testing continue to play foundational roles in annual and perennial species alike. In highly heterozygous and cross-pollinated crops, clonal selection is central, as it allows breeders to perpetuate superior genotypes unchanged. Hybridization has become the backbone of breeding programmes where combining ability is key to expressing heterosis. Controlled pollination techniques have become highly refined, accommodating crop-specific floral biology, pollen viability patterns, and stigma receptivity timelines. In perennial species where one breeding cycle may span a decade, early selection indices, marker-assisted tools, and vegetative propagation have gained exceptional importance for accelerating progress. Mutation breeding provides novel variants in crops where sexual recombination is limited. Polyploidy manipulation enhances biomass, metabolite yield, and stress resilience in both aromatic and medicinal crops. Tissue culture techniques allow large-scale multiplication, pathogen elimination, and preservation of elite lines. Molecular markers enable hybrid verification, genealogy reconstruction, genome diversity analysis, and precision selection.

Because these crops remain in the field for long durations, sustainability is a guiding theme in breeding. The varieties deployed must not only produce high yields and excellent quality but must also withstand diseases and pests with minimal chemical intervention. Durable resistance is essential, especially in humid tropical environments where pathogens proliferate rapidly. Breeding must incorporate tolerance to drought, heat, salinity, storms, and other abiotic stresses that are becoming more frequent under climate change. Architectural modifications, such as compact canopy structure, reduced height, improved root systems, or efficient leaf arrangement, contribute to both resilience and ease of harvesting. Plant ideotypes, designed with a holistic understanding of plant physiology, resource use, and agronomic management, increasingly shape breeding pipelines. Future breeding is being guided by climate scenarios, ecological sensitivities, changing consumer awareness, and the need to reduce environmental footprints.

Breeding Complexity and the Need for Scientific Precision

The improvement of plantation, spice, medicinal, and aromatic crops demands sustained scientific precision because each trait of interest is influenced by a combination of genetic, physiological, environmental, and biochemical factors. The complexities of perennial growth make it necessary to observe plants over multiple years to fully understand their yield potential, flowering stability, stress tolerance, and quality consistency. Breeding decisions require careful evaluation not only of agronomic performance but of how a genotype responds to a spectrum of environmental variables. Diseases and pests often behave unpredictably across seasons, and environmental factors such as shade, humidity, rainfall distribution, soil fertility, and temperature fluctuations can significantly influence crop performance. Such complexity requires breeders to integrate long-term datasets, adopt refined experimental designs, and employ statistical models that capture genotype-by-environment interactions. The demand for varieties that perform uniformly across regions, yet retain adaptability to specific microclimates, underscores the need for diversified breeding populations and multi-location evaluation strategies.

The Spread of Genetic Resources and Their Utilization

The global spread of many of these crops over centuries has resulted in rich reservoirs of genetic variability that serve as the primary foundation of breeding programmes. Coconut, dispersing naturally across oceans, developed distinct genetic pools that adapted to coastal, island, and inland ecosystems. Rubber, originally from the Amazon, was introduced into Asia through a narrow genetic base but later enriched through targeted germplasm collection. Coffee cultivars and wild relatives spread from the Ethiopian highlands to diverse tropical regions, creating variation in flavour, disease tolerance, and altitude adaptation. Tea evolved as separate China and Assam types before hybridization merged their genetic attributes. Spices and medicinal plants maintain valuable variability in phytochemical content, disease resistance, and ecological adaptation across regions of traditional cultivation. Aromatic grasses display remarkable chemotype diversity, with each ecotype presenting unique essential oil compositions shaped by environmental conditions. Utilizing these genetic resources requires a thorough program of exploration, conservation, and characterization. Molecular tools, phenotypic evaluations, and phytochemical profiling collectively help breeders identify elite parents with desirable traits.

Advances in Hybridization Techniques

Hybridization remains central to generating genetic variability and combining elite traits. Achieving successful hybridization in plantation and perennial spices often demands intricate knowledge of floral biology. In coconut, the timing of female flower receptivity and male flower shedding must be aligned precisely, requiring controlled isolation and systematic pollination. Oil palm hybridization requires understanding Dura, Pisifera, and Tenera phenotypic relationships, as commercial production depends heavily on specific shell thickness traits. Tea hybridization often involves synchronizing flowering of genetically distant parents grown under differing environmental conditions. Black pepper breeding demands controlled pollination of minute flowers, careful emasculation or spike isolation, and monitoring of berry set over several months. Medicinal plants often require precise emasculation of small floral structures, and aromatic grasses frequently pose challenges due to protandry, poor seed set, or high sterility. These complexities have led to refined methods of manual pollination, floral manipulation techniques, pollen preservation, and testing of seed viability. The integrity of hybrid progenies is vital to breeding success, making hybrid verification through molecular markers and phenotypic checks a routine part of advanced breeding programmes.

Clonal Selection and Vegetative Propagation

Clonal selection occupies a central place in the improvement of many plantation and aromatic crops. Because superior genotypes in highly heterozygous, cross-pollinated species cannot be fixed through selfing, clonal selection allows the direct preservation of elite individuals. Tea gardens largely rely on vegetative propagation of high-yielding and flavour-rich clones. Rubber plantations depend on bud-grafted elite clones selected for high latex yield, tolerance to diseases, and wind resistance. Cashew improvement frequently utilizes clonal selection to stabilize traits like nut size, kernel weight, and canopy structure. In many aromatic grasses, clonal propagation maintains chemotype purity, ensuring that essential oil composition remains uniform. Medicinal plants such as periwinkle or coleus are often propagated vegetatively to maintain high-metabolite genotypes. Clonal selection requires systematic evaluation of initial plus trees or mother plants, followed by multi-year testing of rootstocks, propagation techniques, and field performance. Issues such as clonal fatigue, somaclonal variation, and disease transmission require breeders to maintain strict clonal hygiene, adopt indexing technologies, and develop tissue-culture-based propagation methods when required.

Mutation Breeding and Polyploidy Induction

When natural variation and hybridization offer limited options, mutation breeding and polyploidy induction become important tools. Many aromatic grasses have benefited from gamma irradiation, which induces beneficial mutations resulting in compact growth, enhanced essential oil concentration, improved leaf morphology, or higher tillering. Mutation breeding has generated novel chemotypes with higher citral, geraniol, or citronellal content. Medicinal plants have seen the development of early-flowering, high-metabolite mutants through mutagenic treatments. Polyploidy induction using colchicine has been successfully used in aromatic grasses to increase biomass, enhance essential oil synthesis, and improve stress tolerance. In certain medicinal crops, polyploids show enhanced root mass, improved metabolite pathways, or greater environmental adaptability. Mutation and polyploidy breeding serve as supplementary strategies that expand the genetic canvas of crops where sexual reproduction is difficult or where breeders aim to introduce novel variation rapidly.

Biotechnological and Molecular Approaches

Modern breeding approaches increasingly rely on biotechnology and molecular genetics to enhance precision and accelerate progress. Molecular markers such as SSRs, RAPDs, SNP arrays, and gene-sequence-based markers enable identification of genetic relationships, assessment of diversity, early selection for resistance traits, and parental verification in hybridization programmes. Marker-assisted selection is particularly valuable in perennial crops where phenotypic evaluation requires many years. Genomic selection, a newer approach, uses genome-wide marker data to predict breeding values, offering high potential for reducing the length of breeding cycles. Transcriptomic and metabolomic studies provide deep insights into pathways governing aroma compounds in spices, secondary metabolites in medicinal plants, flavour attributes in tea and coffee, and essential oil biosynthesis in aromatic grasses. Tissue culture techniques such as micropropagation, somatic embryogenesis, and synthetic seed technology have enabled mass multiplication of elite clones, disease elimination, and maintenance of germplasm free from pathogens. In recalcitrant species where seed viability is low, cryopreservation of embryos, pollen, or in vitro shoots offers long-term security for valuable genetic material.

Quality Improvement and Industrial Relevance

Quality improvement is central to breeding these crops because economic value is often determined by specific biochemical or physical traits. In beverage crops such as tea and coffee, factors like flavour profile, aroma volatiles, polyphenol content, caffeine levels, and cup quality are critical. In spices, qualities like piperine concentration, essential oil richness, capsule size, volatile composition, curcumin content, colour retention, and pungency define market price. In medicinal plants, the concentration and stability of active principles such as alkaloids, glycosides, withanolides, or terpenoids determine therapeutic efficacy. Aromatic grasses derive value from essential oil yield, purity, and composition ratios such as citral balance, geraniol-to-acetate proportion, or menthol content. Breeding programmes increasingly use chromatographic profiling, biochemical assays, and sensory panels to evaluate quality traits alongside agronomic performance. With rising global demand for natural, residue-free, health-promoting products, breeding for quality has become as important as breeding for yield.

Climate Change, Resilience, and Sustainability

Climate change has introduced new dimensions to breeding challenges, particularly in perennial crops that remain in the field for decades. Increasing temperatures, shifting rainfall patterns, prolonged droughts, salinity intrusion in coastal belts, emergence of new pest complexes, and the spread of fungal and viral diseases demand urgent development of climate-resilient varieties. Breeders must address both short-term stress episodes and long-term climatic trends. Traits such as deep root systems, osmotic adjustment, stomatal regulation, thicker leaves, efficient canopy structure, high chlorophyll stability, and robust branching patterns enhance plant resilience. Coastal crops may require salt-tolerant genotypes, while upland crops may need improved drought survival mechanisms. Incorporating resilience traits requires screening germplasm under stress conditions, identifying tolerant landraces or wild relatives, and integrating physiological, genetic, and molecular markers into breeding pipelines. Sustainable agriculture also demands that varieties perform well under reduced input conditions, making nutrient-use efficiency, pest tolerance, and organic compatibility increasingly relevant.

Ideotype Breeding and Future Directions

Ideotype breeding offers a forward-looking approach that designs the "ideal plant type" based on functional attributes tailored to specific agro-ecosystems. For plantation crops, ideotypes may include dwarf or compact palms for high-density cultivation, rubber clones with strong trunks and high latex flow, or tea bushes with optimal leaf arrangement and rapid shoot turnover. For spices, ideotypes may embody robust root systems, improved flowering intensity, and compact architecture suited for intercrops. For medicinal and aromatic crops, ideotypes integrate enhanced metabolite synthesis with resilience, manageable height, consistent quality, and adaptability to varied soils. Ideotype models guide breeders in assembling trait combinations that may not exist naturally but can be constructed through hybridization, selection, and molecular intervention. This approach aligns breeding goals with future production systems that emphasize sustainability, precision agriculture, and climate-smart management.

The improvement of plantation, spice, medicinal, and aromatic crops is both a scientific pursuit and a strategic necessity for global agriculture. These crops embody a rich heritage of human–plant relationships while offering enormous potential for future growth. Their biological complexity demands multidisciplinary breeding approaches that combine classical genetics, evolutionary understanding, biochemical assessment, molecular innovation, physiologic insight, and ecological awareness. The long life span of these crops and their deep integration into landscapes and livelihoods reinforce the importance of developing varieties that are productive, resilient, high-quality, and sustainable.

As climate uncertainty grows and global markets continue to expand, the role of breeding becomes more central than ever. Varieties developed today will shape agricultural landscapes, industry supply chains, and ecological outcomes for decades. The integration of genetic resources, scientific breeding, modern biotechnology, and climate-smart ideotypes will determine the future of these vital crops. The journey of improvement is continuous, demanding innovation, commitment, and deep knowledge. By combining diversity with design, tradition with technology, and resilience with productivity, breeding programmes can ensure that plantation, spice, medicinal, and aromatic crops continue to enrich economies, ecosystems, and societies around the world.

Chapter - 1

Plantation Crops

Objective of Breeding in Plantation Crops

The major objective of breeding plantation crops is the generation of genetic variation and its judicious utilization for developing cultivars of high productivity, superior quality, and wide adaptability. As most plantation crops are perennial and largely cross-pollinated, improvement depends on creating new gene combinations through systematic hybridization followed by long-term selection. The overall goal is to combine high yield with resistance to diseases and pests, adaptability to diverse agro-climatic conditions, and uniform product quality required by domestic and export industries.

1. General Perspective of Plantation-Crop Breeding

Plantation crops form a distinct group of long-lived perennials grown primarily for industrial and commercial produce such as edible or industrial oils, beverages, latex, and high-value export commodities. Coconut, arecanut, oil palm, cashew, coffee, tea, cocoa, and rubber dominate this sector. They are pillars of tropical agriculture, sustaining millions of smallholders and supporting processing industries, foreign exchange, and rural employment. Because stands persist for decades, breeding decisions must be supported by precise objectives, robust experimental design, and reliable propagation systems to ensure uniformity and longevity of performance.

1.1 Economic and Ecological Importance

» **Industrial relevance:** Coconut and oil palm supply edible and industrial oils; coffee, tea, and cocoa underpin global beverage and confectionery markets; rubber provides natural latex vital to automotive, medical, and engineering sectors. Breeding must therefore secure consistent raw-material specifications (e.g., oil profile, cup quality, kernel grade, dry rubber content) that sustain factory throughput and product standards.

» **Export earnings:** Several of these commodities contribute significantly to foreign exchange. Stable, high-performing cultivars with predictable quality enable long-term contracts, reduce price volatility risk, and support geographical indications and quality certifications.

DOI: 10.1201/9781042010332-2

» **Employment generation:** From nursery to processing, plantation value chains offer year-round jobs. Early-bearing, high-yielding cultivars can increase labour demand in harvesting and primary processing while improving household incomes.

» **Environmental services:** Perennial canopies reduce erosion, moderate microclimate, build soil organic carbon, and support biodiversity. Mixed plantations and shade systems promote beneficial insects and pollinators, aiding ecological regulation of pests.

» **Long-term impact of varietal choice:** Given life spans of 20–40+ years, cultivar deployment has multi-decade consequences. Breeding programs must therefore balance near-term gains (precocity, vigour) with traits that sustain mature-phase productivity (e.g., height increment control, chronic-disease tolerance, structural integrity).

1.2 Challenges in Genetic Improvement

» **Perennial habit and longevity:** 10–25 years may elapse from crossing to conclusive evaluation, demanding disciplined records and a staggered pipeline of populations at different stages.

» **Long juvenile phase:** Delayed flowering necessitates early selection indices (morphological, physiological, molecular) to triage seedlings before costly field planting.

» **High heterozygosity:** Extensive segregation is valuable yet difficult to fix via seed; clonal/vegetative propagation becomes essential for deploying superior genotypes.

» **Limited controlled pollination:** Tall canopies, asynchronous flowering, and delicate floral structures require skilled crews, specialized gear, and strict protocols to secure true crosses.

» **Complex reproductive biology:** Monoecy, dioecy, protandry/protogyny, self-(in)compatibility vary by crop and even genotype, demanding crop-specific hybridization methods.

» **Environmental influence:** Yield and quality are highly site- and season-dependent; robust inference requires multi-location, multi-year trials with standardized management.

» **Small effective population size:** Space/time/cost limitations reduce the number of crosses and test trees, increasing genetic drift risk; germplasm conservation and parental refresh are mandatory.

» **Germplasm conservation cost:** Field genebanks and in situ reserves are expensive but indispensable to guard against emerging biotic/abiotic threats.

These constraints justify hybridization, clonal selection, and biotechnological integration to accelerate gain while maintaining genetic breadth.

1.3 General Breeding Objectives

» **Increase productivity per palm/tree/unit area:** Elevate harvest index and partitioning to economic organs while preserving canopy health and structural safety.

» **Improve economic product quality:** Optimize fatty-acid composition in oil crops; refine aroma, flavour, and body in beverages; improve kernel grade and recovery in nut crops; enhance latex properties in rubber.

- **Evolve resistance to major diseases and insect pests:** Build durable resistance by pyramiding qualitative and quantitative mechanisms; reduce inoculum build-up.
- **Develop tolerance to abiotic stresses:** Enhance drought, salinity, heat, wind, waterlogging tolerance through morphological and physiological traits that stabilize yield.
- **Reduce juvenile period and ensure regular bearing:** Select for precocity, stable flowering cycles, and consistent fruit set without compromising lifespan.
- **Evolve dwarf or compact genotypes:** Enable high-density planting and safer harvest via reduced height increment and balanced architecture.
- **Enhance stability and adaptability:** Secure consistent performance across sites, seasons, and management intensities.
- **Suitability for mechanization and processing:** Improve harvest ergonomics, uniform maturity, and raw-material traits that reduce processing losses.
- **Integrate modern molecular and tissue-culture tools:** Apply marker-based identity checks, early selection, and micropropagation for rapid, faithful multiplication.

2. Principles and Approaches of Hybridization in Plantation Crops

Hybridization, deliberate crossing of genetically distinct parents, is the principal engine for creating and assembling favorable alleles in long-lived perennials. Because each cross consumes time and resources and progenies occupy land for years, planning, execution, documentation, and evaluation must be exacting.

2.1 Purpose of Hybridization

- **Creation of genetic variation through recombination:** Generate wide segregation in F_2 and later generations to discover superior recombinants.
- **Combining complementary traits:** Merge high yield × resistance, quality × compact architecture, or precocity × longevity.
- **Exploitation of heterosis:** Capture hybrid vigour for vigour, earliness, resource use efficiency, and yield.
- **Transfer of specific genes:** Introgress donor alleles for resistance, dwarfness, or abiotic tolerance into elite backgrounds via planned backcrossing.
- **Adaptation to new environments and systems:** Evolve types for marginal soils, altered rainfall, heat, or high-density/mechanized systems.

2.2 Basic Genetic Principles

- **Complementary gene action:** Choose parents that contribute distinct favorable alleles to the same trait complex.
- **Segregation and recombination:** Anticipate novel combinations and size populations to make recovery of rare recombinants plausible.

- Heterosis: Recognize dominance/overdominance effects; plan clonal deployment or repeatable seed systems as appropriate.
- **Heritability and selection intensity:** Target high-heritability traits early; delay heavy selection on low-heritability traits until multi-site, multi-year data reduce noise.
- **Genetic diversity of parents:** Moderate to high genetic distance widens variability but may reduce fertility or adaptation if too distant.

2.3 Types of Hybridization Used in Plantation Crops

- **Intervarietal Hybridization:** Crossing between cultivars of the same species (e.g., **Tall × Dwarf coconut**) to combine complementary traits.
- **Interspecific Hybridization:** Crossing related species (e.g., **Elaeis guineensis × E. oleifera**) to transfer special traits such as reduced height increment or improved oil profile.
- **Intergeneric Hybridization:** Rare but valuable for **wide gene introgression**; usually requires **embryo rescue** and careful fertility screening.
- **Recurrent Selection:** Iterative **selection and inter-mating** to enrich favorable alleles for complex, quantitative traits.
- **Backcross Breeding: Targeted introgression** of a specific trait (e.g., disease resistance) while **recovering the recurrent parent** background.
- **Transgressive Breeding:** Designing crosses to combine complementary alleles that produce progeny **exceeding both parents** for key traits.

2.4 Biological Considerations Before Hybridization

- **Study of floral biology:** Map **inflorescence development**, **anthesis**, **stigma receptivity**, **pollen shed**, and **floral longevity** for each parent.
- **Mode of pollination:** Establish **monoecy/dioecy/hermaphroditism, self-(in)compatibility**, and degree of **protandry/protogyny**.
- **Flowering synchronization:** Use **staggered planting, pruning, irrigation pulses**, or **growth regulators** to align flowering.
- **Pollen collection and storage:** Standardize **collection, drying, viability testing**, and **storage**; respect **species-specific viability windows**.
- **Environmental conditions:** Track **temperature/humidity**; schedule pollination in **optimal windows** to maximize fertilization and minimise contamination.

2.5 Detailed Procedure of Hybridization (General for Plantation Crops)

The steps outlined below mirror the systematic approach used in the uploaded reference but adapted for perennial species.

Step 1 – Setting Objectives

The breeder first defines the specific target traits, yield improvement, quality enhancement, resistance, or plant architecture. A clear objective guides parent selection and evaluation. For example, in coconut breeding for root-wilt resistance, one parent must be a resistant donor.

Step 2 – Selection of Parents

Parents are chosen based on contrasting yet complementary traits:

» The female (parent A) is generally a locally adapted, stable variety with known performance.

» The male (parent B) is selected for the specific plus trait to be introduced (e.g., high oil content, disease resistance).

» Geographically diverse parents are preferred to maximize genetic variation.

Step 3 – Evaluation of Parents

Selected parents are grown under uniform conditions and evaluated for phenotypic stability, flowering behavior, and disease reaction. In cross-pollinated trees, ensuring homozygosity is less critical than verifying vigour and compatibility.

Step 4 – Sowing and Flowering Plan

Since flowering periods often differ, staggered sowing or controlled irrigation is used to synchronize anthesis. The usual layout places female parents in central rows and male parents on the periphery to simplify pollen collection and hand pollination.

Step 5 – Emasculation

In bisexual flowers, immature anthers are removed before dehiscence to avoid self-pollination. In monoecious species like coconut or arecanut, male inflorescences on female palms are removed or bagged. The time of emasculation depends on crop species; for example, in coconut, it is done one day before anthesis.

Step 6 – Pollen Collection and Dusting

Pollen is collected from the male parent just as the anthers begin to dehisce. Depending on species, it may be air-dried and stored short-term. Pollination is usually carried out in early morning hours when stigma receptivity is maximum. Fresh pollen is gently brushed or dust-blown onto stigmas.

Step 7 – Labelling and Bagging

Each cross is tagged with a durable label containing the cross number, parentage, and date of pollination. Inflorescences are then covered with cloth or polythene bags to prevent foreign pollen contamination. Bagging is retained until fruit set is visible.

Step 8 – Harvesting and Storage of Hybrid Seeds

Fruits or nuts are harvested at physiological maturity, typically 6 to 12 months after pollination depending on species. Hybrid seeds are collected on a plant basis, cleaned, labelled, and stored under conditions that maintain viability.

Step 9 – Raising and Evaluation of Progeny

Hybrid seeds are germinated and seedlings raised in nurseries before field planting. Initial screening is for vigour and early growth. Detailed yield and quality assessments begin after the juvenile phase ends. Because evaluation takes years, early selection indices based on morphological or molecular markers are increasingly used.

Step 10 – Fixation and Propagation of Desirable Types

Superior hybrids are cloned through vegetative methods such as budding, grafting, or tissue-culture micropropagation to maintain uniformity and accelerate dissemination.

2.6 Modified Hybridization Approaches for Perennial Crops

- » **Recurrent interspecific hybridization:** Stabilize useful interspecific traits through **cyclical crossing and selection.**
- » **Shuttle breeding:** Evaluate segregants across **contrasting environments/seasons** to shorten cycles and strengthen adaptability inference.
- » **Off-season nursery / Rapid generation advancement:** Use **controlled environments** or **alternate locations** to advance generations and **save time.**
- » **Single Seed Descent and Bulk methods:** Advance segregation under **mass conditions** before intensive selection when populations are more **homozygous.**
- » **Controlled clonal hybridization:** Cross selected **clones,** then deploy **F_1 as clones,** maintaining heterosis while ensuring uniform fields.

2.7 Evaluation and Selection in Hybrid Progenies

- » **Field evaluation:** Observe **vigour, earliness, yield attributes, quality,** and **disease/pest tolerance** in **replicated** plots.
- » **Statistical analysis:** Apply **ANOVA/mixed models** to separate genetic effects from environmental noise and to estimate **stability.**
- » **Heritability and genetic advance:** Quantify **selection response** potential; prioritize traits with **higher heritability** for early gain.
- » **Progeny testing:** Advance **elite families/clones** to **multi-location trials** before recommendation.
- » **Molecular assays:** Use **DNA markers** to confirm **hybridity/parentage** and to assist **early selection** for key traits.

2.8 Advantages of Hybridization in Plantation Crops

- » **Broad genetic base** for sustained improvement.
- » **Combination of desirable traits** from divergent parents.
- » **Exploitation of heterosis** for vigour, precocity, and yield.
- » **Introduction of novel qualities** (e.g., oil profile, beverage flavour).

- **Facilitates resistance breeding** against emerging pests/diseases.
- **Foundation for marker-assisted and genomic selection** strategies.

2.9 Limitations and Precautions

- **Long duration** from crossing to conclusive evaluation.
- **Labour-intensive controlled pollination** requiring skilled personnel.
- **Low fruit set/germination** in some species, reducing effective population size.
- **Heterozygosity of hybrids** often precludes seed-based fixation; necessitates **clonal deployment**.
- **Environmental fluctuations** influencing quantitative traits; demand **multi-year, multi-site** validation.

Despite these constraints, **systematic hybridization** remains the most powerful, scalable method for genetic improvement in perennial plantation crops.

2.10 Integration of Modern Tools with Hybridization

- **Marker-Assisted Selection (MAS):** DNA markers enable **identity checks, hybridity confirmation**, and **early selection** of carriers of resistance/quality alleles.
- **Tissue culture and micropropagation: Rapid cloning** of elite hybrids maintains **uniformity** and supplies **large-scale planting material**.
- **Embryo culture and rescue:** Overcome **incompatibility** and **seed recalcitrance** in **wide crosses**, improving recovery of rare hybrids.
- **Polyploidization and chromosome engineering:** Induce **novel variation**, modify **plant stature**, and influence **quality attributes**.
- **Molecular characterization and digital traceability:** Strengthen **germplasm management**, **parentage tracking**, and **field-to-lab data integrity** for long-term programs.

2. COCONUT (*Cocos nucifera* L.)

2.1 Introduction

Coconut (*Cocos nucifera* L.), often referred to as the **"Tree of Life"**, is one of the most economically and socially significant plantation crops in tropical regions. Every part of the palm is of value, food, beverage, fiber, timber, and fuel, making it a unique multipurpose crop. It provides edible oil, copra, coir, and tender nuts for consumption, while also serving as a source of raw material for numerous industries such as confectionery, pharmaceuticals, cosmetics, and biofuels. Globally, coconut cultivation spans over 12 million hectares across 93 countries, with major producers being **Indonesia, the Philippines, India, Sri Lanka, and Thailand.**

In India, coconut occupies around **2.1 million hectares**, contributing substantially to the rural economy, particularly in the states of **Kerala, Tamil Nadu, Karnataka, Andhra Pradesh, and Odisha.** The crop's sustainability lies in its ability to support millions of smallholder farmers, ensuring both nutritional and economic security.

From a breeding standpoint, coconut presents unique challenges: it is a **monocotyledonous, perennial palm, monoecious but cross-pollinated**, and has a **long generation cycle (8–10 years to maturity)**. These biological traits lead to high heterozygosity and low heritability for many yield-related characters, which make **hybridization, selection, and clonal propagation** the most crucial breeding tools. The long productive life (up to 60–80 years) means that every breeding decision has lasting consequences. Hence, coconut breeding demands a **precise combination of long-term genetic planning, field experimentation, and biotechnological support** to generate sustainable improvement.

2.2 Breeding Objectives in Coconut

The ultimate aim of coconut breeding is to evolve cultivars that produce higher and more stable yields, with improved oil quality, disease and pest resistance, and better adaptability to climatic stress and soil conditions. The objectives have evolved over time to meet industrial demand, climate resilience, and consumer preferences.

2.2.1 Major Breeding Objectives

1. **Increase in Copra and Oil Yield per Palm and per Hectare**: The primary breeding goal in coconut is to maximize **copra yield** (dried kernel weight) per palm and per hectare. The copra yield depends on nut number, kernel thickness, and copra weight per nut. By increasing nuts per palm (through high fruit set and reduced immature nut fall) and kernel content, breeders can significantly enhance total oil recovery. Improved **partitioning efficiency**, better **leaf photosynthetic capacity**, and optimized **nut-to-bunch ratio** are physiological targets integrated into selection programs.
2. **Improvement in Oil Quality and Composition**: Coconut oil is unique due to its **high lauric acid (C12:0)** content, which provides excellent oxidative stability and antimicrobial properties. However, modern industrial uses demand **customized oil profiles**, with balanced medium-chain fatty acids (MCFAs) for health-oriented applications. Hence, breeding programs now aim to modify **fatty acid synthesis pathways**, seeking palms with higher **oleic acid (C18:1)** or reduced **saturated fats**, to expand edible and pharmaceutical markets.
3. **Early Bearing and Compact Plant Type**: Traditional tall palms begin bearing after 6–8 years. Dwarf varieties bear earlier (3–4 years) but are short-lived. Breeding seeks **precocious, semi-dwarf palms** that combine early maturity, high yield, and longevity, facilitating **high-density planting (175–200 palms/ha)**. Compact palms also reduce labour for harvesting, improving management efficiency.
4. **Resistance to Major Diseases and Pests**: Major diseases include **root wilt, bud rot (Phytophthora palmivora), stem bleeding (Thielaviopsis paradoxa)**, and **lethal yellowing**. Key insect pests include **rhinoceros beetle (Oryctes rhinoceros), red palm weevil (Rhynchophorus ferrugineus)**, and **mite (Aceria guerreronis)**. Breeding for resistance involves identifying tolerant landraces (e.g., Kalparaksha, Kalpatharu) and incorporating resistance through hybridization or molecular markers.
5. **Tolerance to Abiotic Stresses**: Coconut grows across diverse environments, from saline coastal belts to arid plains. Breeding aims to identify and combine traits for **drought tolerance, salinity tolerance**, and **temperature resilience**. Parameters such as **leaf water potential, osmotic adjustment**, and **root depth** are used as indirect selection criteria.

6. **Improvement in Nut and Kernel Quality:** Industrial demand emphasizes **high kernel recovery (≥ 180 g/nut), uniform nut size,** and **sweet water content.** Breeding for tender-nut types focuses on **water volume, sweetness, total soluble solids (TSS),** and **low titratable acidity.** For copra-type nuts, emphasis is placed on **kernel thickness and oil recovery.**
7. **Adaptability and Stability of Performance:** Given the variability in rainfall, temperature, and soil conditions, breeding aims to evolve **stable genotypes** that perform consistently under varied conditions. Long-term multi-location trials and G×E interaction studies form integral parts of this goal.
8. **Integration of Biotechnological Tools :** Modern breeding integrates **DNA marker technology, genomics, tissue culture,** and **embryo rescue** to shorten breeding cycles and ensure accurate parentage verification, hybrid identity, and early selection of desirable genotypes.

2.3 Genetic Resources and Germplasm in Coconut

2.3.1 Origin and Domestication

Coconut is native to the **Indo-Pacific region,** with two major centres of diversity:

- **Pacific group** – Characterized by large nuts, abundant water, and tall habit.
- **Indian group** – Characterized by smaller nuts, thicker kernels, and drought tolerance.

Natural dispersal by sea currents led to the establishment of coconut across tropical coasts of Asia, Africa, and America. Continuous human selection for tender nuts, copra yield, and adaptability resulted in distinct ecotypes and cultivars.

2.3.2 Types and Races of Coconut

Coconuts are broadly classified into **two genetic races, Tall** and **Dwarf,** and a few natural intermediates.

- **Tall Types (Typica):** Cross-pollinated, highly heterozygous, tall palms (20–30 m), long-lived (up to 80 years). Bearing starts after 6–8 years. They produce large nuts with high copra content. Examples: *West Coast Tall (WCT), East Coast Tall (ECT), Laccadive Ordinary, Andaman Ordinary.*
- **Dwarf Types (Nana):** Predominantly self-pollinated, short stature (6–10 m), early bearing (3–4 years), smaller nuts, often colorful (orange, yellow, green). Examples: *Chowghat Orange Dwarf (COD), Malayan Yellow Dwarf (MYD), Malayan Green Dwarf (MGD).*
- **Intermediate Types (Tall × Dwarf crosses):** Exhibit features of both groups. They combine dwarf's precocity and tall's vigour. Examples: *King Coconut, Kamini.*

2.3.3 Germplasm Resources and Conservation

Coconut improvement depends heavily on maintaining and utilizing its rich genetic base.

- **National Repositories:** The **ICAR–Central Plantation Crops Research Institute (CPCRI),** Kasaragod, Kerala, maintains India's main coconut gene bank, with over **450 accessions** representing 15 countries. Regional stations at **Aliyarnagar (Tamil Nadu), Kidu (Karnataka), and Mohitnagar (West Bengal)** conserve local ecotypes.

- **International Collections:** The **International Coconut Genetic Resources Network (COGENT)** coordinates global conservation efforts. Field Gene Banks exist in Sri Lanka, Indonesia, the Philippines, Papua New Guinea, and Côte d'Ivoire.
- **Conservation Approaches:**
  - **Ex situ:** Field gene banks and in vitro conservation (embryo culture).
  - **In situ:** Community-based conservation of traditional cultivars in farmer fields.
  - **Cryopreservation:** Studies are ongoing for long-term embryo storage.
- **Molecular Characterization:** Molecular markers such as **SSR, RAPD, AFLP, and SNPs** are used for genetic diversity analysis, parentage testing, and varietal identification. Marker data help identify **core collections** for efficient germplasm utilization.

Coconut's breeding success relies on a thorough understanding of its floral biology.

2.4.1 Inflorescence Structure

- The **inflorescence (spadix)** arises in the leaf axil and is enclosed by a double sheath called a **spathe**.
- Each inflorescence contains **40–60 spikelets**, each bearing numerous **male flowers** at the tip and **female flowers** at the base.
- **Male flowers** are small, yellow, and produced in large numbers; **female flowers** are larger, spherical, and fewer in number (10–50 per inflorescence).

2.4.2 Flowering Behaviour

- The palm is **monoecious and protandrous**, male flowers mature and shed pollen before female flowers become receptive, ensuring cross-pollination.
- **Flowering interval:** A new inflorescence emerges every 3–4 weeks.
- **Pollination agents:** Wind and insects (mainly *Apis dorsata* and *Trigona* spp.).
- **Duration of male phase:** 18–20 days; **female phase:** 3–5 days.
- **Anthesis:** Occurs in the early morning, and pollen viability is highest during this period.

2.4.3 Pollen Viability and Storage

Pollen viability is short (2–3 days under ambient conditions). For controlled hybridization, pollen is **collected, air-dried, and stored** at 5–8°C for up to 10 days or at –20°C with silica gel for 3–4 weeks. Viability testing is conducted using acetocarmine staining or germination tests.

2.5 Breeding and Hybridization in Coconut

Coconut breeding has historically evolved through stages, from **mass selection** to **intra- and inter-varietal hybridisation, and is now moving** toward **biotechnological integration**.

2.5.1 Selection Breeding

- **Mass Selection:** Based on the field performance of individual palms within natural populations. Criteria include high nut yield (>100 nuts/palm/year), copra content, and disease tolerance.

» **Progeny Testing:** Offspring of selected palms are evaluated to confirm heritability. Selections like *WCT* and *ECT* emerged from such programs.

2.5.2 Hybridization Breeding

Hybridization exploits **heterosis** by combining complementary traits of different genetic types.

a. **Types of Crosses**

1. **Tall × Tall (T × T):** Combines vigour and stability. Example: *West Coast Tall × East Coast Tall.*
2. **Dwarf × Tall (D × T):** Combines dwarf's precocity and tall's longevity and yield; most widely adopted. Example: *COD × WCT (Kera Sankara).*
3. **Tall × Dwarf (T × D):** Reverse cross; less preferred due to pollination loGIstics.
4. **Dwarf × Dwarf (D × D):** Limited use; primarily for tender-nut markets.

b. **Steps in Controlled Hybridization**

Step 1 – Selection of Parents: Identify vigourous, true-to-type parent palms with contrasting traits.

Step 2 – Emasculation: For female parents, remove or cover male flowers before anthesis to prevent self-pollination.

Step 3 – Bagging: Enclose the entire inflorescence with muslin or polyethylene bags to isolate from external pollen.

Step 4 – Pollen Collection: Collect mature male inflorescences just before anther dehiscence; dry, crush, and sieve to release pollen.

Step 5 – Pollination: Apply pollen uniformly using a brush or air blower during the female receptive period (usually morning).

Step 6 – Labelling and Records: Tag each inflorescence with parentage, date, and cross code; maintain logbooks.

Step 7 – Bag Retention and Aftercare: Keep the bag until fruit set (≈45 days); maintain nutrient and water supply.

Step 8 – Harvesting and Evaluation: Harvest mature nuts (~11–12 months after pollination), germinate, and raise progeny for evaluation.

2.6 Hybrid Seed Production and Evaluation

Hybrid seed production requires **isolated seed gardens** with a **minimum 400 m isolation distance**. Hybrid palms are maintained under uniform management for pollen synchronization. True hybridity is verified morphologically (e.g., petiole colour, leaf base hue, nut shape) and through **molecular assays (SSR-based markers)**. Seedlings are evaluated for **germination rate**, **seedling vigour**, and **true-to-type features** before field planting. Each hybrid generation is tested across **multiple sites** for yield, oil content, and stress resilience.

2.7 Major Achievements in Coconut Breeding

2.7.1 Released Varieties and Hybrids

Type	Name of Variety/ Hybrid	Parentage	Salient Features
Hybrid (D × T)	**Kera Sankara (COD × WCT)**	Chowghat Orange Dwarf × West Coast Tall	Early bearing, 110–120 nuts/palm/year, high copra yield, suitable for tender nut and oil.
Hybrid (D × T)	**Chandrasankara (COD × ECT)**	Orange Dwarf × East Coast Tall	High copra yield, tall adaptability, early bearing.
Hybrid (D × T)	**Kalpa Samrudhi (MYD × WCT)**	Malayan Yellow Dwarf × West Coast Tall	High yield (125 nuts/palm/year), tolerant to drought.
Hybrid (T × T)	**Kalpa Surya (ECT × WCT)**	East Coast Tall × West Coast Tall	Vigourous, drought-tolerant, stable yield across regions.
Tall Selection	**Kera Keralam (ECT selection)**	,	High oil recovery, suited for coastal saline soils.
Dwarf Selection	**Kalpa Jyothi (COD selection)**	,	Early bearing, high tender-nut quality and sweetness.

2.7.2 Contributions of ICAR–CPCRI

- Establishment of a **National Coconut Breeding Programme**.
- Development of **15+ improved varieties and hybrids** adapted to diverse agro-climatic zones.
- Creation of a **molecular database** for 400+ accessions.
- Introduction of **Kalpa Krishi** – a drought-tolerant selection suitable for dryland cultivation.

2.7.3 International Achievements

- In the Philippines, *Laguna Tall* × *Malayan Dwarf* hybrids revolutionized commercial plantations.
- In Sri Lanka, *Sri Lanka Tall* × *Sri Lanka Green Dwarf (SLGD)* hybrids dominate the tender-nut market.
- In Indonesia, *Tenga* × *Malayan Yellow Dwarf* hybrids are popular for high oil yield.

2.8 Modern Advances and Biotechnological Approaches

2.8.1 Marker-Assisted Selection (MAS)

DNA markers enable early hybrid verification, parentage confirmation, and detection of resistance alleles. SSR and SNP markers are routinely used to identify **root wilt-resistant palms** and ensure hybrid purity in seed gardens.

2.8.2 Tissue Culture and Clonal Propagation

Although coconut is recalcitrant to conventional vegetative propagation, recent advances in **somatic**

embryogenesis from immature inflorescences and zygotic embryos show promise. Micropropagation ensures **uniform planting material** and faster dissemination of elite genotypes.

2.8.3 Embryo Rescue and Cryopreservation

Used to recover **interspecific hybrids** (e.g., *Cocos nucifera* × *Cocos weddelliana*) and to conserve embryos of elite varieties. Embryos cultured in vitro yield uniform seedlings with higher survival rates.

2.8.4 Genomics and Transcriptomics

Whole-genome sequencing of coconut has enabled identification of candidate genes for **oil biosynthesis, disease resistance**, and **flowering control**. Transcriptomic data are now used to design **molecular breeding pipelines** for stress-tolerant varieties.

3 – Arecanut (*Areca catechu* L.) and Oil Palm (*Elaeis guineensis* Jacq.)

3.1 Arecanut (*Areca catechu* L.)

3.1.1 Introduction

Arecanut, commonly known as the **betel-nut palm,** is one of the oldest and most socially significant plantation crops of South and Southeast Asia. Belonging to the family **Arecaceae,** it is cultivated mainly for its seeds (nuts) that are chewed with betel leaves as a mild stimulant. India, Bangladesh, Myanmar, Indonesia, and Thailand account for nearly 80 % of the global area under the crop.

In India, arecanut supports more than two million smallholders, forming the economic backbone of the humid tracts of **Karnataka, Kerala, Assam, Meghalaya, and West Bengal.** The palm thrives under humid tropical conditions with well-distributed rainfall and deep loamy soils. Biologically, arecanut is a **monoecious, cross-pollinated perennial** with a single slender stem reaching 15–25 m, bearing a crown of pinnate leaves. Its long generation time, cross-pollination, and high heterozygosity make breeding challenging, yet systematic programs over the past five decades have led to major genetic improvement.

3.1.2 Breeding Objectives in Arecanut

1. **Increase Nut Yield per Palm**
   - Enhance bunch number, nut set, and kernel weight through hybridization and selection.
   - Improve partitioning of assimilates to reproductive organs without compromising palm vigour.
2. **Improve Nut Quality**
   - Increase kernel recovery and total alkaloid content (arecoline ≈ 0.3 – 0.6 %).
   - Standardize biochemical composition for processing grades such as *chali*, *red boiled*, and *white*.
3. **Develop Dwarf and Compact Varieties**
   - Reduce palm height (≤ 12 m) for easier harvesting and higher planting density (up to 1 700 palms ha^{-1}).

4. **Evolve Disease- and Pest-Resistant Genotypes**
   - Resistance to **Kole rot, foot rot (*Phytophthora meadii*)**, and **inflorescence dieback.**
   - Tolerance to **mite** and **spindle bug** infestation.
5. **Adaptation to Stress Environments**
   - Identify genotypes performing well under low-light (mixed cropping) or moisture-stress conditions.
6. **Enhance Regular Bearing**
   - Stabilize year-to-year yield by breeding palms with synchronized flowering and balanced sex ratio.

3.1.3 Genetic Resources and Germplasm

- **Centers of Origin:** The crop is believed to have originated in the **Malay Peninsula**, later spreading to India and the Pacific.
- **Diversity in India:** Indigenous ecotypes like *South Kanara, Mangala, Thirthahalli,* and *Mohitnagar Local* exhibit wide variability in nut size, yield, and disease tolerance.
- **Conservation:** The **ICAR-Central Plantation Crops Research Institute (CPCRI)** maintains a gene bank at Vittal (Karnataka) with > 180 accessions.
- **Evaluation Parameters:** Germplasm is assessed for palm height, number of leaves, nuts per bunch, copra content, and tolerance to foot rot.

3.1.4 Floral Biology and Pollination

- The **inflorescence** (spadix) arises in the leaf axil and is enclosed in a double-sheath spathe.
- Each inflorescence bears **many spikelets**, with **male flowers** toward the tip and **female flowers** toward the base.
- The palm is **monoecious and protandrous**, male flowers release pollen before the female ones become receptive, ensuring cross-pollination.
- **Pollination agents:** mainly wind and insects (*Apis cerana*, *Trigona* spp.).
- **Pollen viability:** 24 – 48 h under ambient conditions; can be extended to 5 days at 4 °C with desiccants.

3.1.5 Breeding and Hybridization Methods

a. Selection Breeding

Initial improvement in arecanut relied on **mass selection** of superior palms within traditional plantations. Selection criteria include:

- ≥ 150 nuts $palm^{-1}$ $year^{-1}$.
- Thick kernel and high dry-matter recovery.
- Resistance to foot rot and mite. Resulting elite selections such as *Mangala* and *Sumangala* became parent materials for later hybridization.

b. Hybridization

Controlled crossing is difficult because of tall stature and asynchronous flowering. Specialized pollination towers and scaffolds are used.

Steps in Controlled Pollination

1. **Parent Selection:** Choose female and male parents with complementary traits (e.g., tall × dwarf).
2. **Bagging:** Enclose unopened spadices of the female parent in muslin bags before male anthesis.
3. **Pollen Collection:** Collect mature male flowers, dry, and store pollen at 4 °C.
4. **Pollination:** Apply pollen during early morning (6–9 a.m.) when stigmas are most receptive.
5. **Labelling and Aftercare:** Tag each inflorescence and retain bags until fruit set (≈ 6 weeks).
6. **Seedling Raising:** Harvest ripe nuts (≈ 12 months), germinate, and evaluate progeny for growth and yield.

c. Backcross and Recurrent Selection

To introgress disease resistance without altering desirable nut quality, resistant donors are backcrossed with elite high-yielding types. Recurrent selection among inter-se crosses has improved *Mangala* and *Sumangala* derivatives.

3.1.6 Released Varieties and Achievements

S. No.	Variety/Hybrid Name	Parentage/Type	Salient Features
1	*Mangala*	Selection from South Kanara Local	Early bearing; 140–160 nuts palm^{-1} yr^{-1}; tolerant to foot rot
2	*Sumangala*	Tall selection	High kernel recovery (25 %); resistant to Kole rot
3	*Sreemangala*	Hybrid (*Sumangala* × *South Kanara Local*)	Precocious (4 years to bearing); 180 nuts palm^{-1} yr^{-1}
4	*Mohitnagar*	Selection from Bengal germplasm	Adapted to humid East India; uniform nuts
5	*Vittal Areca-1*	Recurrent selection population	High yield (≈ 3.5 t chali ha^{-1}); moderate tolerance to foot rot

Achievements include raising national productivity from **1.2 t to > 2.8 t chali ha^{-1}** and establishing hybrid seed-production gardens under CPCRI guidelines.

3.1.7 Modern Advances in Arecanut Breeding

1. **Molecular Characterization:** SSR and AFLP markers delineate genetic relationships among accessions, enabling parent selection.
2. **Marker-Assisted Selection:** QTLs for foot-rot tolerance are under validation for early screening.
3. **Tissue Culture:** Somatic embryogenesis from immature inflorescence segments aids rapid multiplication.

4. **Sex-Ratio Studies:** Understanding male: female flower ratios assists selection for stable bearing.
5. **Digital Phenotyping:** Drones and remote sensing quantify canopy health and yield traits in field trials.

3.2 Oil Palm (*Elaeis guineensis* Jacq.)

3.2.1 Introduction

Oil palm, belonging to the family Arecaceae, is the **highest oil-yielding crop in the world**, producing 4–6 t oil ha^{-1} $year^{-1}$, ten times that of soybean. Native to West Africa, it was introduced to Southeast Asia in the 19th century and now dominates the global vegetable-oil supply.

In India, oil palm cultivation spans **Andhra Pradesh, Telangana, Karnataka, Tamil Nadu, Mizoram, and Kerala,** under irrigated conditions. The crop contributes substantially to India's edible-oil mission by reducing import dependence. Botanically, oil palm is a **monoecious, cross-pollinated perennial** with a single stem (20–25 m tall) bearing large pinnate leaves and unisexual inflorescences. The fruit is a drupe containing mesocarp (oil pulp) and kernel (palm kernel oil). Because of its long life (25–30 years) and cross-pollination, breeding requires meticulous record-keeping and long-term evaluation.

3.2.2 Breeding Objectives in Oil Palm

1. **Increase Fresh Fruit Bunch (FFB) Yield**
   - » Target > 30 t FFB ha^{-1} $year^{-1}$ under irrigated conditions.
   - » Improve bunch number, bunch weight, and oil extraction rate.
2. **Enhance Oil Yield and Quality**
   - » Raise oil-to-bunch ratio to > 25 %.
   - » Modify fatty-acid composition with higher oleic and linoleic acids for edible purposes.
3. **Develop Compact and Early-Bearing Palms**
   - » Reduce trunk height increment (< 45 cm yr^{-1}).
   - » Select for precocity (first harvest within 30 months).
4. **Evolve Disease and Pest Resistance**
   - » Resistance to **Ganoderma basal stem rot, Fusarium wilt**, and **bagworm pests.**
5. **Improve Tolerance to Abiotic Stress**
   - » Select for heat and drought tolerance suitable for semi-arid India.
6. **Maximize Oil Recovery and By-product Use**
   - » Enhance mesocarp oil yield and kernel oil content; reduce shell thickness for better processing efficiency.

3.2.3 Genetic Resources and Germplasm

- » **Centers of Origin:** West and Central Africa (Nigeria, Cameroon, Ghana).

» **Species:** Three species, *E. guineensis* (African oil palm), *E. oleifera* (American oil palm), and *E. odora.*

» **Germplasm Collections:** The Indian Council of Agricultural Research maintains > 120 accessions at ICAR-IIOPR (Pedavegi, Andhra Pradesh).

» **Fruit Forms:**

- **Dura** (thick shell), **Pisifera** (shell-less, female-sterile), and **Tenera** (single thin shell, high oil content). *Tenera* is the commercial hybrid of *Dura* × *Pisifera.*

3.2.4 Floral Biology and Pollination

» Palms bear male and female inflorescences in axils of leaves, produced in successive cycles (≈ 15 days apart).

» Pollination is mainly by insects (*Elaeidobius kamerunicus* weevil) and partly by wind.

» Controlled pollination requires timed bagging of female spadices and artificial dusting of pollen when stigmas are receptive (24 – 48 h window).

3.2.5 Breeding and Hybridization Methods

a. Intra-Specific Hybridization (*E. guineensis*)

» Crosses between selected *Dura* and *Pisifera* parents produce *Tenera* hybrids exhibiting heterosis for yield and oil content.

» **Controlled Pollination Steps:**

1. Isolate female spadix one day before anthesis.
2. Collect pollen from male parent, dry to < 10 % moisture.
3. Dust pollen evenly over stigmas using air blower.
4. Rebag until fruit set (≈ 60 days).
5. Harvest ripe bunches and raise seedlings for evaluation.

b. Inter-Specific Hybridization (*E. guineensis* × *E. oleifera*)

» **Objective:** combine high yield of *E. guineensis* with short stature and unsaturated oil profile of *E. oleifera.*

» Resulting hybrids show slow stem growth and better oil quality but lower yield; used as breeding material for recurrent selection.

c. Selection Breeding

» Mass selection among *Tenera* populations for high bunch number, compact crown, and thinner shell.

» Recurrent selection to improve general combining ability of parent lines.

3.2.6 Released Hybrids and Achievements

S. No.	Hybrid/Variety Name	Parentage/Type	Salient Features
1	*Tenera (D × P)*	Dura × Pisifera	Thin shell (1–2 mm); high oil content (> 25 %); early bearing
2	*IIOPR Tenera Elite 1*	Selected from Pedavegi breeding block	30–32 t FFB ha^{-1}; 4.8 t oil ha^{-1}; moderate drought tolerance
3	*IIOPR Tenera Elite 2*	Improved recurrent selection cycle	High oleic acid (45 %); slow height increment
4	*IIOPR Guineensis × Oleifera*	Interspecific hybrid	Compact palm; enhanced unsaturation in oil

Achievements include raising India's average yield from < 2 t to > 4 t oil ha^{-1} and establishing elite seed-production gardens at Pedavegi and Palode.

3.2.7 Modern Advances in Oil Palm Breeding

1. **Molecular Markers and Genomics**
   - » SNP and microsatellite markers confirm hybrid identity and trace parentage.
   - » DNA markers linked to **shell-thickness (Sh)** gene enable early identification of *Tenera* seedlings, avoiding field culling.
2. **Marker-Assisted Backcrossing**
   - » Introgression of *E. oleifera* genes for slow stem growth and high unsaturation without yield penalty.
3. **Tissue Culture and Clonal Propagation**
   - » Somatic embryogenesis from immature leaf explants produces true-to-type clonal palms of elite *Tenera* hybrids.
   - » Reduces variability and ensures uniform estates.
4. **Genome Sequencing**
   - » Oil palm genome (~1.8 Gb) has been sequenced, revealing gene clusters controlling **oil biosynthesis, height regulation**, and **stress response**.
5. **Phenomic and Digital Breeding Tools**
   - » Use of UAV (drone) imagery for bunch estimation, height monitoring, and canopy health assessment.
   - » Integration of AI algorithms for yield prediction and selection efficiency.

4 – Cashew (*Anacardium occidentale L.*)

4.1 Introduction

Cashew, a tropical tree native to **Brazil**, is one of the most important plantation crops cultivated for its **nut, apple, and by-products**. Belonging to the family **Anacardiaceae**, the species was introduced to India by Portuguese traders in the 16th century. Over the centuries, it has become a major export commodity, providing livelihood to millions of farmers and workers in the nut-processing and value-addition sectors.

India, Vietnam, Côte d'Ivoire, and Tanzania are global leaders in cashew nut production. In India, the crop occupies about **1.2 million ha**, mainly in the **coastal belts of Kerala, Karnataka, Goa, Maharashtra, Odisha, Andhra Pradesh, Tamil Nadu, and West Bengal**. Cashew is a **drought-tolerant, hardy perennial**, capable of thriving on marginal lateritic soils and sloping terrains where few crops succeed.

Botanically, the cashew tree is a **medium-sized evergreen (8 – 12 m high)** with a deep taproot, irregular branching, and a spreading canopy. It produces **bisexual and male flowers** on a **terminal panicle inflorescence**, making the species **predominantly cross-pollinated**, although a proportion of **self-pollination** occurs.

Economically, the kernel, the true nut, is a high-value edible product rich in protein (21 %), fat (47 %), and minerals (iron, magnesium, phosphorus). The **cashew apple**, the swollen pedicel, is a source of juice, syrup, and alcoholic beverages. Cashew nut shell liquid (CNSL) extracted from shells is used in **pharmaceuticals, paints, and polymer industries**. Because cashew is grown mainly by smallholders under rainfed conditions, genetic improvement has focused on yield stability, early bearing, compact canopy, and pest and disease resistance.

4.2 Breeding Objectives in Cashew

4.2.1 Enhancement of Nut Yield and Productivity

- The primary breeding objective is to develop **high-yielding varieties** producing ≥ 10 kg raw nuts palm^{-1} year^{-1} or ≥ 2 t ha^{-1} under rainfed conditions.
- Breeding aims to increase **fruit set percentage, number of nuts per panicle**, and **nut weight** while maintaining kernel quality.
- **Yield components** such as panicle density, bisexual flower ratio, and fruit retention are used as indirect selection indices.

4.2.2 Improvement of Kernel Quality

- High kernel recovery (≥ 30 %) and desirable **grade size (W180 – W240)** are prioritized.
- Breeding also targets **bright, uniform color, low tannin content**, and **crisp texture** suitable for export markets.
- Standardization of biochemical composition (fatty-acid ratio, sugar profile) supports diversification into confectionery and dairy analogues.

4.2.3 Early and Regular Bearing

» Traditional seedlings begin bearing at 4 – 5 years; modern varieties aim for bearing in 2 – 3 years.

» Regular bearing palms with **stable annual production** are favored for commercial orchards and intercropping systems.

4.2.4 Compact Canopy and High-Density Planting

» Selection for **compact or semi-dwarf trees** with strong central leaders allows planting densities up to **400 – 600 trees ha^{-1}**.

» Compact canopy architecture facilitates mechanical harvesting and reduces competition among trees.

4.2.5 Resistance to Major Pests and Diseases

» The **tea mosquito bug (*Helopeltis antonii*)** is the most serious pest, causing up to 80 % yield loss. Breeding seeks field tolerance via thicker epicuticular wax, compact canopy, and synchronized flushing.

» Resistance to **stem and root borer (*Plocaederus ferrugineus*)** and tolerance to **anthracnose (*Colletotrichum gloeosporioides*)** are equally critical.

4.2.6 Adaptation to Abiotic Stress

» Development of varieties tolerant to **moisture stress, salinity**, and **temperature extremes**.

» Physiological traits such as **leaf water potential, stomatal conductance**, and **wax load** guide selection in stress-prone environments.

4.2.7 High Quality Cashew Apple and By-Product Utilization

» Selection of clones with **high-juice yield, TSS > 14 °Brix**, and **low astringency** supports juice and wine industries.

» High-CNSL content shells are targeted for industrial applications.

4.3 Genetic Resources and Germplasm in Cashew

4.3.1 Origin and Distribution

Cashew originates from northeastern Brazil and was dispersed to tropical Asia and Africa in the 16th century. It exhibits wide genetic diversity across its introduced range, forming landraces adapted to local conditions, such as *Vengurla*, *Goa 1*, *Ullal*, and *Bhaskara* in India.

4.3.2 Germplasm Collections and Conservation

» The **ICAR–Directorate of Cashew Research (DCR), Puttur (Karnataka)** maintains the National Cashew Gene Bank with over **500 accessions** representing India and abroad.

» Regional stations (Madakkathara, Vengurla, Bhubaneswar, Madurai) conserve local ecotypes.

» Ex situ conservation in field gene banks is supplemented with **cryostorage of embryos** and **in vitro shoot cultures**.

» Characterization descriptors include plant height, branching pattern, leaf morphology, nut size, shelling percentage, and phenology.

4.3.3 Classification of Cashew Types

Cashew populations are categorized based on growth habit, nut size, and apple characteristics:

S. No.	Type/Class	Characteristic Features
1	**Tall**	Vigourous (8 – 12 m); wide canopy; late bearing; large nuts; long productive life (30 – 40 years)
2	**Dwarf / Semi-Dwarf**	Compact (≤ 6 m); early bearing (2 – 3 years); suitable for high-density planting
3	**High-Juice Apple Type**	Thick, fleshy pedicel; high TSS; used for beverage production
4	**High-CNSL Type**	Thick shell with ≥ 25 % CNSL; used in chemical industry

4.4 Floral Biology and Pollination Behaviour

Cashew trees flower during the dry season. The inflorescence is a **terminal panicle** containing **bisexual and male flowers** in a ratio of 1 : 6 to 1 : 10.

» **Flowering Period:** December – March in peninsular India.

» **Anthesis:** Occurs in early morning (6 – 9 a.m.).

» **Pollination:** Mostly by insects (*Apis cerana*, *Thrips spp.*, *Scelionidae* wasps).

» **Pollen Viability:** About 48 h under ambient conditions.

» **Fertilization:** Occurs 12 – 18 h after pollination; fruit matures in ~60 days.

The reproductive biology is essential for hybridization because **asynchronous flowering** and **low fruit set (≈ 5 %)** complicate controlled pollination.

4.5 Breeding and Hybridization Methods in Cashew

4.5.1 Selection Breeding

Early improvement relied on **clonal selection** of superior trees from heterogeneous seedling populations. Selection criteria include:

» ≥ 8 kg nuts $tree^{-1}$ $year^{-1}$.

» Kernel recovery ≥ 30 %.

» Uniform nut size and apple quality.

» Resistance to pests and diseases. Selected clones are propagated by **softwood grafting**, ensuring true-to-type multiplication.

4.5.2 Hybridization Breeding

a. Objectives of Hybridization

» Combine **high yield** with **compact canopy** and **early bearing**.

- Introduce **Helopeltis tolerance** and **disease resistance** into elite genotypes.
- Improve **kernel and apple quality** simultaneously.

b. Controlled Pollination Procedure

1. Parent Selection

Choose genetically diverse parents based on complementary traits (e.g., Vengurla 4 × Bhaskara).

2. Emasculation

Remove anthers from bisexual flowers at bud stage before dehiscence.

3. Bagging

Cover panicles with fine-mesh muslin bags to prevent foreign pollen contamination.

4. Pollen Collection and Storage

Collect pollen from freshly dehisced male flowers; store at 4 °C in desiccated vials.

5. Pollination

Apply pollen gently using a camel-hair brush when stigmas are glossy and receptive.

6. Tagging and Record Keeping

Label each cross with date, parentage, and cross number.

7. Fruit Harvesting and Seedling Raising

Harvest mature nuts, shell carefully, and germinate kernels in polybags for progeny testing.

c. Clonal Selection from Hybrids

Progenies are evaluated for yield, nut size, kernel recovery, and pest tolerance for at least five years before recommending promising clones for release.

4.6 Released Varieties and Achievements in Cashew

S. No.	Variety/Hybrid Name	Parentage/Type	Salient Features
1	**Vengurla 1**	Selection from Goa local	Early bearing; 9 kg nuts tree^{-1} yr^{-1}; kernel recovery 30 %
2	**Vengurla 4**	Hybrid (Vengurla 1 × Vengurla 3)	Compact canopy; high yield (12 kg nuts tree^{-1} yr^{-1})
3	**BPP 8 (Bhaskara)**	Selection from Andhra Pradesh germplasm	Tolerant to Helopeltis; kernel recovery 31 %
4	**NRCC Selection 2 (Ullal-3)**	Local selection	High apple TSS (14 °Brix); dual purpose nut and juice type
5	**Dhana**	Hybrid (Goa 11 × Vengurla 4)	Regular bearing; suitable for high-density planting
6	**Madakkathara 1**	Recurrent selection line	Resistant to stem and root borer; kernel recovery 32 %

Achievements include:

- Increase of average productivity in India from **400 kg to 1,200 kg raw nuts ha^{-1}**.
- Standardization of **softwood grafting** for mass propagation.
- Establishment of **clonal seed orchards** for true-to-type planting material.

4.7 Modern Advances in Cashew Breeding

1. **Molecular Characterization**
   - RAPD, ISSR, and SSR markers assess genetic diversity and identify duplicates in germplasm.
   - DNA fingerprints for released varieties (Vengurla 4, Bhaskara) ensure varietal authenticity.
2. **Marker-Assisted Selection**
   - Markers linked to **Helopeltis tolerance** and **flowering duration** are under development to shorten selection cycles.
3. **Tissue Culture and Micrografting**
   - In vitro shoot culture of juvenile shoots and **micrografting on seedling rootstocks** aid rapid clonal multiplication.
4. **Embryo Rescue and Hybrid Stabilization**
   - Applied in wide crosses (e.g., Goa 1 × Bhaskara) to recover low-viability hybrids.
5. **Genomic Resources**
   - Draft genome assembly (~350 Mb) enables identification of genes related to flowering and resin biosynthesis.
6. **High-Throughput Phenotyping**
   - Drone-based imaging quantifies canopy volume, flowering intensity, and fruit set, improving precision in selection.

5 – Coffee (*Coffea spp.*)

5.1 Introduction

Coffee, derived from the genus *Coffea* (family **Rubiaceae**), is among the world's most valuable plantation crops, second only to petroleum in international trade value. It serves as a vital export commodity for over 70 tropical countries and provides direct livelihood to more than 25 million farming families globally.

India ranks among the top ten coffee producers, contributing approximately **3 % of global production**, with the majority exported. The principal coffee-growing regions of India include **Karnataka, Kerala, and Tamil Nadu**, collectively known as the **traditional coffee tracts** of southern India, while **Andhra Pradesh, Odisha, and the North-Eastern states** represent expanding non-traditional areas.

The crop is primarily cultivated under **shade-grown, eco-sustainable systems** that conserve biodiversity and support wildlife corridors. In India, coffee is the backbone of hill-economies, generating both domestic employment and export income.

Botanical Overview

The genus *Coffea* includes over **120 species**, but only two are economically significant:

1. **Arabica coffee (*Coffea arabica* L.)**: fine quality beverage; self-pollinated; diploid (2n = 44).
2. **Robusta coffee (*Coffea canephora* Pierre ex Froehner)**: higher yield, disease tolerance, cross-pollinated, diploid (2n = 22).

Arabica originated in the highlands of Ethiopia, while Robusta originated in the Congo Basin. Both species were introduced to India in the 17th–18th centuries.

Economic Importance

- » Coffee forms a **major export earner**, contributing over ₹8,000 crores annually to India's economy.
- » The beans are valued for their **unique aroma, flavour, and caffeine content (0.8–2.5 %)**.
- » Coffee plantations maintain **rich shade-tree ecosystems**, enhancing soil fertility and carbon sequestration.

The **Coffee Board of India** coordinates all activities related to production, processing, breeding, and marketing.

5.2 Breeding Objectives in Coffee

The ultimate goal of coffee breeding is to develop cultivars that combine **high yield, superior cup quality, and resistance to diseases and pests**, along with adaptability to changing climatic conditions.

5.2.1 Enhancement of Yield and Productivity

- » Develop high-yielding, short-statured cultivars capable of producing ≥ 2.5 t clean coffee ha^{-1} under rainfed or irrigated systems.
- » Improve yield components such as **fruit set, berry size, branch productivity**, and **fruit-to-seed conversion ratio.**
- » Incorporate drought and temperature resilience for consistent performance under variable climates.

5.2.2 Improvement of Cup Quality

- » Enhance biochemical composition influencing flavour, aroma, and acidity, especially **chlorogenic acids, sucrose, and volatile aromatics.**
- » Maintain species-specific flavour integrity (Arabica's mild acidity; Robusta's strong body).
- » Avoid off-flavours caused by physiological stress or bean defects.

5.2.3 Resistance to Major Diseases and Pests

- Develop durable resistance to **coffee leaf rust (*Hemileia vastatrix*)**, **berry borer (*Hypothenemus hampei*)**, and **coffee stem borer (*Xylotrechus quadripes*)**.
- Breeding incorporates **vertical and horizontal resistance**, pyramiding genes from wild relatives.

5.2.4 Drought and Abiotic Stress Tolerance

- Selection for **osmotic adjustment, wax deposition**, and **leaf rolling ability**.
- Development of drought-tolerant rootstocks for grafted combinations.

5.2.5 Compact Growth and Early Bearing

- Breeding for compact, low-stature types reduces labour and enables higher-density planting (up to 6,000 plants ha^{-1}).
- Early bearing ensures economic returns within 2–3 years.

5.2.6 Quality Consistency under Climate Variability

- Varieties must maintain bean size, density, and biochemical quality under elevated temperature, CO_2, and rainfall variability.

5.3 Genetic Resources and Germplasm in Coffee

5.3.1 Origin and Distribution

Arabica originated in **Ethiopia**, while *Robusta* originated in **Congo Basin (Central Africa)**. Wild *Coffea* species occur across Africa and Madagascar, offering a valuable gene pool for disease resistance and stress tolerance.

5.3.2 Germplasm Conservation

- The **ICAR–Central Coffee Research Institute (CCRI)** at Chikmagalur, Karnataka, houses the **National Coffee Germplasm Repository** with over **3,500 accessions** of *C. arabica*, *C. canephora*, and allied species.
- Exotic introductions include accessions from Ethiopia, Kenya, Angola, and Costa Rica.
- **Wild relatives conserved:** *C. liberica*, *C. eugenioides*, *C. racemosa*, *C. kapakata*, *C. travancorensis* (endemic to Western Ghats).

5.3.3 Germplasm Characterization

Traits evaluated include:

- Morphological: leaf size, internode length, branching pattern, berry size.
- Agronomic: yield, bearing habit, canopy spread.
- Biochemical: caffeine content, total soluble solids, sugar-acid ratio.
- Resistance: rust, berry borer, stem borer, drought stress.

5.4 Floral Biology and Pollination Behaviour

5.4.1 Flowering Pattern

» Coffee bears **axillary cymes** containing 4–6 white, fragrant flowers.

» **Arabica:** self-pollinated; 95 % autogamy.

» **Robusta:** cross-pollinated; protogynous with insect pollination (*Apis cerana*, *Trigona* spp.).

» Flowering is **triggered by pre-monsoon showers** following a dry spell.

» Anthesis occurs early morning (6 – 9 a.m.), and stigma remains receptive for ~48 h.

5.4.2 Fruit Development

» Fertilization occurs within 24 h after pollination.

» Berries mature in 6 – 8 months, depending on climate.

» Fruit consists of pericarp (outer skin), mucilage, parchment, and two seeds (beans).

5.5 Breeding and Hybridization in Coffee

Breeding strategies differ between **Arabica (self-pollinated)** and **Robusta (cross-pollinated)** species, but both aim to combine yield, quality, and resistance.

5.5.1 Selection Breeding

» **Mass Selection:** Superior trees identified in plantations based on yield, uniform ripening, and disease tolerance.

» **Progeny Testing:** Selected trees evaluated for genetic stability over multiple years.

» **Examples:** *Kent's Arabica*, *Coorg Arabica Selection (Cauvery)*, *Sln.9*, *Sln.13*.

5.5.2 Hybridization Breeding in Arabica

a. Objectives

» Combine *C. arabica*'s beverage quality with resistance from *C. canephora* or *C. liberica*.

» Introduce tolerance to rust, drought, and pests.

b. Steps in Controlled Hybridization

1. **Parent Selection:** Choose genetically contrasting parents; e.g., *S.795* × *HDT* (Hybrid of Timor).
2. **Emasculation:** Remove anthers from flower buds before anthesis.
3. **Bagging:** Cover flowers with butter-paper bags to prevent foreign pollen.
4. **Pollination:** Apply pollen manually during stigma receptivity.
5. **Labelling:** Record date, parentage, cross number.
6. **Fruit Harvesting:** Collect mature berries after 7–8 months; extract seeds for nursery planting.
7. **Evaluation:** Observe progeny for yield, rust tolerance, and bean quality across seasons.

c. **Hybrid Examples**

» ***Sln.9 (Selection 9):*** (*Tafarikela* × *Hibrido de Timor*) – high yield, rust resistant.

» ***Sln.13:*** (*Chandragiri* × *Hibrido de Timor*) – compact, good quality, rust resistant.

5.5.3 Hybridization Breeding in Robusta

a. **Objectives**

» Improve yield, cup quality, and resistance to stem borer and drought.

» Combine robustness with a compact canopy.

b. **Procedure**

1. Select elite female and male parents.
2. Isolate inflorescences by bagging before anthesis.
3. Collect pollen from the male parent and apply it using a camel-hair brush.
4. Maintain an insect-proof bag for 2 weeks post-pollination.
5. Tag and record cross details.

c. **Example Hybrids**

» ***Sln.274:*** Cross between Congensis × Ceylon Hybrid – high yield, drought tolerant.

» ***CxR (Congensis* × *Robusta):*** Early bearing, uniform beans, drought hardy.

5.6 Mutation and Polyploid Breeding

» **Gamma-ray-induced mutations** used to create rust-resistant lines in *C. arabica.*

» Polyploidization (via colchicine) generates autotetraploids with larger leaves and berries, aiding interspecific hybridization.

5.7 Backcross and Recurrent Selection

» Used in *Arabica* to recover beverage quality after introgression of rust-resistance genes from wild species (e.g., *Hibrido de Timor*).

» Recurrent selection cycles enhance general combining ability in *Robusta* populations, improving yield and uniformity.

5.8 Released Varieties and Achievements in Coffee

5.8.1 Arabica Varieties

S. No.	Variety Name	Parentage/Type	Salient Features
1	**Kent's Arabica**	Selection from Typica	First rust-tolerant selection; good cup quality
2	**S.795**	(*Kent* × *S.288*)	High yield; partial rust tolerance; standard commercial type
3	**Sln.9**	(*Tafarikela* × *Hibrido de Timor*)	Compact, high-yielding, rust-resistant

4	**Chandragiri**	Selection from Catimor	High yield; uniform ripening; bold beans
5	**Sln.13**	(*Chandragiri × HDT*)	Excellent cup quality; rust tolerant; compact canopy

5.8.2 Robusta Varieties

S. No.	Variety Name	Parentage/Type	Salient Features
1	**S.274**	(*Congensis × Ceylon Hybrid*)	High yield; drought tolerant
2	**CxR**	(*Congensis × Robusta*)	Compact; uniform bean size; early bearing
3	**Sln.10**	Recurrent selection line	Improved berry set and rust tolerance
4	**Sln.12**	(*CxR × S.274*)	High oil content; balanced flavour; drought tolerant

5.9 Achievements of Coffee Breeding in India

1. **Yield Enhancement**
   - National average yield increased from **400 kg ha^{-1} (1950s)** to **>1100 kg ha^{-1}** through improved cultivars and management.
2. **Rust Resistance**
   - *Hibrido de Timor (HDT)* introduced durable rust resistance into multiple cultivars.
   - Pyramiding of resistance genes (SH1–SH9) achieved in *Sln.9* and *Sln.13.*
3. **Quality Improvement**
   - Varieties like *S.795* and *Sln.13* set global standards for flavour and balance.
   - Specialized selections (e.g., *Bourbon*, *Chandragiri*) cater to premium specialty markets.
4. **Compact and Early-Bearing Types**
   - Semi-dwarf cultivars allow denser planting, improving land-use efficiency.
5. **Expansion to Non-Traditional Areas**
   - Varieties adapted for Eastern Ghats and NE India (*Sln.13, CxR*) expanded coffee beyond the southern hills.
6. **Hybrid Seed Gardens**
   - Established at CCRI, Balehonnur, and regional stations for true-to-type hybrid seed production.

5.10 Modern Advances in Coffee Breeding

5.10.1 Molecular Marker and Genomic Tools

- SSR and SNP markers used for **genetic mapping, varietal identification**, and **rust-resistance gene tracking**.

» Genome sequencing of *C. canephora* and *C. arabica* revealed genes controlling **caffeine biosynthesis**, **flavour pathways**, and **disease response**.

» Marker-assisted selection (MAS) for *H. vastatrix* resistance (SH3 locus) is now integrated into breeding pipelines.

5.10.2 Somatic Embryogenesis and Tissue Culture

» Somatic embryogenesis from leaf and zygotic tissues enables **rapid clonal multiplication** of elite hybrids like *Sln.9*.

» Micropropagation conserves rare genotypes and accelerates field deployment.

5.10.3 Grafting and Rootstock Breeding

» Rust-resistant *Robusta* or *Liberica* used as rootstocks for *Arabica* scions in rust-prone regions.

» Grafting improves soil-borne disease tolerance and drought resilience.

5.10.4 Cytogenetics and Ploidy Manipulation

» Controlled polyploidization allows hybridization between diploid *Robusta* and tetraploid *Arabica.*

» Synthetic hybrids like *Arabusta (Arabica × Canephora)* combine quality with vigour.

5.10.5 Transcriptomics and Bioinformatics

» RNA-seq analysis identifies gene networks associated with **bean maturation**, **flavour compound synthesis**, and **stress adaptation**.

» Differential expression studies link specific genes (e.g., PAL, COMT, CCoAOMT) to aroma precursors.

5.10.6 Smart Phenotyping and AI-Based Selection

» Drone imagery and multispectral sensors evaluate canopy vigour, flowering density, and berry load.

» AI models predict yield, rust incidence, and bean maturity with high accuracy, enabling **data-driven selection**.

5.10.7 Climate-Resilient Breeding Initiatives

» Integrated breeding under the **Coffee Adaptation to Climate Change (CACC)** program develops lines tolerant to heat (>32 °C) and irregular rainfall.

» Quantitative trait loci (QTLs) mapped for **water-use efficiency** and **photosynthetic stability**.

5.11 Integration of Conventional and Molecular Approaches

Modern breeding programs combine field-based selection with molecular precision to enhance efficiency:

1. **Conventional Stage:** Hybridization, progeny evaluation, and multi-location trials.

2. **Molecular Stage:** Genomic screening, marker validation, and trait introgression.
3. **Biotechnological Stage:** In vitro conservation, somatic embryogenesis, and genetic transformation research for rust resistance.
4. **Digital Stage:** GIS mapping and predictive analytics for site-specific varietal recommendations.

6 – Tea (*Camellia sinensis* (L.) O. Kuntze) And Cocoa (*Theobroma cacao* L.)

6.1 TEA (*Camellia sinensis* (L.) O. Kuntze)

6.1.1 Introduction

Tea is the world's most consumed non-alcoholic beverage after water and one of the oldest plantation crops of the humid tropics and subtropics. It belongs to the family **Theaceae**, genus **Camellia**, and is cultivated for its **young, tender shoots ("flush")**, which are processed into black, green, oolong, or white teas.

Globally, tea is grown on more than **5 million ha** in over 40 countries, with **China, India, Kenya, Sri Lanka, Vietnam, and Indonesia** being major producers. In India, the principal tea-growing regions are **Assam, West Bengal, Tamil Nadu, Kerala, and Karnataka**, supporting nearly 1.2 million workers directly and several million indirectly. The tea plant is a **woody evergreen shrub (1 – 3 m tall)** with deep roots and a long productive life (40 – 60 years). Its economic importance stems not only from beverage production but also from employment generation, soil conservation on hilly slopes, and carbon sequestration through perennial canopy cover.

6.1.2 Taxonomy and Botanical Classification

Two main cultivated varieties exist:

S. No.	Variety	Principal Characteristics
1	**Camellia sinensis var. sinensis** (China type)	Small leaves (5 – 8 cm), cold tolerant, suitable for green tea; compact habit, quality flavour.
2	**Camellia sinensis var. assamica** (Assam type)	Large leaves (15 – 25 cm), high yield, low altitude adaptation; used for black tea production.
3	**Assamica × Sinensis hybrids**	Intermediate characters; combine high yield with desirable liquor quality.

6.1.3 Breeding Objectives in Tea

1. **Increase Leaf Yield per Bush**
   - » Select for vigourous shoot growth, high plucking density, and frequent flush cycles.
   - » Improve **shoot replacement rate (SRR)** and **leaf-area index (LAI)**.
2. **Enhance Quality Attributes**
   - » Improve biochemical constituents determining liquor quality: **polyphenols (25 – 30 %), catechins, theaflavins, thearubigins, amino acids**, and **volatile aroma compounds**.

3. **Develop Resistance to Pests and Diseases**
   » Resistance to blister blight (***Exobasidium vexans***), red rust, and root rot.
   » Tolerance to major insects such as tea mosquito bug (***Helopeltis theivora***), red spider mite, and looper caterpillars.
4. **Adaptation to Abiotic Stresses**
   » Tolerance to drought, cold, and high soil acidity; selection for efficient nutrient uptake at low pH.
5. **Morphological Improvement**
   » Compact, erect bush form suited for mechanical plucking; uniform canopy spread.
6. **Clonal Uniformity and Longevity**
   » Stable productivity over decades with resistance to replant failure and canopy degeneration.

6.1.4 Genetic Resources and Germplasm

a. Centers of Origin and Distribution

Tea is indigenous to the **Himalayan foothills of North-East India, China, and Myanmar**. Wild tea relatives occur naturally in Yunnan (China) and Assam (India).

b. Germplasm Collections and Conservation

» The **Tocklai Tea Research Institute (TTRI)**, Jorhat (Assam), maintains the national germplasm with over **3,000 accessions**.
» The **UPASI Tea Research Foundation**, Valparai (Tamil Nadu), conserves South Indian types.
» Characterization uses morphological, biochemical, and molecular markers (SSR, AFLP).
» In vitro conservation and cryopreservation protocols exist for elite clones.

6.1.5 Floral Biology and Pollination

» Tea flowers are **bisexual and self-incompatible**, ensuring cross-pollination.
» Anthesis occurs mainly in early morning hours (4 – 7 a.m.).
» Pollination agents include **bees, flies, and wind**.
» Fertilization takes place within 24 h; seeds mature in 8 – 10 months.
» The long juvenile phase (4 – 5 years) complicates breeding and evaluation.

6.1.6 Breeding and Hybridization Methods

a. Selection Breeding

Initial improvement relied on the selection of elite bushes from seedling populations.
Selection criteria include:

» Shoot density ≥ 60 shoots m^{-2}.
» 35 – 40 flushes $year^{-1}$.

- Polyphenol content ≥ 25 %. Examples: *Tocklai 1, Tocklai 17, UPASI 3, UPASI 9.*

b. Hybridization Breeding

Hybridization combines the high yield of *assamica* types with the quality of *sinensis* types.

Steps in Controlled Crossing

1. Select male and female parents with contrasting traits.
2. Emasculate flower buds one day before anthesis.
3. Bag inflorescences to prevent foreign pollen entry.
4. Collect pollen from the male parent at anthesis and apply it manually to the stigmas of female flowers.
5. Tag cross and record parentage.
6. Harvest seeds after 8 – 10 months and raise progeny in a nursery for field evaluation.

Hybrid Types

- *TV 18* × *UPASI 9* – high yield and flavour.
- *UPASI 3* × *Assam Jat* – vigour and blister blight tolerance.

c. Clonal Selection

Tea's heterozygosity makes clonal propagation essential. Outstanding bushes are cloned through **nodal cuttings** or **micropropagation** to maintain uniformity. Clones show 25 – 40 % yield advantage over seedling progenies.

6.1.7 Released Varieties and Achievements

S. No.	Clone/Variety Name	Origin or Cross	Salient Features
1	**TV 1**	Selection from Assam seedling population	High yield (> 3 000 kg made tea ha^{-1}); moderate blister-blight tolerance
2	**TV 9**	Assamica × Sinensis hybrid	Excellent flavour and cup quality; vigourous growth
3	**UPASI 9**	South Indian selection	Drought tolerant; uniform flush pattern
4	**B/6/61**	Clonal selection from China type	Cold tolerant; suited for high elevations
5	**TV 26**	Tocklai hybrid	Resistance to root rot and leaf blight

Major Achievements

- Replacement of heterogeneous seedling populations by elite clones.
- National average yield increased from 600 kg to 1,800 kg made tea ha^{-1}.
- Development of blister-blight-tolerant hybrids and mechanization-ready bush forms.
- Establishment of DNA fingerprints for varietal authentication.

6.1.8 Modern Advances in Tea Breeding

1. **Molecular Markers and Genomics**
   » SSR and AFLP markers map blister-blight resistance and quality QTLs.
   » Draft genome of *C. sinensis* (~ 3.0 Gb) enables gene-based selection for polyphenol biosynthesis.
2. **Tissue Culture and Micropropagation**
   » Somatic embryogenesis from axillary buds and leaf segments for rapid clonal multiplication.
   » Cryopreservation for germplasm security.
3. **Mutation Breeding**
   » Gamma-ray-induced mutants like *UPASI 27* show higher catechin content.
4. **Marker-Assisted Selection (MAS)**
   » Used for early screening of clones for blister blight resistance and flavour genes.
5. **Phenomic and Digital Breeding Tools**
   » Drone imaging assesses canopy health and flush density.
   » Spectral indices correlate with chlorophyll and polyphenol levels, speeding field screening.

6.2 COCOA (*Theobroma cacao* L.)

6.2.1 Introduction

Cocoa, the source of **chocolate**, belongs to the family **Malvaceae** (formerly Sterculiaceae) and is native to the **humid rainforests of Central and South America**. It is cultivated for its **fermented and dried seeds (cocoa beans)**, which are processed into cocoa butter, powder, and chocolate liquor.

Globally, cocoa occupies about **12 million ha**, mainly in **West Africa (Ghana, Côte d'Ivoire, Nigeria, Cameroon)**, Southeast Asia (Indonesia, Malaysia, Philippines), and Latin America (Brazil, Ecuador). India has emerged as a small but growing producer, cultivating ≈ 100,000 ha under coconut and arecanut intercropping systems in Kerala, Karnataka, Tamil Nadu, and Andhra Pradesh.

Cocoa trees are small evergreens (3 – 8 m height) with glossy leaves and cauliflorous flowers borne on the trunk and older branches. Each fruit (pod) contains 30 – 40 seeds embedded in sweet pulp. The crop is strictly tropical, requiring 20 – 30 °C temperature and 1,500 – 2,500 mm rainfall. Because it is shade-loving, it fits perfectly within multi-storey plantations.

6.2.2 Market Classes and Genetic Groups

S. No.	Group	Breeding Relevance
1	**Criollo**	Superior fine flavour/aroma; low yield; disease-susceptible; niche premium.
2	**Forastero (Amazonian/Amelonado)**	High yield, hardy; coarser flavour; disease tolerance; global workhorse.

3	**Trinitario**	Natural **Criollo × Forastero** hybrid; balances **quality × yield**; foundation for many modern clones.

6.2.3 Breeding Objectives

- » **Increase bean yield and uniformity:** Reduce **pod index** (pods per kg dry beans), improve bean count per pod and 100-bean weight.
- » **Improve butter content and flavour:** Target ≥54% fat, strong flavour precursors, low bitterness/astringency; consistent fermentation performance.
- » **Disease and pest resistance:** Especially **black pod rot, stem canker, vascular streak dieback,** and **mirids.**
- » **Adaptation & shade efficiency:** Productivity under 30–50% shade, tolerance to temporary moisture stress.
- » **Compact crown & precocity:** Suit high-density agroforestry (800–1,000 trees ha^{-1}); fruiting at 24–30 months.
- » **Clonal stability:** High graft success, strong chupon–jorquette framework.

6.2.4 Genetic Resources and Conservation

- » **Repositories: ICAR–CPCRI** (Kasaragod & Vittal) and **ICG-Trinidad** maintain global reference clones (e.g., ICS, T, PA series).
- » **Characterization descriptors:** Pod morphology, ridge form, anthocyanin intensity, bean traits, butter %, disease reaction; **SSR/SNP** fingerprints secure identity.
- » **Conservation:** Field banks, **in vitro** embryos, and **cryopreservation** of zygotic embryos.

6.2.5 Floral Biology and Pollination

- » **Flowers:** Small, pinkish, **cauliflorous**; hermaphroditic yet with **sporophytic self-incompatibility**, promoting outcrossing.
- » **Pollinators:** Tiny midges (*Forcipomyia* spp.); natural fruit set 1–3%.
- » **Pods:** Mature in 5–6 months; 30–40 seeds in mucilage.
- » **Breeding implication:** Low natural set necessitates **hand pollination** for controlled crosses and accurate pedigrees.

6.2.6 Breeding and Hybridization Methodology

a. **Clonal selection and fixation:** Identify elite trees for yield, pod index, butter %, disease tolerance, and bean uniformity; fix by **patch budding/softwood grafting** on robust seedling rootstocks; establish **clonal trials** across sites to verify stability.

b. **Controlled hybridization (Criollo × Forastero → Trinitario-type)**

1. **Parent choice:** Quality donor (Criollo/Trinitario) × vigour/disease-tolerant donor (Forastero).
2. **Emasculation & isolation:** Remove anthers pre-anthesis; bag cushions to prevent stray pollen.

3. **Pollen management:** Collect fresh pollen pre-dawn; test viability; apply with fine brush during stigma receptivity.
4. **Tagging & records:** Cross code, parentage, date; maintain cushion-wise maps.
5. **Raising progenies:** Germinate seeds quickly (recalcitrant); nursery culling for vigour; field evaluation under shade.

c. **Recurrent selection and reciprocal recurrent selection:** Cycle within **Upper Amazon × Trinitario** pools to accumulate favorable alleles for **yield × black-pod tolerance × butter %.**

d. **Backcrossing:** Introgress fine-flavour (Criollo) while recovering **Forastero** field robustness; maintain quality without sacrificing yield.

e. **Marker-assisted hybridity checks:** Use SSR panels to confirm true crosses early, preventing mislabeling in long trials.

6.2.7 Evaluation, Fermentation Quality, and Release

» **Field traits:** Pods tree^{-1} year^{-1}, pod index, cherelle wilt %, chupon vigour, branching angle, canopy efficiency under shade.

» **Bean/processing traits:** 100-bean weight, butter %, cut-test slate %, fermentation index, flavour notes after standard fermentation–drying–roasting protocols.

- **Disease screens:** Hot-spot field nurseries; detached-pod inoculation for **Phytophthora**; score canker lesions.
- **Release:** Only clones/hybrids with **multi-site stability** for yield and fermentation quality, plus disease tolerance, are recommended for farmer deployment.

6.2.8 Released Clones/Hybrids and Achievements

S. No.	Clone/ Hybrid	Parentage/Type	Salient Features
1	VTL C-1	Trinitario selection	Early bearing; pod index ≤ 18; good butter %; field tolerance to black pod.
2	VTL C-2	Upper Amazon derivative	High yield under coconut shade; strong chupon growth; uniform beans.
3	ICG-T ICS 95	International reference clone	Fine flavour; standard in quality trials; moderate disease tolerance.
4	Trinitario Hybrid-A	(Criollo donor × Upper Amazon)	Balanced flavour × yield; low slate %; good fermentability.
5	Forastero Elite-F	Amazonian composite	Hardy; high pod load; acceptable liquor; suits intensive systems.

Program outcomes: Establishment of **true-to-type clonal orchards**, standardized **patch budding** protocols (>85% success), increased smallholder yields through **shade-optimized agroforestry**, and deployment of **black-pod-tolerant** materials in high-rainfall belts.

6.2.9 Modern Tools Integrated with Cocoa Hybridization

» **Marker-assisted selection:** SSR/SNP tags for **black pod** resistance and self-incompatibility alleles accelerate early screening.

» **Genomics & transcriptomics:** Candidate genes for **flavour precursors, fat biosynthesis**, and **defense** guide parent choice and selection mapping.

» **Somatic embryogenesis:** From staminodes/immature zygotic embryos enables rapid clonal multiplication of elite, otherwise difficult-to-root clones.

» **High-throughput phenotyping:** Imaging-based pod counts, canopy vigour indices, and digital cut-test analytics improve objectivity.

» **Quality analytics:** NIR and GC–MS pipelines to predict flavour and butter % from bean spectra, reducing the reliance on time-consuming fermentation–roast cycles during early selection.

6.2.10 Integrated Seed-to-Clone Pipeline (Operational SOP)

1. **Cross design:** Define **yield × quality × disease** targets; choose contrasting but combinable parents; plan reciprocal sets.
2. **Controlled pollination:** Execute emasculation, bagging, timed pollination; maintain stringent field logs.
3. **Nursery & culling:** Germinate immediately; remove weak/aberrant seedlings; barcode trays.
4. **Early tests:** Marker-based hybridity; basic disease screens; preliminary growth vigour scores.
5. **Field testing under shade:** Randomized blocks under standardized shade; record pod load, pod index, disease incidence.
6. **Processing trials:** Standard ferment–dry–roast; conduct **sensory panels** and chemistry (butter %, flavour volatiles).
7. **Clonal fixation:** Bud top performing individuals; establish **clonal trials** across agro-ecologies.
8. **Release & multiplication:** Certify nucleus blocks; distribute budwood; implement **trueness-to-type** audits.

7. Rubber (*Hevea brasiliensis* Muell. Arg.)

7.1 Introduction

Rubber, derived from the **para-rubber tree (*Hevea brasiliensis*)**, is the world's primary source of **natural rubber (NR)** – a critical industrial raw material for tyres, medical devices, footwear, and thousands of engineering products. The tree belongs to the family **Euphorbiaceae** and is native to the **Amazon Basin** of South America. Commercial plantations today are concentrated in the **tropical humid regions of Asia**, especially **Thailand, Indonesia, Malaysia, Vietnam, India, and Sri Lanka**, which together supply more than 90 % of global NR demand.

In India, rubber is primarily grown in **Kerala, Tamil Nadu**, and parts of **Karnataka, Assam, Tripura, and Meghalaya**, covering about **850,000 ha** with an average yield of ~ 1,800 kg ha^{-1} yr^{-1}. Rubber cultivation is not only a major **foreign-exchange earner** but also a **climate-resilient agro-industry** sustaining over a million smallholders, integrating carbon sequestration, erosion control, and livelihood security.

7.2 Botanical and Cytogenetic Background

- **Species:** *Hevea brasiliensis* Muell. Arg.
- **Family:** Euphorbiaceae.
- **Chromosome number:** 2n = 36.
- **Reproduction:** Monoecious, cross-pollinated (entomophilous).
- **Life cycle:** Juvenile phase 5–7 years; economic life 25–30 years.

7.2.1 Morphology

Rubber is a tall, straight tree (20 – 30 m) with a single stem and whorled branching. Leaves are trifoliate, deciduous during a short dry period. The latex is secreted by **laticiferous vessels** distributed in bark phloem, from which latex is tapped by spiral incision.

7.2.2 Reproductive Biology

Inflorescence is a **panicle** with numerous small unisexual flowers.

- **Male: female ratio** ≈ 100: 1.
- **Anthesis** – early morning (04:00 – 08:00 h).
- **Pollination:** By insects (thrips, midges, flies).
- **Fertilization:** Within 24 h of pollination; capsule matures ~ 120 days.
- **Fruit:** 3-seeded capsule; seeds recalcitrant with 6–8 months dormancy.

Knowledge of **flowering synchronization, pollen viability**, and **fruit-set dynamics** is essential for designing hybridization schedules.

7.3 Economic Importance

1. **Industrial relevance:** Natural rubber's high elasticity, tensile strength, and heat dispersion make it irreplaceable in tyres, medical gloves, conveyor belts, and engineering goods.
2. **Socio-economic value:** Employs > 13 million people globally; in India ~ 90 % of production comes from smallholdings.
3. **Environmental services:** Dense canopy enhances **carbon sequestration (≈ 20 t CO_2 ha^{-1} yr^{-1})**, stabilizes slopes, and enriches soil organic matter.
4. **Agroforestry potential:** Intercropping with pineapple, banana, cocoa, and pepper during immature phase diversifies income.

7.4 Breeding Objectives in Rubber

The perennial, cross-pollinated nature demands long-term, multi-trait objectives:

7.4.1 Yield and Latex Productivity

- Increase **latex yield > 2 500 kg ha^{-1} yr^{-1}** through higher **bark thickness, laticifer density, and latex flow duration.**
- Enhance **dry-rubber content (DRC)** > 35 %.
- Combine high yield with low incidence of tapping-panel dryness (TPD).

7.4.2 Timber Value

- Develop dual-purpose clones with high-quality timber (density 0.55 – 0.65 g cm^{-3}) for furniture and plywood.

7.4.3 Disease and Pest Resistance

- Resistance to **South American leaf blight (SALB – *Microcyclus ulei*), powdery mildew, *corynespora* leaf-fall**, and **abnormal leaf-fall.**
- Tolerance to **root rot** and **root-knot nematodes.**

7.4.4 Abiotic Stress Tolerance

- Drought and cold tolerance for expansion to **non-traditional regions** (NE India, Eastern Ghats).
- Salt and water-logging tolerance for coastal lowlands.

7.4.5 Growth and Architecture

- Early girthing for tapping at 5 – 6 years.
- Uniform cylindrical trunks; narrow branching angle for a stable canopy.

7.4.6 Clonal Stability and Latex Quality

- Consistent performance across environments; high **molecular-weight distribution** of rubber polymers for superior elasticity.

7.5 Genetic Resources and Germplasm

7.5.1 Origin and Natural Distribution

Hevea brasiliensis originates from the **Upper Amazon (Brazil–Peru–Bolivia)**. From only **22 seeds** exported to Kew Gardens in 1876, all Asian plantations trace their ancestry, creating a narrow genetic base.

7.5.2 Germplasm Collection and Conservation

S. No.	Repository / Location	Approx. Accessions	Remarks
1	**RRII – Rubber Research Institute of India, Kottayam**	> 1 000 clones	Primary national gene bank – wild & improved clones.
2	**IRRDB – International Rubber Research and Development Board network**	> 5 000 global entries	Exchange of wild Amazonian collections post-1975.
3	**Malaysia (RRIM)**	~ 3 000 accessions	Historic Amazon Expedition 1976 material.
4	**Thailand (RRIT)**	~ 2 500 clones	Germplasm for disease tolerance and timber.

7.5.3 Characterization Traits

- **Morphological:** leaf shape, bark thickness, crown spread.
- **Anatomical:** laticifer number per mm^2 bark, latex-flow rate.
- **Physiological:** photosynthetic rate, stomatal conductance.
- **Biochemical:** dry-rubber content, sucrose / inorganic P ratio in latex.
- **Molecular:** SSR, AFLP, and SNP fingerprints for diversity and clone identification.

7.6 Breeding Approaches and Hybridization Methodology

Because of long generation time (25 – 30 years), **stepwise multi-stage breeding** is followed.

7.6.1 Phase I – Clonal Selection

- Identify elite mother trees in estate populations.
- Clone by **budding (green / brown)** onto uniform rootstock.
- Evaluate yield, girth, latex properties for 10 – 12 years.
- Foundation clones (e.g., *Tjir 1, GT 1, RRII 105*) serve as parents.

7.6.2 Phase II – Hybridization Breeding

a. Objectives

- Combine **high yield × disease tolerance × early girthing**.
- Introduce new alleles from Amazon wild germplasm to widen diversity.

b. Controlled Pollination Procedure

1. **Parent selection:** contrasting performance (e.g., RRII 105 × PB 86).
2. **Bagging:** Immature inflorescences enclosed in muslin bags.
3. **Emasculation:** Manual removal of anthers one day before anthesis.

4. **Pollen collection & storage:** Collected 06:00–07:00 h; desiccated (30 % RH, 4 °C) for short-term use.
5. **Pollination:** Hand-transfer of pollen to receptive stigmas (7–10 a.m.).
6. **Tagging & record:** Cross ID, date, parentage.
7. **Seed harvest:** Capsules mature in ~ 120 days; seeds collected before dehiscence.
8. **Raising progenies:** Germinate immediately in polybags (nonspecific dormancy); evaluate for vigour and early girth.

c. **Seedling Evaluation**

» **Stage I:** Nursery (1 – 2 yrs) – vigour, leaf morphology.

» **Stage II:** Field immature phase (5 – 6 yrs) – girth increment, disease tolerance.

» **Stage III:** Tapping phase (≥ 7 yrs) – latex yield, quality. Superior seedlings are then cloned for comparative testing.

7.6.3 Phase III – Recurrent Selection

Repeated cycles of interspecific and interclonal crossing with periodic selection for combining ability. This broad-based scheme produced India's modern series (*RRII 105*, *RRII 414*, *RRII 430*).

7.7 Interspecific Hybridization

S. No.	Cross Combination	Purpose / Trait Introgressed
1	*H. brasiliensis* × *H. benthamiana*	Leaf-disease tolerance (SALB).
2	*H. brasiliensis* × *H. spruceana*	High latex viscosity and rubber quality.
3	*H. brasiliensis* × *H. guianensis*	Cold tolerance; wider ecological range.
4	Backcrosses to *H. brasiliensis*	Recover yield and bark thickness while retaining disease genes.

Embryo rescue and in vitro culture techniques overcome hybrid inviability and fruit abortion.

7.8 Evaluation of Hybrid Progenies

» Growth traits: Girth increment (> 2 cm yr^{-1}), tree height, bark thickness.

» Latex traits: Yield per tapping, DRC %, viscosity, coagulum character.

» Physiological indices: Sucrose / inorganic P ratio (> 5 indicates high metabolic activity).

» Statistical analysis: ANOVA and heritability estimates guide selection intensity.

7.9 Released Clones and Performance

S. No.	Clone Name	Parentage/Origin	Salient Features
1	**RRII 105**	*Tjir 1 × GT 1*	Benchmark Indian clone; early tapping (5 yrs); DRC > 35 %; wide adaptation.
2	**RRIM 600**	*Tjir 1 × PB 86* (Malaysia)	World standard; high yield; moderate wind resistance.
3	**PB 260**	*PB 5/51 × PB 49*	High yield > 2 500 kg ha^{-1}; good timber quality; Corynespora-tolerant.
4	**RRII 414**	*RRIM 600 × RRII 105*	Superior girth increment; TPD-tolerant; dual purpose (latex + timber).
5	**RRII 430**	*RRII 105 × PB 260*	Drought-hardy; fast growth; high rubber molecular weight.
6	**PB 235**	*PB 5/51 × PB 6/9*	High yield with Corynespora resistance; recommended for Malaysia & India.

7.10 Biochemical and Physiological Markers of Yield

» **Sucrose, inorganic phosphate, thiols, and pH** of latex used as indices of metabolic efficiency.

» High **sucrose: P ratio (> 4)** → strong latex regeneration.

» **High thiol content** → TPD resistance.

» **Clonal diagnosis** through latex enzyme profiling (peroxidase, acid phosphatase).

7.11 Clonal Propagation and Nursery Techniques

1. **Seedling rootstocks** raised from selected open-pollinated seed gardens.
2. **Budding:** Green budding (3–4 months) and brown budding (> 6 months) standardized with > 90 % success.
3. **Polybag plants:** Facilitate year-round planting.
4. **Clonal seed gardens:** Provide uniform rootstock seed supply.

7.12 Molecular and Biotechnological Advances

1. **Molecular Markers and Genomics**

» **Microsatellites (SSR)** and **SNPs** used for mapping yield and TPD resistance QTLs.

» Reference genome (~ 1.4 Gb) assembled by IRRDB consortium facilitates candidate-gene discovery for latex biosynthesis genes (*HbREF, HbSRPP*).

2. **Marker-Assisted Selection (MAS)**

» Linked markers for cold tolerance and drought indices accelerate screening in juvenile phase.

3. **Tissue Culture and Somatic Embryogenesis**
   » **Somatic embryos** from immature anthers or integuments enable rapid propagation of elite clones.
   » **Synthetic seed technology** explored for long-distance germplasm exchange.
4. **Molecular Diagnostics for TPD and Corynespora**
   » PCR kits detect pathogen DNA in asymptomatic leaves, enabling early sanitation.
5. **Genetic Transformation Research**
   » Agrobacterium-mediated transfer of *HbACO* antisense construct down-regulates ethylene oxidase, enhancing latex production. (Experimental stage.)
6. **Phenomic Platforms**
   » High-resolution LiDAR and drone imaging estimate canopy volume and girth increment remotely.

7.13 Achievements of Rubber Breeding

» Yield potential tripled from 600 → > 2 500 kg ha^{-1} yr^{-1}.
» Early-tapping clones reduced immature period from 8 → 5 years.
» Development of TPD-tolerant and Corynespora-resistant clones.
» Dual-purpose latex + timber clones broadened economic returns.
» Introduction of high-latitude tolerant genotypes enabling expansion to NE India and Odisha.
» Integration of **biochemical screening** and **molecular fingerprinting** in selection pipeline.

7.14 Integration of Conventional and Modern Approaches

S. No.	Approach	Contribution to Breeding Efficiency
1	Conventional hybridization	Base for variability creation and recombination.
2	Clonal selection	Rapid fixation and multiplication of elite recombinants.
3	Recurrent intercrossing	Accumulation of additive genetic effects.
4	Marker-assisted screening	Juvenile-stage selection for complex traits.
5	Tissue-culture propagation	Mass multiplication and germplasm exchange.
6	Latex biochemistry profiling	Rapid physiological prediction of yield potential.

Plantation Crops

(Coconut | Arecanut | Oil Palm | Cashew | Coffee | Tea | Cocoa | Rubber)

S. No.	Common Name	Botanical Name (Italicized)	Family	Primary Product / Bioactives	Major Breeding Objectives	Released Varieties / Achievements
1	Coconut	*Cocos nucifera* L.	Arecaceae	Copra oil, lauric acid	High copra yield, dwarf × tall hybrids, disease resistance	*Chowghat Orange Dwarf × West Coast Tall, Kera Sagara, Tiptur Tall*
2	Arecanut	*Areca catechu* L.	Arecaceae	Arecoline alkaloids	High nut yield, uniform fruit size, tolerance to yellow-leaf disease	*VTL-11, Sumangala, Siri, Mangala*
3	Oil Palm	*Elaeis guineensis* Jacq.	Arecaceae	Palm oil (rich in oleic acid)	High oil yield, compact canopy, short stature, tenera hybrids	*Tenera DxP Crosses, Godavari Gold, CARI-OP-1*
4	Cashew	*Anacardium occidentale* L.	Anacardiaceae	Kernel oil, phenolics	High nut yield, early flowering, tea mosquito bug resistance	*BPP-8, NRCC-1, Ullal-3, Vengurla 7*
5	Coffee	*Coffea arabica* L., *C. canephora* Pierre	Rubiaceae	Caffeine, chlorogenic acid	High bean yield, rust and berry-borer resistance, cup quality	*S.795, Chandragiri, Cauvery (Catimor hybrid)*
6	Tea	*Camellia sinensis* (L.) O. Kuntze	Theaceae	Caffeine, catechins	High yield, quality flavour, tolerance to blister blight	*TV-23, UPASI-9, Tocklai Clone TV-29*
7	Cocoa	*Theobroma cacao* L.	Malvaceae	Theobromine, cocoa butter	High bean yield, self-compatibility, black pod resistance	*NC-42, CC-37, VSD-Hyb-2, TNAU-Cocoa-1*
8	Rubber	*Hevea brasiliensis* (Muell.) Arg.	Euphorbiaceae	Latex (rubber hydrocarbons)	High latex volume, wind resistance, disease tolerance	*RRII 105, RRIM 600, PB 217, RRII 430*

Chapter - 2

Spice Crops

2.1 Black Pepper (*Piper nigrum* L.)

2.1.1 Introduction

Black pepper, famously known as the **"King of Spices,"** is one of the most ancient and economically significant spice crops of the world. Belonging to the family ***Piperaceae***, *Piper nigrum* is a **perennial woody climbing vine** cultivated primarily for its berries, which, when dried, yield black peppercorns, a universal flavouring and preservative. The spice owes its unique pungency to **piperine**, an alkaloid, and its aroma to volatile oils dominated by β-caryophyllene and **limonene**.

The **Western Ghats of India**, especially Kerala, are recognized as the **primary center of origin and diversity** for black pepper. Over centuries, the crop spread through maritime trade routes to Southeast Asia, Africa, and the Pacific, becoming a major global commodity. Today, India, Vietnam, Indonesia, Brazil, and Sri Lanka are the principal producers, with Kerala, Karnataka, and Tamil Nadu leading in Indian production.

Pepper thrives under **warm, humid tropical conditions**, preferring well-distributed rainfall (1,500–3,000 mm annually), high relative humidity (≥75%), and mild shade. The vines grow best in soils rich in organic matter with good drainage, often climbing on live standards like *Erythrina indica, Glyricidia sepium*, or *Garuga pinnata.*

The crop's economic value lies not only in its culinary and pharmaceutical uses but also in its **export potential and role in intercropping systems** under coconut, arecanut, and coffee plantations. Given the crop's long productive life (up to 25 years) and perennial nature, genetic improvement requires strategic and long-term breeding approaches that integrate field evaluation, clonal propagation, and modern molecular tools.

DOI: 10.1201/9781042010332-3

2.1.2 Botanical Description and Cytology

» **Taxonomy:**
  - **Kingdom:** Plantae
  - **Family:** Piperaceae
  - **Genus:** *Piper*
  - **Species:** *P. nigrum* L.

» **Chromosome Number:** 2n = 52 (polyploid, with diploid species also occurring within the genus).

» **Growth Habit:** Perennial climbing vine with monopodial growth and adventitious roots arising at nodes, enabling it to cling to supports.

» **Leaves:** Simple, alternate, ovate to cordate, glossy dark green with prominent veins; petioles 2–5 cm long.

» **Inflorescence:** Pendulous spike arising from leaf axils, 6–15 cm long, bearing 50–150 minute bisexual flowers.

» **Fruit:** Drupe (berry), globose, 4–6 mm in diameter, turning red at maturity; drying produces the black wrinkled peppercorn.

» **Seeds:** Single per fruit, endospermous, oily, and rich in alkaloids.

The plant is **perennial, cross-pollinated, and heterozygous**, factors that contribute to wide genetic variability but also make breeding complex and time-consuming.

2.1.3 Economic Importance

1. **Culinary Uses:** Primary flavouring and seasoning spice globally; essential component of curry powders and pickles.
2. **Medicinal Value:** Possesses carminative, antioxidant, antimicrobial, and bioenhancing properties. Piperine enhances the absorption of nutrients and drugs such as curcumin and beta carotene.
3. **Industrial Applications:** Source of essential oil and oleoresin used in perfumery, pharmaceutical formulations, and food processing industries.
4. **Export Revenue:** Constitutes a major share of India's spice exports, contributing significantly to foreign exchange earnings.
5. **Agro-ecological Role:** Grown as an intercrop or mixed crop under plantations; provides perennial income and ecological stability through soil cover and carbon sequestration.

2.1.4 Breeding Objectives

The breeding objectives for *Piper nigrum* are comprehensive, reflecting its dual need for productivity and resilience.

1. **High Yield and Productivity**

   » To develop cultivars with higher spike density, an increased number of berries per spike, and improved fruit set percentage.

- Selection for early bearing (2–3 years) and prolonged fruiting season.
- Traits such as longer spikes (≥12 cm), bold berries (>4.5 mm), and compact bunching are desirable yield indicators.

2. **Resistance to Diseases and Pests**
   - The major disease **Phytophthora foot rot** (*Phytophthora capsici*) causes severe mortality in humid regions.
   - Tolerance to **pollu beetle (*Longitarsus nigripennis*)**, **leaf gall thrips**, and **nematodes** is also targeted.
3. **Improved Quality and Biochemical Composition**
   - Enhance **piperine** content (≥7%), **essential oil** (≥2.5%), and **oleoresin** (>12%).
   - Maintain appealing aroma, pungency, and oil-to-resin ratio for both culinary and industrial grades.
4. **Adaptation and Stability**
   - Develop varieties tolerant to varying altitudes (sea level to 1,500 m), shade conditions, and marginal soils.
   - Select lines that maintain yield under drought-prone or partially shaded environments.
5. **Compact Growth and High-Density Suitability**
   - Shorter internodes, reduced climbing height, and increased lateral branching for easy management and intercropping.
6. **Tolerance to Abiotic Stresses**
   - Drought, high temperature, and salinity tolerance for non-traditional zones.
   - Maintenance of physiological stability under fluctuating moisture and temperature regimes.

2.1.5 Genetic Resources and Germplasm

India harbours the richest diversity of *Piper nigrum*, both cultivated and wild. Germplasm conservation and evaluation are pivotal for continuous breeding advancement.

Major Germplasm Collections

- ICAR–Indian Institute of Spices Research (IISR), Kozhikode: National repository with >2,800 accessions, including wild *Piper* species, landraces, and elite hybrids.
- Regional Research Stations (RRS): Collections at Panniyur (Kerala), Sirsi (Karnataka), and Kottarakkara (Kerala).
- Exotic Accessions: Introduced lines from Sri Lanka, Indonesia, and Malaysia broaden the genetic base.

Notable Landraces and Varieties

S. No.	Name / Type	Distinctive Traits
1	*Karimunda*	High-yielding; standard cultivar; strong aroma, moderate foot rot tolerance.
2	*Kottanadan*	Tolerant to drought and shade; bold berries, lower piperine.
3	*Neelamundi*	Adapted to high elevations; long spikes; moderate yield.
4	*Kalluvally*	Tolerant to root diseases; compact canopy; suitable for mixed cropping.
5	*Panniyur* types	A series of high-yielding hybrids from Kerala; superior piperine content.

Wild Relatives Utilized

Wild species such as *P. attenuatum*, *P. colubrinum*, and *P. mullesua* serve as donors for resistance to **Phytophthora** and drought tolerance. *P. colubrinum* is particularly valuable for **foot rot resistance**, making it an important rootstock source.

Characterization Criteria

» **Morphological:** Spike length, leaf index, branching pattern, and vine vigour.

» **Physiological:** Photosynthetic rate, chlorophyll content, and drought recovery.

» **Biochemical:** Piperine, oil, and oleoresin content.

» **Molecular:** RAPD, ISSR, SSR marker profiling to assess diversity and parentage.

2.1.6 Floral Biology and Pollination

» **Flowering Season:** May to August; occasionally two flushes under humid tropics.

» **Inflorescence:** Pendulous spikes borne opposite leaves, 6–15 cm long.

» **Flowers:** Minute, bisexual, without petals, with short pedicels and inconspicuous calyx.

» **Pollination Mechanism**

- Predominantly **cross-pollinated,** facilitated by wind, gravity, and small insects (thrips and flies).
- **Self-pollination** occurs but results in lower fruit set.

» **Pollen Viability:** Retained for ~48 h; optimum temperature 25–30°C.

» **Fruit Development:** Takes ~180–200 days from pollination to maturity.

» **Seed Biology:** Seeds are recalcitrant; viability drops sharply after 30–40 days.

Understanding flowering behaviour and synchronization between parents is crucial for designing controlled hybridization programs.

2.1.7 Breeding and Hybridization Methods

a. Clonal Selection

Historically, improvement in black pepper began through the **selection of superior clones** from natural populations. The clones were evaluated for yield, spike morphology, and disease tolerance under replicated field trials.

- **Steps in clonal selection:**
    1. Identify elite vines in farmers' fields or germplasm collections.
    2. Propagate through stem cuttings to ensure uniform clones.
    3. Evaluate over multiple seasons for yield and stability.
    4. Standardize for piperine content and resistance traits.
- Notable clones from this approach include *Karimunda*, *Kottanadan*, and *Kalluvally*.

b. Hybridization Breeding

Systematic hybridization at IISR and Kerala Agricultural University since the 1960s revolutionized pepper improvement.

Objectives

- Combine high yield and disease resistance.
- Transfer foot rot resistance from *P. colubrinum*.
- Develop hybrids suitable for diverse ecological regions.

Steps in Controlled Hybridization

1. **Parent Selection:** Based on complementary traits (e.g., *Uthirankotta* for yield × *Cheriyakaniakkadan* for disease tolerance).
2. **Bagging:** Female spikes enclosed before anthesis to prevent contamination.
3. **Pollen Collection:** From male spikes of donor parent; stored at 4°C with desiccant if needed.
4. **Pollination:** Performed during early morning hours when stigma is receptive.
5. **Tagging:** Each cross is labelled with parentage, date, and cross number.
6. **Fruit Harvesting:** Mature berries collected after 6–8 months.
7. **Seed Extraction and Raising:** Germinated in polybags; progeny evaluated for vigour, spike traits, and resistance.

Examples of Successful Hybrids

- *Panniyur-1* (Uthirankotta × Cheriyakaniakkadan) – high yield, bold berries.
- *Panniyur-2* – early maturing, tolerant to foot rot.
- *IISR Thevam* and *IISR Shakthi* – high piperine and disease resistance.

c. Mutation and Polyploid Breeding

- **Gamma irradiation** induces dwarfness, compact growth, and yield enhancement.
- **Colchicine treatments** are used experimentally to produce polyploids with thicker leaves and spikes.

d. Biotechnological Interventions

- **Somatic embryogenesis** and **organogenesis** from leaf explants used for large-scale micropropagation of elite hybrids.
- **In vitro screening** for Phytophthora resistance by culture filtrate methods.
- **Molecular markers (SSR, RAPD)** applied for early hybrid confirmation and diversity analysis.

2.1.8 Released Varieties and Achievements

S. No.	Variety / Hybrid	Parentage / Type	Distinct Features
1	**Panniyur-1**	*Uthirankotta* × *Cheriyakaniakkadan*	High yielding; long spikes; 7% piperine; adaptable across regions.
2	**Panniyur-2**	Hybrid	Early maturing; bold berries; tolerant to foot rot.
3	**Panniyur-5**	Hybrid selection	Resistant to Phytophthora; compact canopy; suited for high-density systems.
4	**IISR Thevam**	Clonal selection	Tolerant to foot rot; vigourous; high oil content.
5	**IISR Shakthi**	Hybrid	High piperine (>8%); bold berries; disease tolerance.
6	**IISR Girimunda**	Selection from *Karimunda* × *Neelamundi*	Drought hardy; adaptable to hill slopes; uniform spike size.
7	**IISR Malabar Excel**	Elite clonal selection	High oleoresin yield; good blending quality; recommended for export.

Key Achievements

- Yield enhancement from <1 t to 3–4 t dry berries ha^{-1} under improved varieties.
- Development of **Phytophthora-tolerant lines** and **drought-resilient hybrids**.
- Standardization of micropropagation for large-scale clonal planting.
- Integration of molecular marker systems for parental authentication.

2.1.9 Molecular, Biochemical, and Breeding Advances

1. Marker-Assisted Selection (MAS)

- Molecular markers linked with foot rot resistance loci are used for early-stage screening.

2. Transcriptomics and Genomics

- Gene expression studies identified candidate genes for piperine biosynthesis and stress tolerance.

- Draft genome assembly (≈500 Mb) facilitates gene-based breeding.

3. **Somatic Embryogenesis and Synthetic Seeds**
   - Used for germplasm exchange and long-term conservation of elite lines.
4. **Cryopreservation**
   - Tissue and pollen banks established for secure conservation of elite materials.
5. **Biochemical Screening**
   - High-performance liquid chromatography (HPLC) is used to quantify piperine, chavicine, and essential oil fractions for selection accuracy.

2.2 Cardamom (*Elettaria cardamomum* Maton)

2.2.1 Introduction

Cardamom, celebrated as the **"Queen of Spices,"** is the world's third-most-valuable spice after saffron and vanilla. It belongs to the family **Zingiberaceae** and is native to the **evergreen forests of the Western Ghats** in southern India. The crop's economic product is the **capsule-like fruit** enclosing aromatic seeds rich in **volatile oils** (5 – 8 %) responsible for its characteristic fragrance and flavour.

The spice is cultivated principally in **India, Guatemala, Sri Lanka, Tanzania, and Papua New Guinea,** with India being both a major producer and consumer. Within India, commercial cultivation spans the **Cardamom Hills of Kerala, the Coorg region of Karnataka, and the Anamalai ranges of Tamil Nadu**.

Cardamom thrives under **humid tropical climates** between 600 – 1500 m asl, preferring 18 – 28 °C temperature, 1500 – 2500 mm evenly distributed rainfall, and filtered shade beneath evergreen canopy species. Its **perennial rhizomatous habit, cross-pollination**, and **prolonged juvenile period** make breeding a slow and complex process requiring long-term evaluation. The spice commands premium prices because of its unique aroma contributed by **1,8-cineole, α-terpinyl acetate,** and other monoterpenes. Global demand for natural flavouring agents and nutraceutical extracts has further enhanced the crop's commercial importance.

2.2.2 Botanical Description and Cytology

- **Family:** Zingiberaceae
- **Genus:** *Elettaria*
- **Species:** *E. cardamomum* Maton
- **Chromosome number:** 2n = 48
- **Growth habit:** Perennial, rhizomatous, clump-forming herb 2–3 m tall with leafy pseudostems arising from underground rhizomes.
- **Leaves:** Simple, lanceolate, alternate, glabrous; sheathing base enclosing pseudostem.
- **Inflorescence:** Racemose panicle or spike emerging from the rhizome base; 0.3 – 1.0 m long; bearing numerous bisexual, zygomorphic flowers.

- **Flower:** Tubular calyx; white to pale green corolla; lip violet-streaked; **self-incompatible,** promoting cross-pollination.
- **Fruit:** Trilocular capsule, 1 – 2 cm long, trilobed, green when immature, straw-coloured on drying.
- **Seeds:** 15 – 20 per capsule, angular, black, aromatic; each enclosed in mucilaginous aril rich in essential oils.

Cytologically, *E. cardamomum* is considered allopolyploid with high heterozygosity. Vegetative propagation maintains uniformity, but genetic recombination requires controlled crossing.

2.2.3 Economic Importance

1. **Flavouring and Culinary Uses:** An integral component of traditional Indian and Middle Eastern cuisines, confectionery, and bakery products.
2. **Pharmaceutical and Nutraceutical Applications:** Essential oils act as carminative, expectorant, and antioxidant; seed extracts exhibit antimicrobial and anti-inflammatory activities.
3. **Industrial Utilization:** Used in flavouring soft drinks, perfumery, cosmetics, and oral-care formulations.
4. **Export Revenue:** India and Guatemala together contribute nearly 80 % of global trade; cardamom remains one of India's highest foreign-exchange earners among spices.
5. **Agro-ecological Benefits:** Under-story crop in evergreen forests, conserving soil and biodiversity while providing perennial income to smallholders.

2.2.4 Breeding Objectives

1. **Increased Yield and Productivity**
   - Development of clones with **more panicles per clump (≥ 15), greater capsule set (≥ 60 %),** and **bold seed weight**.
   - Early tillering and quick clump establishment.
2. **Resistance to Diseases and Pests**
   - Tolerance to **capsule rot** (*Phytophthora meadii*), **rhizome rot,** and **leaf blotch** (*Colletotrichum gleosporioides*).
   - Resistance to **thrips (*Sciothrips cardamomi*), shoot borer (*Conogethes punctiferalis*),** and **root grub (*Basilepta fulvicorne*).**
3. **Improved Quality**
   - Higher volatile-oil content (≥ 8 %), enriched α-**terpinyl acetate and 1,8-cineole** concentration, and low fiber content.
   - Uniform green colour retention upon curing.
4. **Adaptation to Environmental Stress**
   - Drought tolerance and shade adaptation for marginal and secondary forest zones.

5. **Compact Growth and Clonal Uniformity**
   » Short pseudostems, narrow leaves, high clump density for mechanization and ease of harvest.
6. **Extended Bearing Period and Perennial Stability**
   » Continuous flushing and panicle emergence across seasons to ensure year-round production.

2.2.5 Genetic Resources and Germplasm

a. Centers of Origin and Distribution

The primary center of origin of *E. cardamomum* lies in the **Western Ghats of India**, particularly in the Cardamom Hills of Kerala. Secondary centers exist in **Sri Lanka** and **Southeast Asia**.

b. Germplasm Repositories

S. No.	Institution / Location	Approx. Accessions	Remarks
1	**ICAR-Indian Institute of Spices Research (IISR), Kozhikode**	> 550	National gene bank; maintains wild and cultivated types.
2	**IISR Regional Station, Appangala (Karnataka)**	≈ 350	High-altitude germplasm and breeding materials.
3	**Karnataka and Kerala Agricultural Universities**	≈ 200	Working collections for selection and hybridization.

c. Diversity within Cultivated Types

S. No.	Type / Local Name	Salient Features
1	**Malabar type**	Prostrate panicles; short plants; suitable for low altitudes.
2	**Mysore type**	Erect panicles; vigourous clumps; suited for hills and high rainfall.
3	**Vazhukka type**	Natural hybrid between Malabar and Mysore; semi-erect habit; superior yield and quality.

d. Wild Relatives and Utility

Wild *Elettaria* species and allied genera (*Amomum*, *Aframomum*) provide genes for pest resistance, drought tolerance, and aromatic diversity. Controlled introgression from these taxa is limited due to strong reproductive barriers, yet molecular tools now allow partial gene transfer or marker association studies.

2.2.6 Floral Biology and Pollination

Cardamom flowers are protandrous and highly attractive to insects, making natural pollination dependent on bees and flies.

» **Flowering Season:** April – October (depending on rainfall).

» **Floral Structure:** Bisexual, zygomorphic, with a conspicuous labellum aiding pollinator landing.

» **Pollination Mechanism**

- **Entomophilous**, mainly by **honeybees (*Apis cerana, A. dorsata*)** and **stingless bees (*Trigona* spp.)**.
- **Self-incompatibility** prevents self-fertilization; cross-pollination rate > 90 %.

» **Fertilization and Fruit Development:** Occurs within 72 h of pollination; capsules mature in 100–120 days.

» **Seed Viability:** Short-lived; storage requires 12 °C and 40 % RH to maintain viability for 6 months.

Pollination ecology is vital for hybridization programs; in controlled breeding, artificial pollination substitutes natural bee activity.

2.2.7 Breeding and Hybridization Methods

a. Clonal Selection

The earliest and most practical method for improving cardamom has been the identification of superior clones from existing plantations.

Steps in Clonal Selection

1. **Survey and Selection:** Locate high-yielding, disease-free clumps with > 1.5 kg dry capsules $clump^{-1}$ yr^{-1}.
2. **Vegetative Propagation:** Separate and plant tillers in nursery beds; propagate through suckers to maintain genetic purity.
3. **Evaluation:** Monitor for three to five years for yield, panicle length, capsule set, and oil content.
4. **Stabilization and Release:** Selected clones were multiplied and released after multi-location trials.

Notable Selections: *CL 37*, *CL 666*, *Green Gold*, *Njeduveliyan*, and *Appangala 1*.

b. Hybridization Breeding

Because the species is cross-pollinated and heterozygous, controlled crosses enable recombination for yield and resistance.

Objectives

» Combine Malabar type's quality with Mysore type's vigour.

» Develop hybrids resistant to capsule rot and thrips.

Procedure for Controlled Hybridization:

1. **Parent Selection:** Identify contrasting parents for desired traits.
2. **Bagging:** Inflorescences enclosed with butter-paper bags before anthesis.
3. **Emasculation:** Remove anthers just before dehiscence using fine forceps.
4. **Pollen Collection:** From male flowers on the day of anthesis; stored at cool temperature.

5. **Pollination:** Apply fresh pollen to stigmas in the early morning; re-bag immediately.
6. **Tagging and Record:** Parentage, date of pollination, and cross number noted.
7. **Seed Harvesting:** Collect mature capsules after 4 months; extract and germinate seeds.
8. **Progeny Evaluation:** Seedlings assessed for vigour, panicle traits, yield, and disease resistance.

Examples of Hybrids

» *Appangala-1* (Malabar × Mysore) – early bearing, high yield, resistant to capsule rot.

» *Appangala-2* (Vazhukka × Mysore) – bold capsules, 8 % oil content.

c. Mutation Breeding

Physical (mutagenic radiation) and chemical mutagens have produced semi-dwarf types with compact growth and enhanced oil yield.

d. Polyploid Breeding

Artificial tetraploids show larger capsules and thicker pericarp but reduced fertility; research is continuing to stabilize yield traits.

e. Biotechnological Interventions

» **Micropropagation:** In vitro multiplication from nodal segments and rhizome buds ensures rapid clonal production of elite genotypes.

» **Somaclonal Variation:** Used to generate novel variants with altered oil profiles.

» **Molecular Marker Assisted Selection:** RAPD and ISSR markers help distinguish Malabar, Mysore, and Vazhukka types and identify hybrid progenies early.

» **In Vitro Disease Screening:** Culture-filtrate tests used for evaluating resistance to *Phytophthora* and *Colletotrichum.*

2.2.8 Released Varieties and Achievements

S. No.	Variety / Clone / Hybrid	Parentage / Origin	Salient Features
1	**Appangala-1**	(Malabar × Mysore)	High yield (500 kg ha^{-1} dry capsules); tolerant to capsule rot.
2	**Appangala-2**	(Vazhukka × Mysore)	Bold capsules; 8 % essential oil; thrips tolerant.
3	**Green Gold**	Clonal selection	Uniform panicles; long capsules; superior colour retention.
4	**CL 666**	Selection	High yield and leaf blotch tolerance; good curing quality.
5	**Njallani Green Gold**	Farmers' selection (Kerala)	Early bearing; drought resilient; bright green capsules.
6	**IISR Avinash**	Breeding line	Stable yield; moderate capsule rot tolerance; high α-terpinyl acetate.

Key Breeding Achievements

- » Replacement of heterogeneous seedling populations with clonal cultivars yielding > 3 t ha^{-1}.
- » Identification of capsule-rot-resistant germplasm.
- » Establishment of micropropagation protocols for mass clonal production.
- » Initiation of molecular marker work for quality and resistance genes.

2.2.9 Biochemical and Molecular Advances

1. **Essential-Oil Chemoprofiling:** Quantitative GC–MS analysis of monoterpenes identifies chemotypes rich in α-terpinyl acetate and 1,8-cineole; these profiles serve as biochemical markers in selection.
2. **DNA Fingerprinting:** RAPD, ISSR, and SSR markers delineate population structure and confirm varietal identity.
3. **Gene Expression Studies:** Transcriptomic analyses reveal key genes involved in terpene biosynthesis and disease-response pathways.
4. **Cryopreservation and Germplasm Conservation:** Encapsulation-dehydration techniques preserve shoot tips for long-term storage without genetic alteration.
5. **Tissue Culture Applications:** Somatic embryogenesis from rhizome buds allows production of disease-free plantlets for rejuvenating old plantations.

2.3 Ginger (*Zingiber officinale* Roscoe)

2.3.1 Introduction

Ginger is one of humanity's oldest cultivated spice crops and among the most versatile rhizomatous herbs in the tropics. Belonging to the family **Zingiberaceae**, it is widely used as a spice, condiment, and medicinal herb because of its characteristic flavour, aroma, and pungency imparted by **essential oils** and **oleoresins**. *Zingiber officinale* is believed to have originated in the Indo-Malayan region and was domesticated in the Indian subcontinent more than 3,000 years ago. Today, it is grown across the humid tropics of **India, China, Nigeria, Indonesia, Thailand, and Nepal**, together contributing over 70 % of global production.

India ranks first in area and production, cultivating ginger on roughly **1.6 lakh ha** with an average yield of **4 t ha^{-1} of dry rhizome equivalent**. Major producing states include **Kerala, Karnataka, Meghalaya, Odisha, and Sikkim**. The crop forms a crucial part of the smallholder spice economy and serves as a model species for vegetative propagation and disease-management research.

Ginger's economic organ is the **underground rhizome**, which stores volatile oil (1–3 %), oleoresin (5–8 %), starch, and fiber. The pungency arises from **gingerols** and **shogaols**, while the aroma derives from **zingiberene, citral, camphene**, and related terpenes. Apart from culinary uses, ginger occupies an important place in traditional medicine (Ayurveda, Unani, Chinese) as a carminative, stimulant, anti-nausea, and anti-inflammatory agent. Because the crop is **sterile, highly heterozygous, and vegetatively propagated**, conventional breeding faces exceptional challenges; nevertheless, careful clonal selection, mutation induction, and biotechnology-based improvement have yielded significant progress.

2.3.2 Botanical Description and Cytology

- **Taxonomic position**
  - *Kingdom* Plantae
  - *Family* Zingiberaceae
  - *Genus Zingiber*
  - *Species Z. officinale* Roscoe
- **Chromosome number:** 2n = 22 (triploid nature in most cultivars results in sterility).
- **Growth habit:** Perennial herbaceous plant with underground branched rhizomes and aerial pseudostems (60–120 cm).
- **Leaves:** Simple, alternate, lanceolate, with entire margins and sheathing bases.
- **Inflorescence:** A compact spike arising either terminally or laterally on a separate shoot; bracts imbricate, green or purplish.
- **Flowers:** Bisexual, zygomorphic; calyx tubular; corolla yellow-green with a purple lip streaked with yellow.
- **Fruit:** Rarely formed under cultivation because of sterility.

Cytologically, most commercial clones are **triploid (3n = 33)** or aneuploid, accounting for their absence of viable seed and dependence on rhizome propagation. This condition restricts sexual recombination but allows perpetuation of superior clones through vegetative means.

2.3.3 Economic Importance

1. **Culinary and Processing Uses**
   - Fresh or dried rhizomes flavour curries, bakery goods, beverages, and pickles.
   - Dried and powdered ginger forms a key ingredient in spice blends.
   - Processed products include ginger oil, oleoresin, candy, syrup, beer, and carbonated drinks.
2. **Pharmaceutical and Nutraceutical Value**
   - Active principles (*[6]-gingerol*, *[6]-shogaol*) exhibit anti-inflammatory, antioxidant, and anti-emetic properties.
   - Used in treating motion sickness, indigestion, arthritis, and metabolic disorders.
3. **Industrial Applications**
   - Essential oil utilized in perfumery and confectionery industries.
   - Oleoresin is valued in flavour technology for encapsulated seasoning formulations.

4. **Export Significance**
   - India exports dried ginger, oil, and oleoresin; principal markets are the Middle East, Europe, and North America.
5. **Agro-ecological Role**
   - Fits well in intercropping systems under coconut, arecanut, and shade trees; contributes to soil conservation through ground cover.

2.3.4 Breeding Objectives

Because of the vegetatively propagated, sterile nature of *Z. officinale*, the objectives emphasize **selection, disease tolerance, quality improvement**, and **adaptation** rather than hybrid creation.

1. **High Yield and Productivity**
   - Selection for **bold, well-branched rhizomes**, high *rhizome index* (ratio of rhizome weight to plant weight), and quick maturity (8 months).
   - Increase in fresh-rhizome yield ≥ 25 t ha^{-1} and dry recovery ≥ 20 %.
2. **Resistance to Diseases and Pests**
   - **Major diseases: rhizome rot** (*Pythium aphanidermatum*), **bacterial wilt** (*Ralstonia solanacearum*), and **leaf spot** (*Phyllosticta zingiberi*).
   - Insect pests: **shoot borer (*Conogethes punctiferalis*), rhizome scale**, and **nematodes**.
   - Breeding for tolerance to these biotic stresses is of paramount importance.
3. **Improved Quality and Biochemical Composition**
   - Enhance **oleoresin (>8 %), essential-oil (>2.5 %)**, and **gingerol content (>1 %)**.
   - Select for low fiber (<5 %) and high dry-matter recovery.
4. **Adaptation and Abiotic-Stress Tolerance**
   - Varieties tolerant to high rainfall, waterlogging, or drought for varied agro-ecological zones.
   - Suitability for high-altitude temperate hills (Sikkim type) and coastal plains (Kerala type).
5. **Morphological Improvement**
   - Compact plant habit, erect pseudostems, and uniform maturity are desirable for mechanized harvesting.

2.3.5 Genetic Resources and Germplasm

a. Centers of Origin and Diversity

Primary center lie in the **Indo-Malayan region**; secondary centers extend across Southeast Asia, China, and the Pacific Islands. India houses tremendous intra-specific variability represented by local landraces and farmer-selected clones.

b. Major Germplasm Holdings

S. No.	Institution / Location	Approx. Accessions	Remarks
1	ICAR–Indian Institute of Spices Research (IISR), Kozhikode	> 700	National gene bank; disease-indexed clones and wild relatives.
2	IISR Regional Station, Appangala (Karnataka)	≈ 250	High-altitude and disease-resistant accessions.
3	State Agricultural Universities (Kerala & Meghalaya)	≈ 150	Local cultivars maintained for evaluation.

c. Important Cultivars / Landraces

S. No.	Cultivar / Landrace	Distinctive Traits
1	*Nadroga* (Fiji type)	Large rhizomes, mild flavour, suitable for processing.
2	*Rio-de-Janeiro*	High yield, 2.5 % oil; popular in Kerala.
3	*Himachali*	Adapted to cool hills; strong aroma, fibrous.
4	*Thingpui* (Mizo type)	Resistant to rot; good keeping quality.
5	*China Ginger*	Light coloured rhizomes; high oleoresin; low fiber.

d. Wild Relatives

Zingiber zerumbet, *Z. cassumunar*, *Z. rubens*, and *Z. montanum* serve as reservoirs of resistance to pathogens and abiotic stress; they also contribute unique aroma and metabolite profiles for introgression through biotechnology.

2.3.6 Floral Biology and Reproduction

- **Flowering:** Seasonal (May–August) in subtropics; occurs sporadically under shade.
- **Inflorescence:** Compact spike with 30–50 flowers.
- **Flower:** Bisexual, zygomorphic, protandrous.
- **Pollination:** Insect-pollinated (bees); however, most commercial clones are male-sterile or non-flowering.
- **Fertilization:** Extremely rare; viable seed formation is almost absent.

Consequently, genetic recombination through sexual means is limited, and vegetative propagation through rhizome bits (each with a viable bud) remains the only practical method of perpetuation. Nevertheless, the study of floral biology assists in occasional hybridization programs using fertile diploids.

2.3.7 Breeding and Improvement Methods

a. Clonal Selection

The principal strategy in ginger improvement.

Steps

1. **Field Survey:** Identify high-yielding, disease-tolerant clumps in farmers' fields.
2. **Rhizome Sampling:** Select healthy, bold rhizomes free from internal necrosis.
3. **Propagation:** Plant in disease-free nursery beds; maintain clones separately.
4. **Evaluation:** Monitor over 3–4 seasons for yield, quality, and resistance.
5. **Stabilization:** Superior clones multiplied for multi-location testing.

Examples: *Varada, Rio-de-Janeiro, Thingpui, IISR-Maran, Suruchi, Rejatha.*

b. Mutation Breeding

Induced mutagenesis using gamma irradiation (5–10 kR) or chemical mutagens (EMS 0.2–0.4 %) generates somatic variants with altered pigmentation, compact growth, or disease tolerance. Mutant selections such as *Himgiri* exhibit enhanced rot tolerance.

c. Polyploid Breeding

Colchicine-induced autotetraploids have been explored for larger rhizomes and higher dry-matter content, though fertility and field vigour remain constraints.

d. Biotechnological Approaches

1. **Micropropagation:** Shoot-tip and bud-meristem culture enable rapid multiplication of pathogen-free clones.
2. **Somaclonal Variation:** Tissue-culture-derived lines screened for resistance to *Pythium* and *Ralstonia.*
3. **In vitro Pathogen Indexing:** ELISA and PCR used to confirm disease-free status before distribution.
4. **Molecular Markers:** RAPD and AFLP profiling assist in clonal identification and genetic-diversity assessment.
5. **Cryopreservation:** Long-term storage of shoot tips using vitrification methods for germplasm conservation.

2.3.8 Major Diseases and Breeding for Resistance

S. No.	Disease / Causal Agent	Symptoms	Resistant Sources / Measures
1	Rhizome rot – *Pythium aphanidermatum*	Yellowing, soft rot, foul odour	*Tingpui, China Ginger* shows tolerance; biocontrol using *Trichoderma.*
2	Bacterial wilt – *Ralstonia solanacearum*	Sudden wilting, vascular browning	*IISR Varada, Maran* tolerant; selection in endemic soils.
3	Leaf spot – *Phyllosticta zingiberi*	Brown spots, blighting	High-altitude types show resistance.
4	Soft rot complex	Multiple pathogens	Soil-solarization and rhizome treatment integrated with resistant clone use.

Breeding integrates phenotypic screening under natural epiphytotics and molecular tagging of tolerant clones.

2.3.9 Released Varieties and Performance

S. No.	Variety / Clone	Parentage / Origin	Salient Features
1	**IISR Varada**	Selection from Kerala landrace	High yield (25 t ha^{-1}); bacterial-wilt tolerant; 1.9 % oil.
2	**IISR Maran**	Clonal selection	Rhizome-rot tolerant; oleoresin 9 %; bright lemon colour.
3	**Suruchi**	Farmers' selection (Odisha)	Compact plant; drought-tolerant; strong aroma.
4	**Rejatha**	Mutation selection	High dry-recovery (22 %); bold rhizomes; early maturing.
5	**Thingpui**	Mizoram landrace	Disease-tolerant; good keeping quality.
6	**Himgiri**	Gamma-ray mutant	Adapted to temperate zones; compact growth.

Adoption of these varieties has improved yield stability and disease management in diverse agro-ecological zones.

2.3.10 Quality Evaluation and Biochemical Screening

- **Essential-Oil Analysis:** GC–MS profiling quantifies major components (zingiberene 25–35 %, camphene 15 %, citral 10 %).
- **Oleoresin Estimation:** Solvent extraction followed by spectrophotometric evaluation.
- **Pungency Index:** Determined by HPLC quantification of *[6]-gingerol* and *[6]-shogaol.*
- **Colour and Fiber Content:** Visual grading and gravimetric methods for export quality.
- **Molecular Chemotyping:** Fingerprinting links chemical phenotype with genetic identity to assist breeding for specific flavour niches.

2.3.11 Advances in Molecular and Genomic Studies

1. **Genetic Diversity:** RAPD and ISSR analyses reveal wide heterogeneity among Indian clones, confirming multiple domestication events.
2. **Gene Expression:** RT-PCR studies identify up-regulated genes in the shikimate and phenylpropanoid pathways responsible for gingerol synthesis.
3. **Molecular Diagnostics:** PCR-based assays for *Ralstonia* and *Pythium* enable rapid screening of propagules.
4. **Proteomic Mapping:** Protein markers linked with stress-response enzymes (peroxidase, superoxide dismutase) correlate with disease tolerance.
5. **Bioinformatics Resources:** De novo transcriptome assemblies (> 45,000 unigenes) available for marker discovery.

2.3.12 Propagation and Nursery Management

- **Seed Rhizome Selection:** Use disease-free, medium-sized pieces (20–25 g with 1 bud).
- **Treatment:** Hot-water treatment (50 °C for 10 min) followed by fungicide dip.
- **Planting:** In raised beds with organic mulch and partial shade.
- **Micropropagation:** Rooted plantlets from tissue culture hardened in mist chambers and field-transferred with > 95 % survival.

These techniques ensure clonal fidelity and phytosanitary safety of planting materials used in breeding trials.

2.3.13 Integration of Conventional and Modern Approaches

S. No.	Approach	Contribution to Improvement
1	Clonal selection	Identification of elite high-yielding, disease-free clones.
2	Mutation breeding	Induction of novel variability in otherwise clonal crop.
3	Tissue culture	Rapid propagation and disease elimination.
4	Molecular diagnostics	Pathogen indexing and true-to-type verification.
5	Chemoprofiling	Direct selection for flavour and oleoresin traits.
6	Integrated disease screening	Combines phenotypic and molecular tests for durable resistance.

2.4 Turmeric (*Curcuma longa* L.)

2.4.1 Introduction

Turmeric, often hailed as the **"Golden Spice of India,"** is one of the most ancient, sacred, and economically valuable spices in human civilization. It belongs to the family **Zingiberaceae**, the same family as ginger, and is cultivated for its **rhizomes** that yield a brilliant yellow pigment and a characteristic earthy aroma. The spice has been a cornerstone of Asian cuisine, medicine, and ritual for more than 4,000 years and has recently attained global attention as a nutraceutical and natural colourant.

The species *Curcuma longa* L. is believed to have originated in the **Indo-Malayan region**, probably in India or Indonesia, and later spread across Southeast Asia and the tropics of Africa, Central America, and the Caribbean. India remains the largest producer, consumer, and exporter, contributing over **75 % of global production.** Major growing states include **Telangana, Tamil Nadu, Karnataka, Maharashtra, Odisha, and Meghalaya**, occupying roughly 3 lakh ha with an average productivity of 6–7 t ha^{-1} dry rhizome equivalent.

The economic portion, the **underground rhizome**, contains **curcuminoids** (2–6 %), **volatile oil** (3–5 %), **starch** (30–50 %), and fiber. The major colouring and bioactive constituent is **curcumin,** accompanied by demethoxy- and bis-demethoxy-curcumin. The essential oil, dominated by **turmerone, zingiberene,** and **ar-turmerone**, contributes aroma and antimicrobial activity. Because turmeric is largely a **sterile triploid** and propagated vegetatively, genetic improvement depends on **clonal**

selection, induced variation, and biotechnology rather than classical hybridization. Nevertheless, systematic breeding has produced numerous high-yielding, disease-tolerant, and quality-rich cultivars sustaining India's leadership in the spice trade.

2.4.2 Botanical Description and Cytology

- **Family:** Zingiberaceae
- **Genus:** *Curcuma*
- **Species:** *C. longa* L.
- **Chromosome number:** 2n = 63 (Triploid)
- **Growth habit:** Perennial, rhizomatous herb 60–100 cm tall with pseudostems formed by sheathing leaf bases.
- **Leaves:** Oblong-lanceolate, alternate, with entire margins and prominent midrib; arranged in two ranks.
- **Inflorescence:** A terminal spike borne on a separate peduncle emerging before the foliage; bracts are bright green or tinged pink.
- **Flowers:** Bisexual, zygomorphic, yellowish-white; calyx tubular; corolla with three lobes; lip pale yellow with red streaks.
- **Fruit:** Rare capsule; seeds seldom viable due to sterility.

Cytologically, *C. longa* is a **polyploid complex** derived from inter-specific hybridization among diploid species such as *C. aromatica*, *C. zedoaria*, and *C. mangga*. Sterility ensures clonal fidelity but limits sexual recombination. Vegetative propagation through rhizome bits ensures uniformity but also demands rigorous disease indexing.

2.4.3 Economic Importance

1. **Culinary Uses** – Turmeric powder is a universal colouring and flavouring ingredient in curries, pickles, sauces, and processed foods; curcumin acts as a natural substitute for synthetic dyes.
2. **Medicinal and Nutraceutical Applications** – Curcumin and related polyphenols exhibit antioxidant, anti-inflammatory, hepatoprotective, and anticancer properties; turmeric is integral to Ayurvedic formulations.
3. **Industrial Uses** – Essential oil finds applications in perfumery, cosmetics, and pharmaceuticals; curcumin is used in textiles and food colourants (E100).
4. **Export Revenue** – India exports over 1 lakh t annually of dry turmeric and value-added products such as oleoresin and curcumin extract.
5. **Agro-ecological Value** – Turmeric suits multi-storey cropping systems with coconut, arecanut, and fruit trees, providing soil cover and additional income.

2.4.4 Breeding Objectives

1. **Higher Yield and Rhizome Quality**
   - » Selection for bold, branched rhizomes with high *rhizome index* and dry-recovery ≥ 20 %.
   - » Early maturity (7–8 months) and uniform bulking.
2. **Enhanced Curcumin and Oil Content**
   - » Curcumin ≥ 6 %, volatile oil ≥ 5 %, and intense yellow pigmentation.
3. **Resistance to Major Diseases and Pests**
   - » Tolerance to **rhizome rot** (*Pythium aphanidermatum*), **leaf blotch** (*Taphrina maculans*), and **bacterial wilt** (*Ralstonia solanacearum*).
   - » Resistance to **shoot borer** (*Conogethes punctiferalis*) and **nematodes.**
4. **Adaptation and Stress Tolerance**
   - » Clones suitable for both irrigated plains and rain-fed hill slopes; drought and salinity tolerance.
5. **Morphological and Physiological Improvement**
   - » Compact canopy, erect pseudostems, vigorous root system for nutrient use efficiency.
6. **Uniformity and Stability**
   - » True-to-type clonal lines with stable performance across seasons and regions.

2.4.5 Genetic Resources and Germplasm

a. Centers of Origin and Distribution

The *Curcuma* genus comprises over 80 species distributed from India to the Pacific. The primary center of diversity is the **Indian subcontinent**, particularly Kerala, Tamil Nadu, and the Eastern Himalayas. Secondary diversity extends to **Thailand, Indonesia, and southern China**.

b. Germplasm Collections

S. No.	Institution / Location	Accessions	Remarks
1	ICAR–IISR, Kozhikode	> 1 000	National repository; includes wild and cultivated *Curcuma* species.
2	IISR Regional Station, Appangala	≈ 400	High-altitude and disease-tolerant types.
3	State Agricultural Universities (Kerala, Telangana, Odisha)	≈ 300	Local landraces and farmers' clones.

c. Important Landraces and Cultivars

S. No.	Name / Origin	Salient Features
1	*Alleppey Finger* (Kerala)	Bright orange rhizome; curcumin > 6 %; high oleoresin.
2	*Erode local* (Tamil Nadu)	Early maturity; golden colour; good for powder.
3	*Suvarna*	High yield; rhizome-rot resistant; 3 % oil.

4	*Suguna*	Resistant to leaf blotch; curcumin 6 %; compact clump.
5	*Roma* (Odisha)	Bold rhizomes; adapted to rain-fed uplands.
6	*Megha Turmeric-1* (Meghalaya)	Drought hardy; long rhizomes; 5 % curcumin.

d. Wild Relatives

Species such as *C. aromatica*, *C. zedoaria*, and *C. mangga* serve as genetic reservoirs for aroma, pigment, and disease resistance; inter-specific hybrids and molecular introgression are being explored.

2.4.6 Floral Biology and Reproduction

» **Flowering season:** June – September.

» **Inflorescence:** Cylindrical spike 15–20 cm with 40–60 flowers.

» **Pollination:** Entomophilous (bees, flies), but self-incompatibility and triploidy cause sterility.

» **Fruit set:** Extremely rare; asexual propagation predominates.

Despite sterility, floral studies are essential for understanding evolutionary relationships and for limited hybridization using fertile diploids of *Curcuma aromatica*.

2.4.7 Breeding and Improvement Methods

a. Clonal Selection

Primary route for developing improved cultivars.

Procedure

1. Survey traditional fields and select high-yielding, disease-free clumps.
2. Propagate selected rhizomes and evaluate in replicated trials.
3. Analyze for curcumin, oil, and dry recovery.
4. Stabilize elite clones over three crop cycles before release.

Successful selections: *Suvarna*, *Suguna*, *Sudarsana*, *Prabha*, *Pratibha*, *Roma*, *Rajendra Sonia.*

b. Mutation Breeding

Gamma irradiation (5–10 kR) and chemical mutagens (EMS) were used to create variation in sterile clones. Mutants with compact canopies, high pigment, and disease tolerance were identified.

c. Polyploid and Somatic Hybridization

Somatic fusion of protoplasts from *C. longa* and *C. aromatica* yielded hybrids combining aroma and colour. Tetraploids show vigorous growth but reduced dry recovery.

d. Biotechnological Interventions

1. **Micropropagation:** Shoot-tip and rhizome-bud culture allow rapid multiplication of disease-free plantlets; hardened in polyhouses.
2. **Somaclonal Variation:** In vitro-derived lines screened for *Pythium* and *Ralstonia* tolerance.
3. **Molecular Diagnostics:** PCR and ELISA tests ensure pathogen-free stock.

4. **Marker Assisted Selection:** RAPD and ISSR markers linked with high curcumin content were identified.
5. **Cryopreservation:** Encapsulated buds stored in liquid nitrogen for long-term germplasm conservation.

2.4.8 Major Diseases and Resistance Breeding

S. No.	Disease / Causal Agent	Symptoms	Resistant Sources / Management
1	Rhizome rot – *Pythium aphanidermatum*	Wilting, soft rot of rhizomes	*Suguna, Sudarsana, Roma* show tolerance. Use *Trichoderma* as biocontrol.
2	Leaf blotch – *Taphrina maculans*	Brown necrotic spots	*Suvarna, Suguna* resistant.
3	Bacterial wilt – *Ralstonia solanacearum*	Sudden yellowing, vascular browning	Field screening in endemic areas used to identify tolerant lines.
4	Nematodes (*Meloidogyne* spp.)	Root galls, poor growth	Integrated approach using *Paecilomyces lilacinus* and resistant clones.

2.4.9 Released Varieties and Achievements

S. No.	Variety / Clone	Parentage / Origin	Salient Features
1	**Suvarna**	Clonal selection from 'Duggirala'	High yield (30 t ha^{-1}); tolerant to leaf blotch; curcumin 6 %.
2	**Suguna**	Kerala selection	Resistant to rhizome rot; bright colour; 5 % curcumin.
3	**Sudarsana**	IISR selection	High oil (6 %); early maturity; compact clump.
4	**Prabha**	High-yield line from Alleppey	Curcumin > 6 %; disease tolerant.
5	**Roma**	Odisha selection	Adapted to rain-fed uplands; stable yield.
6	**Megha Turmeric-1**	ICAR RC NEH, Meghalaya	Drought tolerant; curcumin 5 %; bright yellow rhizome.

Achievements – Productivity doubled over traditional clones; standardized tissue-culture propagation; identified molecular markers for quality traits; established integrated disease-management protocols linked with resistant varieties.

2.4.10 Quality Evaluation and Biochemical Screening

» **Curcumin Determination:** Spectrophotometric quantification at 425 nm after solvent extraction.

» **Essential Oil Analysis:** GC–MS identification of turmerone, zingiberene, β-sesquiphellandrene.

» **Oleoresin Content:** Evaluated by solvent recovery and residue weight (> 12 %).

- **Colour Value:** Measured as E 1%1cm at 425 nm; export grade > 150.
- **Molecular Chemotyping:** Correlates curcumin biosynthetic genes (*CURS1–3, DCS*) with phenotypic pigmentation.

2.4.11 Molecular and Genomic Advances

1. **Genetic Diversity Analysis:** RAPD, ISSR, and AFLP markers reveal broad variation among Indian landraces.
2. **Gene Expression Studies:** Transcriptomic profiling identified key enzymes (curcumin synthase, diketide-CoA synthase).
3. **Functional Genomics:** Overexpression of curcumin pathway genes in model systems clarifies metabolic flux.
4. **Marker Assisted Selection:** SSR markers linked to high-curcumin and rot tolerance facilitate early screening.
5. **Proteomics and Metabolomics:** Identify stress-responsive proteins and secondary metabolites under biotic pressure.
6. **Genome Resources:** Draft transcriptome (> 60,000 unigenes) available for gene mining and SNP discovery.

2.4.12 Propagation and Nursery Practices

- **Seed Rhizome Selection:** Medium size (25 g) pieces with one healthy bud; taken from disease-checked stock.
- **Treatment:** Hot-water dip (52 °C for 10 min) + fungicide before planting.
- **Planting:** 30 × 25 cm spacing in raised beds enriched with organic manure.
- **Mulching:** Essential for moisture conservation and weed control.
- **Micropropagation:** Shoot-tip culture produces uniform, disease-free planting material with >95 % field survival.

These standardized propagation systems underpin breeding programs by ensuring true-to-type and pathogen-free material.

2.4.13 Integration of Breeding Approaches

S. No.	Approach	Contribution to Improvement
1	Clonal selection	Base of most high-yielding cultivars.
2	Mutation breeding	Created compact, high-pigment variants.
3	Somatic hybridization	Combined colour and aroma traits from related species.
4	Tissue culture	Rapid multiplication and pathogen elimination.
5	Molecular marker analysis	Linked curcumin and resistance traits.
6	Integrated field screening	Ensured stability and adaptability of released varieties.

2.5 Nutmeg (*Myristica fragrans* Houtt.)

2.5.1 Introduction

Nutmeg, often referred to as the **"aromatic gem of the tropics,"** is a highly valued spice tree known for its distinct fragrance, warm flavour, and multifaceted uses in culinary, medicinal, and industrial applications. It belongs to the family **Myristicaceae**, which includes about 100 species distributed throughout Southeast Asia, the Pacific Islands, and tropical America. The economic products, **nutmeg** and **mace**, are obtained from the **seed and aril**, respectively, of the same fruit. *Myristica fragrans*, native to the **Moluccas (Spice Islands) of Indonesia**, is the only species cultivated commercially for spice production.

Nutmeg was one of the first spices that drove the early maritime spice trade linking Asia with Europe. The crop was introduced to India in the 18th century and is now widely cultivated in the **humid tropics of Kerala, Karnataka, Tamil Nadu, and the Andaman–Nicobar Islands.** It grows well under partial shade and is often intercropped with coconut, arecanut, and coffee, forming a vital component of spice-based agroforestry systems. The tree is a **large evergreen dioecious species** reaching up to 20 m in height, bearing yellow drupaceous fruits. The dried kernel is the nutmeg, while the dried, crimson-red aril surrounding the seed constitutes mace, a separate high-value spice. Both are rich in **essential oils (7–15 %)** containing **myristicin**, **sabinene**, **terpenes**, and **elemicin**, responsible for the characteristic aroma and flavour.

2.5.2 Botanical Description and Cytology

- **Family:** Myristicaceae
- **Genus:** *Myristica*
- **Species:** *M. fragrans* Houtt.
- **Chromosome number:** 2n = 46
- **Growth Habit:** Large, evergreen, dioecious tree with dense canopy and straight trunk.
- **Leaves:** Simple, alternate, elliptic-lanceolate, glossy dark green, entire margins; 10–15 cm long.
- **Flowers:** Small, yellowish-white, unisexual; male flowers in axillary racemes; female solitary or few together.
- **Fruit:** Fleshy, yellow, globose drupe (~5–7 cm diameter) that splits open at maturity, revealing the bright crimson aril (mace) surrounding the seed (nutmeg).
- **Seed:** Hard, oval, oily kernel with reticulated testa.

Cytologically, nutmeg is diploid and cross-pollinated, with a high degree of heterozygosity. Its dioecious nature (separate male and female trees) makes early sex identification one of the primary breeding challenges. Vegetative propagation ensures true-to-type reproduction but delays seedling sex determination.

2.5.3 Economic Importance

1. **Spice and Condiment Use** Nutmeg and mace are premium flavouring agents in bakery, confectionery, sauces, and beverages. Mace imparts a delicate aroma, while nutmeg gives warm, spicy notes.
2. **Medicinal Value** The essential oils possess carminative, stimulant, and antimicrobial properties. Active compounds such as myristicin and elemicin exhibit analgesic and anti-inflammatory effects.
3. **Industrial Applications** Oils and oleoresins are used in perfumery, toothpaste, pharmaceutical balms, and cosmetic formulations.
4. **Economic and Export Significance** India exports over 2,000 t annually; value-added products like nutmeg butter and oil fetch premium prices.
5. **Agroforestry Role** Long-lived tree crop suited for intercropping under coconut and arecanut; contributes to carbon sequestration and sustainable income for smallholders.

2.5.4 Breeding Objectives

Given the tree's long juvenile phase, dioecious nature, and perennial habit, breeding objectives emphasize **yield stability, early sex identification, disease tolerance, and quality enhancement.**

1. **Increase Yield and Fruit Set**
   » Development of high-yielding female trees producing > 3 000 fruits annually.
   » Selection for high fruit retention and uniform maturation.
2. **Early Sex Identification and Balanced Sex Ratio**
   » Techniques to determine sex at seedling or juvenile stage to optimize orchard composition (1 male: 10–15 females).
3. **Improved Quality**
   » Enhancement of essential-oil content (>12 %), myristicin concentration, and uniform aroma in both nutmeg and mace.
4. **Resistance to Diseases and Pests**
   » Tolerance to **fruit rot** (*Colletotrichum gloeosporioides*), **dieback**, and **leaf spot** (*Phyllosticta myristicae*).
   » Resistance to **root-knot nematodes** and **borers**.
5. **Compact Growth and Early Bearing**
   » Dwarf or semi-dwarf canopy types for high-density planting and easier harvesting.
6. **Adaptation and Environmental Tolerance**
   » Varieties suitable for both coastal and hilly humid tropics with waterlogging tolerance and wind firmness.

2.5.5 Genetic Resources and Germplasm

a. Centers of Origin and Distribution

The **Moluccas Islands (Indonesia)** are the primary center of origin. India serves as a secondary diversification center, particularly in Kerala and Karnataka, where farmer selections have produced distinct ecotypes.

b. Germplasm Collections

S. No.	Institute / Location	Accessions	Remarks
1	**ICAR–Indian Institute of Spices Research (IISR), Kozhikode**	> 200	National repository including male, female, and wild types.
2	**Regional Station, Appangala (Karnataka)**	100	Germplasm from Western Ghats maintained for evaluation.
3	**Kerala Agricultural University (KAU), Vellanikkara**	80	Field gene bank emphasizing yield and oil quality.

c. Important Landraces / Selections

S. No.	Cultivar / Line	Distinct Features
1	*Arikkulam local*	High fruit set; compact canopy; 10–12 % oil.
2	*IISR Vishwashree*	Clonal female line; high mace yield.
3	*IISR Keralashree*	Resistant to fruit rot; suitable for shade.
4	*IISR Nutmeg-1*	High myristicin (7 %); uniform bearing.
5	*Wayanadan type*	Large fruit, thick aril, moderate oil.

2.5.6 Floral Biology and Pollination

Nutmeg exhibits **dioecy**, with distinct male and female trees.

- » **Flowering Period:** January – May, varying by locality.
- » **Pollination:** By small insects, mainly **thysanopterans and flies**; occasionally by wind.
- » **Male Flowers:** Arranged in clusters (panicles) with numerous stamens; produce abundant pollen.
- » **Female Flowers:** Solitary or few per node; ovary superior, single locule with one ovule.
- » **Fruit Development:** 7 – 9 months after pollination; fruit splits open at maturity revealing mace and nutmeg.
- » **Seed Viability:** Short-lived, retaining viability only for 4 – 6 weeks.

2.5.7 Breeding and Hybridization Methods

a. Selection Breeding

Steps

1. **Field Survey:** Identify superior female trees with high yield, bold fruits, and thick aril.
2. **Vegetative Propagation:** Grafting or softwood grafting to maintain elite genotypes.

3. **Evaluation:** Assess yield, oil content, and disease resistance across seasons.
4. **Stabilization:** Clonal lines multiplied and recommended for release.

Notable selections include *IISR Vishwashree*, *Keralashree*, and *IISR Nutmeg-1*.

b. Controlled Hybridization

Hybridization is limited because of dioecy and long juvenile phase but occasionally performed using grafted parental lines.

Procedure

1. Identify synchronously flowering male and female trees.
2. Bag female flowers before anthesis.
3. Collect fresh pollen from male trees and apply manually.
4. Tag pollinated flowers; record parentage.
5. Harvest mature fruits; extract and germinate seeds.
6. Raise seedlings for early sex determination and selection.

The progeny are screened for vigour, female sex ratio, and yield potential.

c. Sex Identification and Early Selection

One of the greatest challenges in nutmeg breeding is **sex determination**. Sex can only be identified after 5–8 years under natural conditions, when flowering begins. Recent advances include:

» **Morphological indicators:** Slightly broader leaves and thicker stems in females, though unreliable.

» **Biochemical markers:** Differential peroxidase and esterase enzyme profiles between sexes.

» **Molecular markers:** DNA markers (RAPD, ISSR) linked to male-specific bands used for early seedling screening, enabling culling of excess males in nurseries.

d. Mutation Breeding

Physical and chemical mutagens have been used experimentally to induce compact growth and early bearing, though large-scale success is limited due to long generation time.

e. Clonal Propagation Techniques

» **Softwood grafting** on six-month-old seedlings is the standard method for clonal multiplication.

» **Approach grafting** and **air layering** are also successful in high-humidity conditions.

» Grafted plants bear within 4 – 5 years, halving the juvenile period.

f. Biotechnological Interventions

1. **Micropropagation:** Shoot-tip culture standardized for rapid multiplication of elite female genotypes.
2. **Somatic Embryogenesis:** Callus cultures induced from immature embryos for regeneration.
3. **Molecular Characterization:** SSR and AFLP markers used for genetic diversity and sex-linked marker development.

4. **Tissue Culture Pathogen Indexing:** Ensures disease-free planting stock for large-scale propagation.

2.5.8 Major Diseases and Breeding for Resistance

S. No.	Disease / Causal Organism	Symptoms	Resistant Sources / Control
1	Fruit rot – *Colletotrichum gloeosporioides*	Premature fruit fall; aril decay	*IISR Keralashree* tolerant; sanitation + copper sprays.
2	Leaf spot – *Phyllosticta myristicae*	Brown necrotic lesions	Moderate tolerance in *Wayanadan* type.
3	Dieback – *Botryodiplodia theobromae*	Branch drying	Pruning and biocontrol with *Trichoderma*.
4	Root-knot nematode – *Meloidogyne incognita*	Root galls	Resistant rootstocks identified among wild *Myristica* spp.

Screening for disease resistance is integrated into clonal evaluation trials under endemic conditions.

2.5.9 Released Varieties and Performance

S. No.	Variety / Clone	Origin / Type	Distinct Features
1	**IISR Vishwashree**	Clonal female line from Kerala	High yield (2 800 fruits tree^{-1}); bold nutmeg, thick mace, 12 % oil.
2	**IISR Keralashree**	Selection from Wayanadan local	Resistant to fruit rot; high mace recovery; compact canopy.
3	**IISR Nutmeg-1**	Selection from farmer's field	High myristicin (7 %); stable yield; early bearing.
4	**Arikkulam Local**	Traditional clone	Good flavour; 10 % oil; moderate yield.

2.5.10 Quality Evaluation and Biochemical Screening

» **Essential Oil Estimation:** Steam distillation followed by GC–MS quantification of myristicin, sabinene, safrole, and elemicin.

» **Mace Colour and Recovery:** Measured by chromameter (L*, a*, b* values) and aril weight ratio.

» **Nutmeg Butter Content:** Soxhlet extraction; ideal range 30 – 40 %.

» **Spice Volatility Index:** Ratio of volatile to fixed oil; indicator of aroma intensity.

» **Molecular Chemotyping:** Correlates chemical composition with genotype, ensuring stability in breeding lines.

2.5.11 Advances in Molecular and Genomic Research

1. **Molecular Sexing Markers:** RAPD primers (e.g., OPA-12, OPC-2) produce male-specific bands; conversion to SCAR markers allows accurate early sex diagnosis.

2. **Genetic Diversity Assessment:** SSR and AFLP markers reveal moderate diversity among Indian accessions, reflecting narrow base from few introductions.
3. **Gene Expression Studies:** RT-PCR analysis identifies differential expression of floral meristem genes in male vs. female buds.
4. **Transcriptome Sequencing:** Initial de novo assembly identifies genes related to essential-oil biosynthesis and dioecy regulation.
5. **Biotechnological Prospecting:** Somatic embryogenesis coupled with molecular fingerprinting aids conservation of elite female clones.

2.5.12 Propagation and Nursery Practices

» **Seed Propagation:** Fresh seeds sown immediately after harvest; seedlings transplanted after 12 months.

» **Grafting:** Softwood grafting is preferred for maintaining elite female genotypes.

» **Shade and Mulching:** Young plants require 50 % shade and organic mulch for moisture retention.

» **Spacing:** 8 × 8 m under monocrop; 10 × 10 m in intercrop systems.

» **Fertilization:** Balanced N:P:K (50:50:50 g plant^{-1}) with organic manures promotes early growth.

These practices are integrated into breeding trials to ensure uniform evaluation and orchard establishment.

2.5.13 Integration of Conventional and Modern Approaches

S. No.	Approach	Contribution to Nutmeg Improvement
1	Selection breeding	Identification of elite female trees for yield and oil quality.
2	Early sex identification	RAPD/SCAR markers enable seedling-level screening.
3	Clonal propagation	Softwood grafting ensures uniform orchards and early bearing.
4	Molecular chemotyping	Standardizes flavour and oil composition.
5	Tissue culture	Mass multiplication and germplasm conservation.
6	Disease-resistance screening	Integration of field and molecular evaluation for durable tolerance.

2.6 Cinnamon (*Cinnamomum verum* J. Presl)

2.6.1 Introduction

Cinnamon, known as the **"true bark spice of the East"**, is one of the oldest recorded spices in human history. Belonging to the family **Lauraceae**, *Cinnamomum verum* J. Presl (syn. *C. zeylanicum* Blume) is native to the **wet evergreen forests of Sri Lanka and the Western Ghats of India.** The dried inner bark of young shoots yields the well-known aromatic spice "cinnamon," while related species such as *C. cassia* (Chinese cassia) produce similar but coarser barks marketed separately.

Cinnamon's fragrance and flavour come from its **essential oils (1–4 %)**, rich in **cinnamaldehyde, eugenol**, and **linalool**. The spice is valued for culinary, medicinal, and industrial purposes. India, Sri Lanka, Indonesia, and Seychelles dominate global production; within India, the crop is cultivated in **Kerala, Tamil Nadu, Karnataka, and Meghalaya**, often under mixed plantations or agroforestry systems with coconut, arecanut, and clove.

The tree is a **medium-sized, evergreen perennial** reaching 8–10 m. It thrives in warm, humid tropics (22–30 °C, 1,250–2,500 mm annual rainfall) on well-drained loams and lateritic soils. Improvement programs in India have targeted high-oil bark types, uniform leaf aroma for cinnamon leaf oil, and resistance to leaf-spot and die-back diseases.

2.6.2 Botanical Description and Cytology

- **Family:** Lauraceae
- **Genus:** *Cinnamomum*
- **Species:** *C. verum* J. Presl
- **Chromosome number:** 2n = 24
- **Growth habit:** Evergreen tree with multiple stems; coppices freely after cutting.
- **Leaves:** Opposite or sub-opposite, simple, ovate-lanceolate, 8–18 cm long; three prominent parallel veins from the base; aromatic when crushed.
- **Flowers:** Small, bisexual, yellowish-white, borne in terminal or axillary panicles.
- **Fruit:** One-seeded berry (drupe) on a cup-shaped receptacle.

Cytological studies reveal a diploid genome with moderate heterozygosity. The tree's ability to coppice facilitates vegetative perpetuation of superior types, which is central to clonal breeding and maintenance of varietal purity.

2.6.3 Economic Importance

1. **Culinary Uses**
    - Bark powder and quills flavour confectionery, bakery goods, and curries.
    - Cinnamon leaf oil substitutes for clove oil in flavour and fragrance industries.
2. **Pharmaceutical and Nutraceutical Value**
    - Cinnamaldehyde and eugenol exhibit antimicrobial, antifungal, and insulin-sensitizing properties.
    - Used in Ayurveda and modern medicine for digestive and circulatory disorders.
3. **Industrial Applications**
    - Essential oils in perfumery, toothpaste, soaps, and cosmetics.
    - Extracts serve as natural preservatives and antioxidants.

4. **Export and Economic Value**
   - Sri Lanka leads global exports, while India's share is increasing through high-oil clones from IISR.
   - Leaf-oil industry provides additional rural employment.
5. **Agro-ecological Benefits**
   - Suited for contour planting on hill slopes; its deep root system controls erosion and supports soil carbon enrichment.

2.6.4 Breeding Objectives

Because cinnamon is a **perennial, cross-pollinated species**, breeding programs emphasize clonal selection and quality improvement.

1. **Enhanced Yield and Bark Quality**
   - Increased bark thickness, smoothness, and ease of peeling.
   - Uniform quill formation and high recovery (≥ 50 % peelable bark).
2. **High Essential-Oil Content and Improved Composition**
   - Cinnamaldehyde ≥ 65 % in bark oil, eugenol ≥ 80 % in leaf oil.
3. **Resistance to Diseases and Pests**
   - Tolerance to **leaf spot** (*Phyllosticta cinnamomi*), **shoot die-back** (*Diplodia* spp.), and **scale insects**.
4. **Adaptation and Growth Habit**
   - Compact, multi-stemmed types suitable for coppice harvest cycles of 2–3 years.
   - Broad adaptability to mid-elevation humid tropics.
5. **Improved Regeneration and Uniformity**
   - Clones with high coppicing ability and synchronized sprouting after harvest.
6. **Leaf-Oil Type Development**
   - Selection of lines rich in eugenol or cinnamyl acetate for diversified oil industry.

2.6.5 Genetic Resources and Germplasm

a. Centers of Diversity

Primary center: Sri Lanka; secondary diversification: Southern India and Seychelles.

b. Germplasm Repositories

S. No.	Institute / Location	Accessions Maintained	Remarks
1	ICAR–IISR (Kozhikode)	250 +	National gene bank; bark- and leaf-oil chemotypes catalogued.
2	Appangala Regional Station (Karnataka)	≈ 120	High-altitude and high-oil lines.

3	State Agricultural Universities (Kerala, Tamil Nadu)	≈ 80	Local farmer selections.

c. Important Clones and Landraces

S. No.	Clone / Landrace	Distinctive Traits
1	*Nithyashree*	High-oil bark (3.2 %); compact canopy.
2	*Dhanasree*	Leaf-oil rich in eugenol (82 %); early coppicing.
3	*Navashree*	Disease-tolerant; large leaves; rapid growth.
4	*Konkan Tej*	Suited for coastal laterite; good coppicing.
5	*Sri Lanka Type*	Superior quill quality; standard export grade.

2.6.6 Floral Biology and Pollination

- **Flowering Season:** February – April.
- **Inflorescence:** Panicles of small bisexual flowers.
- **Pollination:** Mainly entomophilous (bees and flies), partly self-compatible.
- **Fruiting:** Single-seeded berries ripen by September – October.
- **Seed Viability:** Short; germinates within 2 weeks.

Controlled pollination is feasible because of bisexual flowers; however, natural cross-pollination ensures genetic diversity used in selection programs.

2.6.7 Breeding and Hybridization Methods

a. Clonal Selection

The backbone of cinnamon improvement.

Procedure

1. Identify plus trees with superior bark yield, smooth peel, and high oil content.
2. Propagate vegetatively by cuttings or air-layering.
3. Evaluate for bark yield (t ha^{-1} $year^{-1}$), oil composition, and coppicing vigour.
4. Multiply elite clones for regional testing.

Notable clones: *Nithyashree*, *Dhanasree*, *Navashree*, *Konkan Tej*.

b. Controlled Hybridization

Used experimentally to combine high-bark and high-oil traits.

Steps

1. Bag selected inflorescences before anthesis.
2. Apply pollen from the desired male parent using a brush.

3. Tag and record crosses.
4. Collect seeds and raise seedlings for evaluation.

F_1 hybrids show heterosis for growth and leaf area; selected hybrids are cloned for stability.

c. Vegetative Propagation

Stem cuttings from semi-hardwood shoots root readily under mist; air-layering and grafting on wild *Cinnamomum camphora* rootstocks are also successful. Such methods preserve elite genotypes.

d. Mutation Breeding

Gamma irradiation (2–5 kR) was used to induce dwarfism and early coppicing; mutant lines were screened for oil composition changes.

e. Biotechnological Approaches

» **Micropropagation:** Nodal and shoot-tip culture enable mass propagation and virus-free stocks.

» **Somaclonal Variation:** In vitro variants screened for leaf-spot resistance.

» **Molecular Markers:** RAPD and ISSR used for clonal identification and chemotype characterization.

» **Cryopreservation:** Shoot-tips stored in liquid nitrogen for germplasm security.

2.6.8 Major Diseases and Resistance Breeding

S. No.	Disease / Causal Agent	Symptoms	Resistant Sources / Management
1	Leaf spot – *Phyllosticta cinnamomi*	Brown patches on leaves	*Navashree* tolerant; biocontrol with *Trichoderma*.
2	Shoot die-back – *Diplodia sp.*	Tip dieback of shoots	Pruning + fungicidal spray; select vigourous clones.
3	Root rot – *Pythium sp.*	Wilting, coppice death	Resistant rootstocks and soil drainage.
4	Scale insect & mite	Yellowing, reduced growth	Natural predators + oil sprays; tolerant clones used.

2.6.9 Released Varieties and Performance

S. No.	Variety / Clone	Origin / Type	Distinct Features
1	**Nithyashree**	IISR selection	High bark yield (1.5 t ha^{-1}); oil 3 %; cinnamaldehyde 70 %.
2	**Dhanasree**	Leaf-oil type clone	Eugenol 80 %; fast regrowth; disease tolerant.
3	**Navashree**	High-altitude clone	Robust growth; leaf-spot tolerant.
4	**Konkan Tej**	Coastal selection	Adapted to lateritic soils; early coppicing.
5	**Sri Lanka Type (standard)**	Imported reference line	Premium quill quality for export blend.

2.6.10 Quality Evaluation and Biochemical Screening

» **Bark-Oil Analysis:** Steam distillation followed by GC–MS to quantify cinnamaldehyde, eugenol, and linalool.

» **Leaf-Oil Analysis:** GC profiling of eugenol and cinnamyl acetate ratios.

» **Quill Texture:** Measured for fiber content and curl uniformity.

» **Cinnamaldehyde Index:** Indicator of flavour strength; values > 65 % preferred.

» **Molecular Chemotyping:** DNA barcodes correlated with chemical profiles for traceability and quality control.

2.6.11 Molecular and Genomic Advances

1. **Genetic Diversity Assessment:** RAPD and SSR markers confirm moderate diversity among Indian clones.
2. **Gene Expression:** Transcriptomic studies identify key enzymes in phenylpropanoid and cinnamaldehyde pathways (PAL, 4CL, CAD).
3. **Molecular Identification of Chemotypes:** Marker-trait association links specific alleles to high cinnamaldehyde content.
4. **Marker-Assisted Selection:** Pilot screening for oil composition and disease resistance underway at IISR.
5. **In vitro Mutagenesis and Screening:** Cell cultures exposed to UV and chemical mutagens to generate variants with enhanced oil biosynthesis.

2.6.12 Propagation and Nursery Management

» **Seed Propagation:** Seeds sown fresh in nurseries; viability 1–2 months.

» **Vegetative Propagation:** Semi-hardwood cuttings root under mist (70–80 % success); air-layering and grafting are used for elite clones.

» **Planting Density:** 1.5 × 1.5 m (≈ 4400 plants ha^{-1}) for bark harvest rotation.

» **Coppicing:** Main harvest 2–3 years after planting; shoots cut at 15 cm above ground to regenerate.

» **Fertilization and Mulching:** Organic mulch promotes soil moisture and microbial activity for better regrowth.

2.6.13 Integration of Breeding Approaches

S. No.	Approach	Contribution to Improvement
1	Clonal selection	Main strategy for yield and quality enhancement.
2	Controlled hybridization	Introduces novel oil combinations and vigour.
3	Mutation breeding	Dwarf and early-coppicing lines.

4	Micropropagation	Rapid multiplication of elite clones.
5	Molecular marker analysis	Chemotype identification and genetic fidelity.
6	Integrated field screening	Ensures disease tolerance and site adaptability

2.7 Clove (*Syzygium aromaticum* L.)

2.7.1 Introduction

Clove, the **"Queen of Spices,"** is one of the most valuable and universally recognized aromatic tree spices. It belongs to the family **Myrtaceae**, genus *Syzygium*, and is native to the **Moluccas (Spice Islands of Indonesia)**. The dried, unopened flower buds constitute the commercial product known as **clove**. Beyond its culinary importance, it is valued for its **essential oil, pharmaceutical**, and **industrial uses**.

Clove was introduced into India in the late 18th century and is now widely cultivated in **Kerala, Tamil Nadu, Karnataka, and the Andaman–Nicobar Islands**. Globally, Indonesia, Madagascar, Tanzania, and Sri Lanka are major producers. The tree is **evergreen, medium-sized (10–15 m tall)**, with dense foliage and glossy leaves, thriving best in humid tropics (20–30 °C) with 1500–2500 mm annual rainfall.

The spice derives its powerful aroma and flavour primarily from **eugenol**, which constitutes 70–85 % of the essential oil in buds and 90–95 % in leaves. Clove oil is widely used in flavourings, perfumery, and dentistry. Its bark and leaves yield additional by-products, making the crop multifunctional and economically sustainable for spice-based agroforestry.

2.7.2 Botanical Description and Cytology

- **Family:** Myrtaceae
- **Genus:** *Syzygium*
- **Species:** *S. aromaticum* L. (syn. *Eugenia caryophyllata* Thunb.)
- **Chromosome number:** 2n = 22
- **Habit:** Evergreen tree with conical canopy; young flushes pinkish-red.
- **Leaves:** Opposite, elliptic-lanceolate, 8–12 cm, thick, aromatic; dotted with oil glands.
- **Inflorescence:** Terminal or axillary cymes bearing numerous buds.
- **Flower buds:** Four-lobed calyx with prominent tube (hypanthium) turning crimson on maturity; buds harvested just before opening.
- **Flowers:** Bisexual, actinomorphic, cream-coloured with numerous stamens.
- **Fruit:** One-seeded berry, purple when ripe.

Cytologically, clove is diploid and predominantly **cross-pollinated**, though partial self-compatibility occurs. Its longevity (over 75 years) and slow juvenile phase (7–8 years to bearing) make improvement a long-term process.

2.7.3 Economic Importance

1. **Culinary and Flavouring Uses**
   - Dried flower buds and powder flavour bakery, confectionery, pickles, sauces, and spice blends.
   - Clove is an essential ingredient in Indian garam masala and global cuisines.
2. **Essential Oil and Industrial Uses**
   - Clove bud oil (15–20 %) rich in eugenol used in perfumery, toothpaste, soaps, and pharmaceutical preparations.
   - Leaf oil (yield > 2 %) is a commercial substitute for bud oil in flavour industry.
   - Clove stem oil serves as a raw material for synthetic vanillin.
3. **Medicinal and Pharmaceutical Applications**
   - Eugenol exhibits analgesic, antiseptic, and anesthetic properties; used in dental care.
   - Extracts show antioxidant, antiviral, and anti-inflammatory activities.
4. **Economic and Export Value**
   - India imports cloves to meet domestic demand, yet expansion of high-yielding plantations has boosted self-sufficiency.
   - The value-added clove oil trade contributes significantly to the spice economy.
5. **Agro-ecological and Environmental Value**
   - Shade-loving species ideal for intercropping under coconut, arecanut, and nutmeg; excellent windbreak.
   - Long-lived trees contribute to **carbon sequestration** and biodiversity conservation in homestead gardens.

2.7.4 Breeding Objectives

1. **High Yield and Early Bearing**
   - Increase flower-bud production per tree; select genotypes that start bearing by 5–6 years instead of 8–10.
   - Enhance the percentage of fruit set and synchronized flowering.
2. **Improved Oil Yield and Composition**
   - Bud oil ≥ 18 %, leaf oil ≥ 3 %, eugenol ≥ 85 %, with balanced β-caryophyllene for aroma.
3. **Resistance to Diseases and Pests**
   - Tolerance to **seedling wilt** (*Cylindrocladium quinqueseptatum*), **leaf spot** (*Colletotrichum gloeosporioides*), and **stem borer**.
4. **Adaptation and Growth Habit**
   - Compact, spreading canopy types for dense planting (6 × 6 m).

» Drought- and wind-tolerant lines for coastal belts.

5. **Quality Stability**

» Consistent oil composition and quill (bud) size across environments.

6. **Propagation Efficiency**

» Selection for good coppicing and high rooting in cuttings to facilitate vegetative multiplication.

2.7.5 Genetic Resources and Germplasm

a. **Centers of Origin and Distribution**

The **Moluccas Islands** are the primary center of origin; secondary diversity exists in **Sri Lanka, Madagascar, Zanzibar, and India.**

b. **Germplasm Holdings**

S. No.	Institute / Location	Accessions Maintained	Remarks
1	ICAR–Indian Institute of Spices Research (IISR), Kozhikode	120 +	National repository; morpho-chemical cataloguing.
2	Appangala Regional Station (Karnataka)	80	High-altitude types and disease-tolerant clones.
3	State Agricultural Universities (Kerala & Tamil Nadu)	≈ 50	Local farmer selections and progenies.

c. **Important Landraces / Clones**

S. No.	Clone / Landrace	Distinctive Features
1	*IISR Madagaskari*	High bud-oil yield (20 %); eugenol > 85 %.
2	*Kaniyambetta Local*	Early bearing; large buds; adaptable to shade.
3	*Karamoola Selection*	Resistant to leaf spot; compact canopy.
4	*Amini Island Type*	Coastal adaptation; thick peduncle; uniform flowering.
5	*IISR Kalpaka*	High oil recovery; excellent flavour stability.

2.7.6 Floral Biology and Pollination

» **Flowering Season:** December – March (Kerala); extended to May in hills.

» **Inflorescence:** Terminal cymose panicles with 40–100 buds.

» **Flower:** Bisexual; protandrous, encouraging cross-pollination.

» **Pollination:** Mainly by small bees and flies; occasional selfing observed.

» **Fruit Development:** Takes 6 months; fruit turns purplish-red when ripe.

» **Seed Viability:** Short; viable for 4–6 weeks; immediate sowing essential.

Flower biology aids synchronization of controlled pollination and selection of early-flowering genotypes.

2.7.7 Breeding and Hybridization Methods

a. Clonal Selection

The principal method used in India is due to cross-pollinated, perennial nature.

Procedure

1. Identify high-yielding mother trees (> 3 kg dried buds tree^{-1} year^{-1}).
2. Collect bud samples for oil analysis.
3. Propagate superior clones via grafting or cuttings.
4. Evaluate across multiple environments for yield and oil stability.
5. Multiply elite clones for release.

Released clones: *IISR Kalpaka, IISR Madagaskari, Karamoola Selection.*

b. Controlled Hybridization

Limited use because of long generation time. Conducted by manual transfer of pollen from selected donors to emasculated buds before anthesis.

Steps

1. Select synchronously flowering parents.
2. Bag buds before anthesis.
3. Hand-pollinate using fine brush.
4. Tag crosses and record parentage.
5. Harvest mature fruits, sow seeds, and evaluate progeny after 6–8 years.

Hybrid progenies are assessed for vigour, early bearing, and oil quality.

c. Mutation and Polyploid Breeding

Gamma irradiation and colchicine treatments on seedlings and shoot cultures have produced dwarf and early-flowering mutants, though few are stabilized.

d. Vegetative Propagation

- **Softwood grafting** and **air-layering** standard for elite clone multiplication.
- **Cuttings** treated with IBA (3 000 ppm) yield > 70 % rooting.
- Vegetatively propagated plants bear 2–3 years earlier than seedlings.

e. Biotechnological Approaches

1. **Micropropagation:** Nodal culture for large-scale production of disease-free plants.
2. **Molecular Characterization:** RAPD, ISSR, and SSR markers assess diversity and confirm clonal fidelity.
3. **Somaclonal Variation:** In-vitro lines screened for leaf-spot resistance.
4. **Cryopreservation:** Encapsulated shoot-tips stored for long-term germplasm conservation.

2.7.8 Major Diseases and Breeding for Resistance

S. No.	Disease / Causal Agent	Symptoms	Resistant Sources / Management
1	Seedling wilt – *Cylindrocladium quinqueseptatum*	Damping-off, seedling death	*IISR Kalpaka* tolerant; soil solarization + biocontrol.
2	Leaf spot – *Colletotrichum gloeosporioides*	Brown necrotic lesions	*Karamoola Selection* shows tolerance.
3	Stem borer	Girdling, branch dieback	Use tolerant clones; mechanical removal.
4	Scale insect	Leaf yellowing	Neem-oil spray and biological control.

Field screening in endemic zones combined with tissue-culture indexing ensures disease-free propagation.

2.7.9 Released Varieties and Performance

S. No.	Variety / Clone	Origin / Type	Salient Features
1	**IISR Kalpaka**	Selection from Kerala local	High oil (20 %); eugenol 85 %; tolerant to wilt.
2	**IISR Madagaskari**	Exotic introduction stabilized in India	Early bearing; bold buds; 18 % oil.
3	**Karamoola Selection**	Farmer selection	Disease tolerant; compact crown.
4	**Amini Island Type**	Coastal ecotype	Salt-tolerant; uniform flowering.

2.7.10 Quality Evaluation and Biochemical Screening

- » **Essential-Oil Estimation:** Steam-distillation of buds and leaves followed by GC–MS analysis of eugenol, β-caryophyllene, humulene.
- » **Oil Yield Index:** Weight of oil per 100 g dry bud; export standard ≥ 18 %.
- » **Volatile-Oil Stability:** Evaluated through accelerated aging tests.
- » **Flavour Index:** Ratio of eugenol : β-caryophyllene (ideal = 6 : 1).
- » **Molecular Chemotyping:** DNA fingerprints correlated with chemical profiles to ensure varietal authenticity.

2.7.11 Molecular and Genomic Advances

- » **Diversity Assessment:** RAPD and ISSR markers show moderate diversity among Indian and Indonesian accessions.
- » **Marker-Trait Association:** Specific RAPD bands linked with high eugenol content and disease tolerance.
- » **Gene Expression Studies:** Identification of key genes (EOMT, CFAT) regulating eugenol biosynthesis.

- **Transcriptomic Resources:** Partial EST libraries developed at IISR for future molecular breeding.
- **Marker-Assisted Clonal Verification:** DNA fingerprinting now used routinely for clone authentication in nurseries.

2.7.12 Propagation and Nursery Management

- **Seed Propagation:** Fresh seeds sown immediately; germination 70–80 %.
- **Vegetative Multiplication:** Grafting or cuttings preferred for elite clones.
- **Nursery Care:** Partial shade, high humidity, and organic mulch essential.
- **Planting Density:** 6 × 6 m spacing; intercropping with coconut or nutmeg recommended.
- **Manuring:** Organic compost + balanced N:P:K (50:50:75 g plant^{-1}) for early growth.

Such practices ensure uniform, vigourous material for breeding and production plantations.

2.8 Garcinia Species (*Garcinia indica*, *G. cambogia*, *G. gummi-gutta*, etc.)

2.8.1 Introduction

The genus *Garcinia* (family **Clusiaceae**, formerly **Guttiferae**) comprises over 200 species distributed across the **humid tropics of Asia, Africa, and Polynesia**. Among them, *Garcinia indica* Choisy (kokum), *G. cambogia* (syn. *G. gummi-gutta* Roxb.), and *G. xanthochymus* are economically important as spice, condiment, and nutraceutical crops. These species are native to the **Western Ghats and Northeast India**, representing a vital component of traditional food systems and indigenous medicine.

Garcinia indica, popularly known as **kokum**, is endemic to the Western Ghats of India, particularly **Goa, Karnataka, Kerala, and Maharashtra**. Its dried rind is used as a natural acidulant and colourant in curries, beverages, and pickles. The fruit is rich in **anthocyanins** and **hydroxycitric acid (HCA)**, while its seeds yield **kokum butter**, a solid fat comparable to cocoa butter, extensively used in confectionery and cosmetics.

Garcinia cambogia and *G. gummi-gutta*, distributed across southern India and Sri Lanka, are valued for their sour fruits and high HCA content, which has gained global attention as a potential anti-obesity agent. Owing to these nutraceutical and industrial attributes, *Garcinia* species are now recognized as promising underutilized spice crops requiring systematic breeding and genetic improvement.

2.8.2 Botanical Description and Cytology

- **Family:** Clusiaceae (Guttiferae)
- **Genus:** *Garcinia*
- **Important species:** *G. indica*, *G. cambogia*, *G. gummi-gutta*, *G. xanthochymus*
- **Chromosome number:** 2n = 44 (diploid)
- **Growth habit:** Evergreen, medium-sized trees (6–15 m) with dense canopy and straight bole.
- **Leaves:** Opposite, oblong-lanceolate, thick, glossy, exuding yellow resin when cut.
- **Flowers:** Unisexual or polygamous; borne in fascicles; female flowers larger than male.

- **Fruit:** Round berry with thick, fleshy pericarp (rind) and 4–8 seeds embedded in acidic pulp.
- **Seed:** Large, oily, enclosed in mucilaginous aril; source of kokum butter.

Cytologically, most *Garcinia* species are **diploid, cross-pollinated**, and exhibit moderate heterozygosity. Some are **dioecious or polygamous**, posing breeding challenges due to difficulty in sex identification at the seedling stage.

2.8.3 Economic Importance

1. **Culinary Uses**
   - Dried rind of *G. indica* and *G. cambogia* used as souring agents in curries and beverages, replacing tamarind.
   - *Kokum sharbat* and *sol kadhi* are popular traditional drinks in western India.
2. **Industrial and Pharmaceutical Applications**
   - Fruits contain **hydroxycitric acid (HCA)** (10–30 %), a citric-acid derivative that inhibits lipogenesis and appetite.
   - **Kokum butter** (from seeds) is a high-quality solid fat used in confectionery, chocolate, and cosmetics; melting point 38–40 °C, similar to cocoa butter.
   - Extracts serve as natural colourants and acidulants in the food industry.
3. **Medicinal and Nutraceutical Value**
   - Traditional medicine uses rind for digestive and anti-inflammatory purposes.
   - Modern pharmacological studies confirm antioxidant, anti-obesity, hepatoprotective, and cardiotonic effects of HCA and polyphenols.
4. **Environmental and Socio-economic Role**
   - *Garcinia* trees are integral to homestead and agroforestry systems in the Western Ghats.
   - The species contributes to **carbon sequestration, soil stabilization, and biodiversity conservation.**
   - Fruit collection provides seasonal income to rural communities.

2.8.4 Breeding Objectives

Given their perennial and partially dioecious nature, improvement programs for *Garcinia* focus on selection, clonal propagation, and utilization of natural diversity.

1. **Increased Fruit Yield and Quality**
   - Selection for high fruit-bearing female trees with regular bearing and large fruit size (≥ 80 g).
   - Uniform maturation and high rind-to-seed ratio.
2. **Improved Biochemical Composition**
   - Higher HCA content (≥ 25 %), balanced acidity, and anthocyanin pigmentation for processing quality.

3. **High Butter Recovery and Oil Quality**
   » Seed oil ≥ 40 %; high stearic and oleic acid composition; solidification point close to cocoa butter.

4. **Disease and Pest Resistance**
   » Tolerance to **die-back** (*Phytophthora palmivora*), **leaf spot**, and **fruit borer**.

5. **Environmental Adaptability**
   » Drought-tolerant and shade-enduring genotypes suited for mixed plantation systems.

6. **Reduced Juvenile Period and Early Sex Identification**
   » Development of early-bearing clones and molecular markers for sex determination at seedling stage.

2.8.5 Genetic Resources and Germplasm

a. Centers of Diversity

The **Western Ghats of India** represent the primary diversity center for *G. indica* and *G. gummi-gutta*, while *G. cambogia* occurs widely in Kerala, Tamil Nadu, and Sri Lanka.

b. Germplasm Holdings

S. No.	Institute / Location	Accessions Maintained	Remarks
1	ICAR–Indian Institute of Spices Research (IISR), Kozhikode	200 +	National repository; accessions of *G. indica*, *G. cambogia*, *G. gummi-gutta*.
2	KFRI (Kerala Forest Research Institute), Peechi	≈ 150	Wild collections and morphological descriptors.
3	Goa University / Dapoli (MPKV)	≈ 120	Landraces from Konkan region with variable rind colour and HCA content.

c. Notable Accessions and Landraces

S. No.	Clone / Landrace	Distinctive Traits
1	*IISR Kokum-1*	High fruit yield (5 t ha^{-1}); HCA 28 %.
2	*IISR Kokum-2*	Early bearing; deep purple rind; high anthocyanin.
3	*Konkan Ratna*	Large fruit (80 g); butter yield 40 %; tolerant to drought.
4	*Goa Local Red*	Uniform colouration; popular beverage type.
5	*Malabar Type*	High pulp recovery; regular bearer; adapted to shade.

2.8.6 Floral Biology and Pollination

» **Flowering Season:** December – February in *G. indica*; February – April in *G. cambogia*.

» **Sexual System:** Predominantly dioecious; some trees polygamous with both male and hermaphrodite flowers.

» **Inflorescence:** Axillary clusters of 5–10 flowers.

» **Pollination:** Entomophilous (small bees and flies).

» **Fruit Set:** 50–70 % in female trees; parthenocarpy rare.

» **Fruit Maturity:** 5–6 months after flowering.

» **Seed Viability:** 2–3 months under ambient conditions.

2.8.7 Breeding and Hybridization Methods

a. Selection Breeding

Since sexual reproduction is limited by dioecy and long juvenility, **selection of superior female trees** forms the cornerstone of improvement.

Steps

1. **Field Survey:** Identify elite female trees based on yield, fruit size, rind thickness, and biochemical quality.
2. **Clonal Propagation:** Propagate through softwood grafting or air-layering.
3. **Evaluation:** Analyze fruit yield, HCA content, and anthocyanin profile over multiple seasons.
4. **Stabilization:** Multiply superior clones for release and distribution.

Released selections: *IISR Kokum-1, IISR Kokum-2, Konkan Ratna.*

b. Controlled Hybridization

Experimental hybridization between *G. indica* and *G. gummi-gutta* aims to combine high pigment with high HCA.

Procedure

1. Select synchronously flowering male and female trees.
2. Bag female inflorescences before anthesis.
3. Hand-pollinate with fresh pollen using camel-hair brush.
4. Label, record, and harvest fruits after 5 months.
5. Germinate seeds and evaluate progeny for sex, vigour, and fruit quality.

c. Sex Identification Techniques

Since sex can only be distinguished after 5–7 years by flowering, **early diagnostic tools** are vital.

» **Morphological:** Slightly broader leaves and thicker stems in females (not reliable).

» **Biochemical:** Differential peroxidase and esterase profiles between sexes.

» **Molecular:** RAPD and ISSR markers linked to male-specific loci allow seedling-stage sexing with >90 % accuracy.

d. Mutation and Polyploid Breeding

Mutagenesis using gamma rays (2–6 kR) has produced dwarf, precocious variants in *G. indica*. Polyploid induction (4n) by colchicine resulted in larger fruits but lower seed viability.

e. **Vegetative Propagation**

» **Softwood grafting** on 8-month-old rootstocks is the most reliable technique.

» **Air-layering** successful during monsoon; rooting with IBA (2,000 ppm).

» Vegetative plants bear within 4–5 years, compared with 7–9 years in seedlings.

f. **Biotechnological Interventions**

1. **Micropropagation:** Shoot-tip and nodal culture established for rapid clonal multiplication.
2. **Embryo Rescue:** Applied to interspecific hybrids with post-zygotic incompatibility.
3. **Molecular Characterization:** SSR and ISSR markers used for diversity analysis and identification of chemotypes.
4. **In vitro Germplasm Conservation:** Encapsulated buds stored under cryogenic conditions.
5. **Somaclonal Variation Screening:** Tissue-culture variants analyzed for enhanced pigment and HCA levels.

2.8.8 Major Diseases and Breeding for Resistance

S. No.	Disease / Causal Agent	Symptoms	Resistant Sources / Management
1	Dieback – *Phytophthora palmivora*	Twig blight, branch dieback	*IISR Kokum-2* and *Malabar Type* tolerant.
2	Leaf spot – *Colletotrichum gloeosporioides*	Brown necrotic lesions	*Konkan Ratna* resistant.
3	Fruit rot – *Phytophthora nicotianae*	Premature fruit drop	Field sanitation and biocontrol with *Trichoderma*.
4	Fruit borer – *Conogethes punctiferalis*	Bore holes in fruits	Integrated pest management with pheromone traps.

Screening under endemic conditions and molecular tagging of tolerant genotypes form integral parts of breeding programs.

2.8.9 Released Varieties and Performance

S. No.	Variety / Clone	Origin / Type	Distinct Features
1	**IISR Kokum-1**	Selection from Western Ghats	High yield (5 t ha^{-1}); HCA 28 %; bright red rind.
2	**IISR Kokum-2**	Elite selection from Goa	Early bearing; anthocyanin 120 mg 100 g^{-1}; disease tolerant.
3	**Konkan Ratna**	Selection from Dapoli (MPKV)	Large fruits (80 g); butter yield 40 %; drought hardy.
4	**Malabar Type**	Local farmer clone	Adapted to shade; stable bearing; moderate HCA.
5	**Goa Local Red**	Traditional cultivar	High pigment; popular for beverages.

2.8.10 Quality Evaluation and Biochemical Screening

- **HCA Estimation:** Titrimetric and HPLC methods quantify hydroxycitric acid; values 10–30 %.
- **Anthocyanin Content:** Spectrophotometric assay (pH-differential method); correlates with rind colour intensity.
- **Butter Quality:** Seed oil profile analyzed by GC–MS for stearic, palmitic, oleic acids; melting range 38–40 °C.
- **Antioxidant Capacity:** DPPH and FRAP assays; high in purple-rind types.
- **Molecular Chemotyping:** DNA markers correlated with HCA and pigment profiles to define "chemovarieties."

2.8.11 Molecular and Genomic Advances

1. **Genetic Diversity:** RAPD, ISSR, and SRAP analyses reveal wide variability among *G. indica* accessions in Western Ghats.
2. **Molecular Sexing:** SCAR markers derived from male-specific RAPD bands enable seedling-stage sex identification.
3. **Gene Expression Studies:** qRT-PCR profiling identifies upregulation of citrate lyase and hydroxycitric acid synthase genes in high-HCA lines.
4. **Genome Resources:** De novo transcriptome sequencing (> 60 000 unigenes) provides molecular insight into anthocyanin and HCA biosynthesis pathways.
5. **Metabolomics:** LC–MS-based metabolite fingerprinting distinguishes chemotypes by pigment and organic-acid profiles.

2.8.12 Propagation and Nursery Practices

- **Seed Propagation:** Seeds sown fresh; viability 2–3 months; germination 70–80 %.
- **Grafting:** Softwood grafting on 1-year-old rootstocks ensures uniform orchards.
- **Air-Layering:** Performed during monsoon with IBA 2 000 ppm.
- **Planting Density:** 8 × 8 m under monocrop; 10 × 10 m in agroforestry.
- **Nursery Management:** Partial shade, organic mulch, and regular watering; seedlings ready in 10–12 months.

2.9 Coriander (*Coriandrum sativum* L.)

2.9.1 Introduction

Coriandrum sativum L., commonly known as **coriander**, **cilantro**, or **Chinese parsley**, is one of the most ancient and widely cultivated spice and herb plants. Belonging to the family **Apiaceae (Umbelliferae)**, it is grown both for its **fruits (dry seeds)**, used as a spice, and for its **leaves**, which are used fresh as a culinary herb. The crop is native to the **Mediterranean and Middle Eastern regions** but has become naturalized across the world, particularly in **India, Egypt, Morocco, and Mexico**, which are now major producers and exporters.

Coriander is unique in that both its vegetative and reproductive parts are economically valuable. The **green leaves** (cilantro) are rich in **vitamin C, carotenoids, and antioxidants**, while the **dry fruits (coriander seeds)** yield **coriander essential oil (0.1–1.5 %)**, containing **linalool (60–70 %)**, which gives the spice its distinctive aroma. The crop serves both **culinary and industrial purposes**; the oil is used in perfumery, flavourings, and pharmaceuticals.

India accounts for over **70 % of global production**, with major growing states being **Rajasthan, Madhya Pradesh, Gujarat, Tamil Nadu, and Andhra Pradesh**. It is cultivated in both **rabi (winter)** and **kharif (monsoon)** seasons under irrigated and rainfed conditions. Despite its wide cultivation, productivity is often constrained by drought stress, pest infestations (aphids, cutworms), and diseases like **stem gall** and **powdery mildew**. Therefore, genetic improvement programs have focused on developing **high-yielding, disease-resistant, early-maturing, and high-oil-content varieties** suitable for diverse agro-climatic zones.

2.9.2 Botanical Description and Cytology

- **Family:** Apiaceae (Umbelliferae)
- **Genus:** *Coriandrum*
- **Species:** *C. sativum* L.
- **Chromosome number:** 2n = 22
- **Growth habit:** Annual herb, 30–80 cm tall, glabrous, aromatic.
- **Root:** Taproot with secondary laterals.
- **Stem:** Slender, erect, branched, green with fine grooves.
- **Leaves:** Lower leaves broad, lobed, similar to parsley; upper leaves finely dissected.
- **Inflorescence:** Compound umbel with 5–7 rays.
- **Flowers:** Small, white or pale pink; hermaphrodite and protandrous.
- **Fruit:** A schizocarp, nearly spherical, aromatic, splitting into two mericarps at maturity.

Cytologically, *C. sativum* exhibits moderate genetic variability and self-pollination tendency (~60 %), though cross-pollination (up to 40 %) occurs through insect vectors, providing an opportunity for recombination breeding.

2.9.3 Economic Importance

1. **Culinary and Industrial Uses**
   - Dried fruits (spice) flavour curries, pickles, and bakery products.
   - Fresh leaves serve as a herb and garnish in global cuisines.
   - Essential oil and oleoresin extracted from seeds used in perfumery, soap, and flavour industries.
2. **Medicinal and Nutraceutical Value**
   - Seeds possess **carminative, antispasmodic, and hypolipidemic** properties.
   - Oil used in traditional and modern medicine to aid digestion, lower cholesterol, and control microbial infections.

3. **Export and Economic Value**
   » India exports over 40,000 tonnes annually, accounting for nearly 60 % of global trade.
   » Essential oil fetches premium value in international markets.
4. **Agro-ecological Significance**
   » Short-duration crop suitable for intercropping and multiple cropping systems.
   » Tolerates marginal soils and residual moisture, supporting dryland agriculture.

2.9.4 Breeding Objectives

1. **High Yield and Early Maturity**
   » To develop short-duration cultivars (90–110 days) suitable for double cropping.
   » Improve seed yield (> 1.5 t ha^{-1}) and uniform ripening.
2. **Improved Quality and Oil Composition**
   » Increase essential-oil content (≥ 1.0 %) and linalool proportion (≥ 70 %).
   » Enhance flavour and aroma stability.
3. **Resistance to Major Diseases and Pests**
   » Tolerance to **stem gall** (*Protomyces macrosporus*) and **powdery mildew** (*Erysiphe polygoni*).
   » Resistance to aphids and cutworms.
4. **Adaptation and Agronomic Efficiency**
   » Tolerance to drought, temperature fluctuations, and salinity.
   » Suitability for both irrigated and rainfed systems.
5. **Morphological and Seed Characteristics**
   » Compact plants with uniform height for mechanical harvesting.
   » Bold, light-coloured seeds preferred for spice trade.
6. **Dual-purpose Varieties**
   » Selections with tender, aromatic leaves and high seed yield for combined herb-spice use.

2.9.5 Genetic Resources and Germplasm

a. Centers of Diversity

Coriander originated in the **eastern Mediterranean and Near East**, spreading through Asia Minor and South Asia. India holds secondary diversity, with numerous landraces adapted to local environments.

b. Germplasm Holdings

S. No.	Institute / Location	Accessions Maintained	Remarks
1	ICAR–National Bureau of Plant Genetic Resources (NBPGR), New Delhi	> 800	Global base collection; diverse origins.
2	ICAR–National Research Centre on Seed Spices (NRCSS), Ajmer	> 500	Active field gene bank for coriander, fennel, cumin.
3	State Agricultural Universities (Rajasthan, MP, TN)	≈ 300	Landraces and breeding lines for regional adaptation.

c. Important Landraces and Germplasm Accessions

S. No.	Accession / Local Name	Traits of Importance
1	*RCr-41*	High yield; powdery mildew tolerance.
2	*CO-3*	Early maturity; dual-purpose type.
3	*CS-287*	High oil (1.2 %); rich in linalool.
4	*Sindhu Local*	Drought tolerant; bold seeds.
5	*Ajmer Local*	Aromatic; suitable for dryland cultivation.

2.9.6 Floral Biology and Pollination

- **Flowering Season:** 35–50 days after sowing; lasts 15–20 days.
- **Pollination:** Mixed, self-pollinated but also insect-assisted (bees, syrphid flies).
- **Floral Structure:** Five petals, five stamens, bifid style; protandry reduces selfing.
- **Seed Set:** High under mild temperatures (20–25 °C).
- **Fruiting:** Fruits mature 30–40 days after flowering.

Knowledge of floral biology aids controlled hybridization and breeding design for combining ability studies.

2.9.7 Breeding and Hybridization Methods

a. Introduction and Evaluation

Introduced lines from Turkey, Bulgaria, and Egypt evaluated for yield and oil quality. Some serve as donors for linalool and bold-seed traits.

b. Selection Breeding

Mass and pure-line selection remain the primary breeding strategies because of the mixed mating system.

Procedure

1. Collect seed from superior plants showing desired traits.
2. Evaluate progenies in replicated trials for 3–4 seasons.
3. Advance uniform, stable lines for regional testing.

Example: Selections from *Ajmer Local* led to development of *Ajmer Coriander-1 (ACr-1)* and *ACr-2* with high yield and quality.

c. Hybridization

Employed to create recombinants combining high yield and oil content.

Steps

1. Select parents differing in yield and oil traits.
2. Emasculate immature florets (early morning) and bag umbels.
3. Pollinate manually with fresh pollen; tag crosses.
4. Grow F_1 plants; self or backcross depending on objective.
5. Evaluate segregating generations under field conditions.

Hybrid progenies show moderate heterosis for yield and earliness.

d. Mutation Breeding

Gamma irradiation (100–300 Gy) applied to seeds has produced mutants with compact growth, early maturity, and bold seeds. *NRCSS mutants M-15* and *M-27* exhibit improved yield and powdery mildew tolerance.

e. Polyploidy Breeding

Artificial tetraploids induced by colchicine (0.2 %) resulted in large seeds and thicker pericarps but reduced fertility; mainly of experimental interest.

f. Molecular and Biotechnological Approaches

1. **DNA Fingerprinting:** RAPD, ISSR, and AFLP used to classify accessions and detect duplicates.
2. **Marker–Trait Associations:** QTLs linked to essential-oil yield and powdery mildew resistance identified.
3. **In vitro Regeneration:** Shoot-tip and callus culture developed for clonal propagation and mutagenesis.
4. **Micropropagation:** Protocols standardized using MS medium + BAP (1.0 mg L^{-1}) for rapid multiplication.

2.9.8 Major Diseases and Resistance Breeding

S. No.	Disease / Causal Agent	Symptoms	Resistant Sources / Management
1	Stem gall – *Protomyces macrosporus*	Swellings on stems, branches	*Rcr-41* tolerant; crop rotation; seed treatment.
2	Powdery mildew – *Erysiphe polygoni*	White powdery growth on leaves	*CS-287*, *ACr-2* tolerant; sulfur spray effective.
3	Wilt – *Fusarium oxysporum*	Yellowing, vascular browning	*CO-3* tolerant; resistant source in progress.
4	Aphid infestation	Curling leaves, stunted growth	Resistant lines with dense trichomes preferred.

Resistance breeding through recurrent selection and incorporation of molecular markers ensures stability across locations.

2.9.9 Released Varieties and Performance

S. No.	Variety / Line	Origin / Type	Key Features
1	**RCr-41**	NRCSS, Ajmer	Powdery mildew tolerance; 1.3 % oil.
2	**CO-3**	TNAU, Coimbatore	Early maturing (95 days); dual-purpose.
3	**CS-287**	NRCSS selection	High oil (1.4 %); bold seeds.
4	**GCr-2**	Gujarat	High yield; drought tolerance.
5	**ACr-1 (Ajmer Coriander-1)**	Selection from Ajmer local	Uniform ripening; strong aroma.

2.9.10 Quality Evaluation and Biochemical Screening

- **Oil Content:** Determined by steam distillation; commercial standard 0.8–1.2 %.
- **Oil Composition:** GC–MS profiling shows linalool (60–70 %), γ-terpinene (8–10 %), and geraniol (3–5 %).
- **Aroma Index:** Calculated from linalool/terpinene ratio; ideal > 6:1.
- **Seed Morphometry:** Bold seeds (diameter ≥ 3 mm) preferred for export.
- **Antioxidant Assays:** DPPH and FRAP methods confirm high radical-scavenging activity in leafy types.

2.9.11 Molecular and Genomic Advances

1. **Genetic Diversity Studies:** SSR and ISSR markers reveal three major clusters corresponding to Indian, European, and Mediterranean ecotypes.
2. **Transcriptomics:** EST libraries identified genes for linalool synthase and terpene pathway enzymes.
3. **Marker-Assisted Selection:** Markers linked with powdery mildew tolerance and high-oil QTLs in development.
4. **Genome Size and Cytogenetics:** 2C DNA content ~4.5 pg; genome sequencing initiatives underway.
5. **Metabolomics:** LC–MS-based profiling distinguishes chemotypes with distinct aroma signatures.

2.9.12 Propagation and Crop Improvement Practices

- **Seed Propagation:** Split (decorticated) seeds sown directly in field; 8–10 kg ha^{-1} seeding rate.
- **Seed Treatment:** Fungicide + bio-inoculant enhances germination and early vigour.
- **Spacing:** 30 × 15 cm; thinning at 15 days for uniform stand.

» **Fertilization:** 30:40:30 kg N:P:K ha^{-1} recommended.

» **Harvesting:** Umbels cut when fruits turn brown; dried and threshed gently.

» **Seed Storage:** Dry to 9 % moisture; viability up to one year.

Breeding lines are evaluated under standard agronomic conditions for multi-location adaptability before varietal release.

2.10 Cumin (*Cuminum cyminum* L.)

2.10.1 Introduction

Cuminum cyminum L., commonly known as **cumin**, is one of the most ancient seed spices cultivated in the semi-arid regions of the Old World. It belongs to the family **Apiaceae (Umbelliferae)** and is widely valued for its **aromatic seeds** used in culinary, pharmaceutical, and industrial applications. The crop is native to the **Eastern Mediterranean and Irano-Turanian regions** but has adapted remarkably to dry subtropical climates, particularly in **India, Iran, and North Africa**.

India ranks as the **world's largest producer, consumer, and exporter** of cumin, contributing over 70 % of global output. Major producing states include **Gujarat, Rajasthan, and Uttar Pradesh**, where it is grown mainly as a **rabi crop under residual moisture** after the monsoon. Iran, Syria, and Turkey are other important producers.

The cumin seed, containing **2.5–4.0 % essential oil**, is valued for its warm, slightly bitter flavour and distinctive aroma derived mainly from **cuminaldehyde (25–35 %)**, γ-terpinene, **p-cymene**, and β-pinene. It is a key ingredient in spice blends such as curry powder and garam masala, and its oleoresin is extensively used in the food, beverage, and perfumery industries. Because the crop is **high-value but stress-prone**, breeding programs emphasize **disease resistance (especially wilt), drought tolerance**, **uniform maturity**, and **high essential-oil yield** to support export-grade production and sustainability in marginal environments.

2.10.2 Botanical Description and Cytology

» **Family:** Apiaceae (Umbelliferae)

» **Genus:** *Cuminum*

» **Species:** *C. cyminum* L.

» **Chromosome number:** 2n = 14

» **Growth habit:** Annual herb, 20–50 cm tall, aromatic, glabrous to sparsely hairy.

» **Root:** Slender taproot with lateral branches adapted to light soils.

» **Stem:** Erect, finely ridged, much-branched, green to glaucous.

» **Leaves:** Alternate, pinnatisect to bipinnatisect with filiform lobes; sessile and soft.

» **Inflorescence:** Compound umbel with 3–5 umbellules bearing 10–20 flowers each.

» **Flowers:** Small, white to pinkish, bisexual, pentamerous, and protandrous.

» **Fruit:** A laterally compressed schizocarp (4–6 mm long) breaking into two mericarp halves on ripening.

Cumin is largely **self-pollinated (≈ 80 %)**, with a minor degree of cross-pollination by insects. The diploid genome is small and uniform, facilitating selection breeding. Cytogenetic studies have revealed stable chromosomal behaviour with little structural variation, though intraspecific diversity exists for aroma profile and maturity duration.

2.10.3 Economic Importance

1. **Culinary and Industrial Uses**
   - Dried fruits (seeds) used whole or ground as spice in cuisines worldwide.
   - Essential oil used in soups, pickles, sausages, cheeses, and liqueurs.
   - Oleoresin and extracts applied in flavour, pharmaceutical, and cosmetic industries.
2. **Medicinal and Nutraceutical Value**
   - Acts as a digestive stimulant and carminative.
   - Exhibits antioxidant, antimicrobial, and hypoglycemic properties.
   - Cumin extracts used in traditional Ayurvedic and Unani systems of medicine.
3. **Economic and Export Significance**
   - India exports > 200 000 t annually to over 90 countries.
   - High market value and foreign exchange contribution make it a strategic seed spice.
4. **Agro-ecological Benefits**
   - Ideal for dryland farming due to low water requirement (250–300 mm).
   - Acts as a soil binder and fits well in rotation with wheat and mustard.

2.10.4 Breeding Objectives

1. **Higher Seed Yield and Uniform Maturity**
   - Develop varieties yielding > 1.5 t ha^{-1} with synchronized ripening for mechanized harvest.
2. **Improved Essential-Oil Content and Composition**
   - Enhance oil content to ≥ 4 % and cuminaldehyde proportion ≥ 35 %.
3. **Resistance to Diseases and Pests**
   - Breed tolerant lines to **Fusarium wilt, Alternaria blight**, and aphids.
4. **Abiotic-Stress Tolerance**
   - Drought and salinity resilience for arid zones of Rajasthan and Gujarat.
5. **Plant Type Improvement**
   - Compact, semi-erect plants with stiff stems and short umbel height.
6. **Early Maturity and Stability**
   - Short duration (95–110 days) to escape terminal heat stress.

2.10.5 Genetic Resources and Germplasm

a. Centers of Diversity

Primary center – Eastern Mediterranean and Iran; Secondary center – India (Punjab, Rajasthan, Gujarat), where diverse ecotypes exist for maturity, colour, and aroma.

b. Germplasm Collections

S. No.	Institute / Location	Accessions Maintained	Remarks
1	ICAR–NRCSS (Ajmer)	> 500	Core collection for seed spices.
2	NBPGR (New Delhi & Jodhpur)	> 400	Global repository; wild relatives.
3	SAUs (Gujarat, Rajasthan)	≈ 250	Regional landraces and breeding lines.

c. Notable Accessions / Lines

S. No.	Accession / Line	Distinctive Traits
1	*GC-4*	High oil (4 %); wilt-tolerant.
2	*RZ-19*	Early maturing; uniform ripening.
3	*GCr-1*	High seed yield; compact plant habit.
4	*EC 37120*	High cuminaldehyde content (38 %).
5	*Local Rajasthan Type*	Drought resilient; bold seed.

2.10.6 Floral Biology and Pollination

- **Flowering:** Occurs 40–45 days after sowing; continues for 20 days.
- **Inflorescence:** Compound umbel with 5–7 umbellules; each umbellule bears 10–20 small flowers.
- **Flower:** Bisexual; protandrous – stamens dehisce before stigma receptivity reducing selfing.
- **Pollination:** Predominantly self-pollinated but ≈ 20 % cross-pollination by bees and flies.
- **Seed set:** Enhanced under low humidity and moderate temperature (20–25 °C).

2.10.7 Breeding and Hybridization Methods

a. Mass and Pure-Line Selection

Primary approach used to improve locally adapted populations.

Procedure

1. Survey traditional fields and identify superior plants.
2. Bulk or individual plant selections advanced for stability and uniformity.
3. Evaluate for yield, maturity, and disease reaction over seasons.

Resultant Varieties: *GC-4*, *RZ-19*, *GCr-1* and *RZ-209*.

b. Hybridization

Used to combine high yield and wilt resistance.

Steps

1. Select parents contrasting for target traits.
2. Emasculate florets in the evening before anthesis.
3. Pollinate next morning with fresh pollen from male parent.
4. Bag and tag crosses; harvest mature fruits.
5. Advance F_2 and subsequent generations for selection of stable recombinants.

Controlled hybridization enabled the development of lines combining high oil and early maturity.

c. Recurrent Selection

Applied to maintain genetic diversity and improve polygenic traits like yield and drought tolerance.

d. Mutation Breeding

Physical (mutagenic rays 100–400 Gy) and chemical (EMS 0.2 %) mutagens used to generate variability. Mutants with erect plant type and enhanced oil content selected.

e. Molecular Breeding and Biotechnological Approaches

1. **Molecular Markers:** SSR and SRAP markers identify diverse genotypes and tag wilt resistance genes.
2. **Gene Expression:** Studies on terpene synthase genes associated with cuminaldehyde biosynthesis.
3. **In vitro Regeneration:** Callus culture from hypocotyl explants using BAP and NAA; used for mutagenesis.
4. **Micropropagation:** Efficient shoot-tip culture established for rapid multiplication and disease-free stock.

2.10.8 Major Diseases and Resistance Breeding

S. No.	Disease / Causal Agent	Symptoms	Resistant Sources / Management
1	Wilt – *Fusarium oxysporum f.sp. cumini*	Yellowing, vascular browning, plant death	*GC-4*, *RZ-19*, and *MCC 95* tolerant; soil solarization + biocontrol.
2	Alternaria blight – *A. burnsii*	Dark lesions on leaves and umbels	*GCr-1* shows tolerance.
3	Powdery mildew – *Erysiphe polygoni*	White powdery growth on foliage	Moderate resistance in *RZ-209*.
4	Aphid and cutworm infestation	Leaf curling and defoliation	Dense-foliage lines less susceptible.

Resistance sources are incorporated through hybridization and marker-assisted selection.

2.10.9 Released Varieties and Performance

S. No.	Variety / Line	Origin / Type	Distinctive Features
1	**GC-4**	SDAU, Gujarat	High yield (1.6 t ha^{-1}); oil 4 %; wilt tolerant.
2	**RZ-19**	ARS, Rajasthan	Early maturity (95 days); uniform ripening.
3	**RZ-209**	NRCSS, Ajmer	Blight tolerant; high cuminaldehyde.
4	**GCr-1**	Gujarat	Drought hardy; bold seeds.
5	**MCC 95**	Mutation line	Compact plant; resistant to wilt.

2.10.10 Quality Evaluation and Biochemical Screening

» **Essential Oil:** Steam distillation of ground seeds; content 2.5–4 %.

» **Oil Composition:** GC–MS reveals cuminaldehyde (25–35 %), p-cymene (15–20 %), γ-terpinene (10 %), and β-pinene (5 %).

» **Aroma Index:** Cuminaldehyde to p-cymene ratio > 2.5 considered superior.

» **Oleoresin:** Yield 8–10 %; standard for export flavour industries.

» **Antioxidant Activity:** Evaluated by DPPH and ABTS assays; correlated with phenolic content.

2.11 Fennel (*Foeniculum vulgare* Mill.)

2.11.1 Introduction

Foeniculum vulgare Mill., commonly known as **fennel**, is a highly aromatic and flavourful seed spice belonging to the family **Apiaceae (Umbelliferae)**. It is one of the oldest cultivated plants used for both **culinary and medicinal purposes**, valued for its **sweet, anise-like flavour** that comes primarily from the volatile oil constituent **anethole**. Fennel is grown extensively across the **Mediterranean region, the Middle East, and South Asia**, with **India** and **Egypt** being the leading producers.

In India, fennel is known as *saunf* and is grown mainly in **Gujarat, Rajasthan, Uttar Pradesh, and Haryana** as a **rabi crop** under irrigated conditions. The dried fruits (commonly called seeds) and tender leaves are used as condiments and flavouring agents in a wide range of foods, beverages, and pharmaceutical preparations. The crop has significant export potential due to its high essential-oil yield (1.0–3.5 %) and distinctive sweet aroma.

Fennel holds additional importance because of its **therapeutic and nutraceutical properties**, the seeds act as carminative, digestive, and antioxidant agents, while the essential oil is used in perfumery and medicine. Because the plant is **hardy and adaptable**, it fits well in dryland and irrigated crop rotations, contributing to income diversification for small and marginal farmers.

2.11.2 Botanical Description and Cytology

» **Family:** Apiaceae (Umbelliferae)

» **Genus:** *Foeniculum*

- **Species:** *F. vulgare* Mill.
- **Chromosome number:** 2n = 22
- **Growth habit:** Erect, glabrous, biennial or short-lived perennial herb, 1–2 m tall.
- **Root:** Strong taproot with secondary branches.
- **Stem:** Cylindrical, hollow, finely grooved, highly branched.
- **Leaves:** Alternate, 3- to 4-pinnate, finely dissected filiform segments giving a feathery appearance.
- **Inflorescence:** Compound umbel bearing 10–30 umbellules, each with 20–40 small yellow flowers.
- **Flowers:** Bisexual, pentamerous, actinomorphic; epigynous ovary with two carpels.
- **Fruit:** Schizocarp splitting into two curved mericarps ("fennel seeds").
- **Aroma:** Distinct sweet fragrance due to anethole, fenchone, and estragole.

The species shows considerable cytological uniformity, being a diploid (2n = 22). However, **chemotypes and ecotypes** vary widely in oil composition, plant stature, and maturity period, providing a broad genetic base for selection.

2.11.3 Economic Importance

1. **Culinary and Industrial Uses**
   - Dried seeds used whole or powdered in bakery goods, pickles, and spice blends.
   - Essential oil and oleoresin employed in flavouring candies, liquors, toothpaste, and soaps.
   - Tender leaves and stems used as garnishes and salads.

2. **Medicinal and Nutraceutical Value**
   - Seeds possess carminative, galactagogue, and antioxidant properties.
   - Fennel tea used for digestive ailments and infant colic.
   - Essential oil components (anethole, fenchone) exhibit antimicrobial and estrogenic activities.

3. **Export Potential and Economic Significance**
   - India exports > 25 000 t annually, commanding a major share of the global spice trade.
   - Essential-oil export fetches premium value due to high purity (> 90 % anethole).

4. **Agro-ecological and Environmental Advantages**
   - Suited for semi-arid regions with low rainfall (350–500 mm).
   - Deep roots aid soil binding and water extraction.
   - Ideal for crop rotation with cumin, wheat, and chickpea.

2.11.4 Breeding Objectives

1. **High Seed Yield and Early Maturity**
   » Develop varieties yielding ≥ 2.0 t ha^{-1} within 150–160 days.
2. **Improved Oil Content and Composition**
   » Enhance essential-oil content ≥ 2.5 %; anethole ≥ 80 %, fenchone ≤ 10 %.
3. **Resistance to Major Diseases and Pests**
   » Wilt (*Fusarium oxysporum*), powdery mildew (*Erysiphe polygoni*), and aphid resistance.
4. **Adaptation and Plant Type**
   » Semi-dwarf, erect plants with synchronized maturity for mechanical harvest.
5. **Dual-purpose Varieties**
   » Aromatic foliage for greens plus high seed and oil yield.

2.11.5 Genetic Resources and Germplasm

a. Centers of Diversity

Primary center – Mediterranean region; secondary center – India and Egypt, where diverse landraces exist.

b. Germplasm Collections

S. No.	Institute / Location	Accessions Maintained	Remarks
1	ICAR–NRCSS (Ajmer)	> 500	Active collection for seed spices in India.
2	NBPGR (New Delhi & Jodhpur)	≈ 400	Base and core collections for diversity analysis.
3	State Agricultural Universities (Rajasthan, Gujarat)	≈ 200	Landraces and breeding lines.

2.11.6 Floral Biology and Pollination

» **Flowering Time:** 45–60 days after sowing; duration 15–25 days.

» **Inflorescence:** Compound umbel with 15–30 umbellules each bearing 20–40 flowers.

» **Flower:** Bisexual, protandrous; 5 sepals, 5 petals, 5 stamens, inferior ovary.

» **Pollination:** Mainly entomophilous (insect-pollinated) with bees and flies as major agents; cross-pollination ≈ 60 %.

» **Seed Set:** Improved under moderate temperature (18–25 °C).

Knowledge of floral biology helps synchronize pollination for controlled crossing and heterosis breeding.

2.11.7 Breeding and Hybridization Methods

a. Mass and Pure-line Selection

Because of its cross-pollinated nature, mass selection followed by family selection is most effective.

Steps

1. Survey fields for superior plants with large umbels and bold seeds.
2. Harvest separately; evaluate progenies for yield, oil content, and maturity.
3. Select uniform, stable lines for release.

Examples: *GF-1*, *GF-2*, *RF-125* developed through this approach.

b. Hybridization

Aimed at combining high oil with disease resistance.

Procedure

1. Emasculate immature flowers before anthesis.
2. Pollinate the next morning with fresh pollen.
3. Bag and label umbels; harvest mature fruits.
4. Advance segregating generations to isolate stable recombinants.

c. Recurrent Selection

Used to accumulate additive genes for polygenic traits like yield and oil composition.

d. Mutation Breeding

Gamma irradiation (100–300 Gy) produced early maturing mutants with compact plants and enhanced oil content (> 3 %).

e. Biotechnological and Molecular Approaches

1. **Molecular Markers:** RAPD and SSR used to analyze diversity and confirm varietal purity.
2. **Marker-Trait Linkage:** Markers identified for high anethole and wilt resistance genes.
3. **In vitro Regeneration:** Callus from hypocotyl explants cultured on MS medium + BAP (1 mg L^{-1}).
4. **Micropropagation:** Shoot-tip culture for rapid multiplication and pathogen-free stocks.
5. **Somaclonal Variation:** In vitro lines screened for improved oil composition.

2.11.8 Major Diseases and Resistance Breeding

S. No.	Disease / Causal Agent	Symptoms	Resistant Sources / Management
1	Wilt – *Fusarium oxysporum f.sp. foeniculi*	Sudden wilting, vascular browning	*HF-33*, *RF-125* tolerant; crop rotation + biocontrol.
2	Powdery mildew – *Erysiphe polygoni*	White powdery growth on leaves	*GF-1* shows tolerance; sulfur spray effective.
3	Alternaria leaf blight – *A. alternata*	Brown necrotic lesions on leaves	Moderate resistance in *GF-2*.
4	Aphids (*Hyadaphis foeniculi*)	Yellowing, distorted leaves	Dense-foliage lines less susceptible; neem-based biocontrol.

2.11.9 Released Varieties and Performance

S. No.	Variety / Line	Origin / Type	Salient Features
1	**GF-1**	SDAU (Gujarat)	High yield (2.0 t ha^{-1}); oil 2.5 %; powdery mildew tolerant.
2	**GF-2**	SDAU (Gujarat)	Bold seeds; compact plant habit.
3	**RF-125**	ARS Rajasthan	Oil 2.8 %; anethole 83 %; wilt tolerant.
4	**HF-33**	Haryana	Early maturity (145 days); wilt resistant.
5	**Ajmer Fennel-1 (AF-1)**	NRCSS Ajmer	Dual-purpose; good green foliage and seed yield.

2.11.10 Quality Evaluation and Biochemical Screening

- **Essential Oil:** Determined by steam distillation; standard range 1–3.5 %.
- **Oil Composition:** GC–MS reveals anethole (70–85 %), fenchone (5–10 %), estragole (3–5 %).
- **Flavour Index:** Anethole/fenchone ratio > 8 considered superior.
- **Oleoresin:** Yield 7–9 %; used in flavour industry.
- **Antioxidant Assays:** DPPH and FRAP show high activity in green-seeded types.

2.11.11 Molecular and Genomic Advances

1. **Genetic Diversity:** SSR and AFLP analyses identified three distinct chemotypic groups (anethole-rich, fenchone-rich, mixed).
2. **Marker-Trait Association:** QTLs linked with oil content and wilt resistance mapped.
3. **Transcriptomics:** Genes in phenylpropanoid pathway (phenylalanine ammonia-lyase, O-methyltransferase) characterized.
4. **Marker-Assisted Selection:** Used for introgressing resistance into elite high-oil lines.
5. **Molecular Diagnostics:** DNA fingerprinting employed for varietal identification and seed purity.

2.11.12 Propagation and Crop Improvement Practices

- **Seed Propagation:** Split (decorticated) fruits direct-sown at 12–15 kg ha^{-1}.
- **Spacing:** 30 × 20 cm for irrigated fields.
- **Seed Treatment:** Thiram 2 g kg^{-1} seed + Trichoderma bio-inoculant.
- **Fertilization:** 30:40:20 kg N:P:K ha^{-1}; organic manure improves oil content.
- **Irrigation:** Four light irrigations at branching, flowering, and seed-filling stages.
- **Harvesting:** When umbels turn straw-yellow; dried and threshed carefully to retain aroma.

These protocols ensure true-to-type seed stocks for breeding and commercial production.

2.12 Fenugreek (*Trigonella foenum-graecum* L.)

2.12.1 Introduction

Trigonella foenum-graecum L., commonly known as **fenugreek**, is an annual leguminous spice crop belonging to the family **Fabaceae (Leguminosae)**. It is cultivated for its **seeds**, which possess a distinct bittersweet flavour and characteristic aroma owing to the presence of **sotolone, diosgenin,** and **trigonelline.** The young **leaves** are used as a leafy vegetable (*methi*), and the dried foliage (*kasuri methi*) serves as a flavouring agent.

Originating in the **Mediterranean and West Asian** region, fenugreek is now cultivated throughout South Asia, North Africa, and southern Europe. India leads global production, with major growing states being **Rajasthan, Gujarat, Madhya Pradesh, and Uttar Pradesh.**

Fenugreek is unique among spices for its dual role as a **pulse and condiment**, its ability to **fix atmospheric nitrogen**, and its value in **nutraceutical and pharmaceutical industries.**

2.12.2 Botanical Description and Cytology

- **Family:** Fabaceae
- **Genus:** *Trigonella*
- **Species:** *T. foenum-graecum* L.
- **Chromosome number:** 2n = 16
- **Growth habit:** Annual, erect herb (30–60 cm), branched, glabrous.
- **Root:** Deep taproot with profuse nodulation.
- **Stem:** Hollow, green, angular.
- **Leaves:** Trifoliate, obovate leaflets 2–3 cm, pale green.
- **Flowers:** Papilionaceous, solitary or in axillary pairs, pale yellow.
- **Fruit:** Linear pods 10–15 cm long bearing 10–20 seeds.
- **Seed:** Hard, yellow-brown, cuboid, aromatic.

Fenugreek is **self-pollinated,** though 2–3 % out-crossing occurs via bees. The diploid genome shows modest genetic variability, sufficient for recurrent and mutational breeding.

2.12.3 Economic Importance

1. **Culinary Use:** Seeds flavour pickles, curries, and bakery products; leaves used as green vegetable.
2. **Medicinal Value:** Contains saponins (diosgenin), alkaloids (trigonelline), and fiber (> 40 %), showing hypoglycemic and hypocholesterolemic effects.
3. **Industrial Applications:** Diosgenin used in steroid drug synthesis; gum as stabilizer in foods and cosmetics.
4. **Agronomic Advantages:** Nitrogen-fixing legume improves soil fertility; short duration (90–110 days).

2.12.4 Breeding Objectives

1. **High seed and green-leaf yield** (> 1.5 t ha^{-1} seed; > 20 t ha^{-1} green leaves).
2. **Early maturity** (80–90 days) for rabi systems.
3. **Resistance** to powdery mildew (*Erysiphe polygoni*), leaf spot (*Alternaria* spp.), and aphids.
4. **High diosgenin and protein content.**
5. **Tolerant to salinity and frost.**

2.12.5 Genetic Resources and Germplasm

S. No.	Institute / Location	Accessions Maintained	Remarks
1	ICAR–NRCSS (Ajmer)	> 450	Active field gene bank.
2	NBPGR (New Delhi)	≈ 600	Global collection of *T. foenum-graecum*.
3	SAUs (Rajasthan, TN, MP)	≈ 300	Local landraces and dual-purpose types.

2.12.6 Floral Biology and Pollination

» **Flowering:** 30–40 days after sowing; protogynous tendency favors selfing.

» **Pollination:** Predominantly autogamous; occasional insect visitation.

» **Fruit Development:** Pods mature 30 days post-anthesis.

2.12.7 Breeding and Hybridization Methods

a. Mass and Pure-line Selection

Field selection of vigourous plants with bold seeds and disease tolerance followed by progeny testing. Varieties like *AFg-3* and *RMt-303* originated this way.

b. Hybridization

Employed to combine early maturity and disease resistance.

» Emasculation in pre-anthesis buds.

» Manual pollination next morning.

» Segregants selected in F_2 for seed weight and pod length.

c. Mutation Breeding

Gamma irradiation (200–400 Gy) generated compact, early, and high-diosgenin mutants.

d. Biotechnological Approaches

» Tissue culture and somatic embryogenesis for clonal propagation.

» Molecular markers (RAPD, SSR) used for diversity and disease-resistance mapping.

2.12.8 Major Diseases and Resistance

S. No.	Disease / Causal Agent	Symptoms	Resistant Sources / Lines
1	Powdery mildew – *Erysiphe polygoni*	White patches on leaves	*AFg-3, RMt-305.*
2	Leaf spot – *Alternaria tenuissima*	Circular necrotic spots	*IC-370243* tolerant.
3	Root rot – *Rhizoctonia solani*	Damping-off of seedlings	Biocontrol + resistant lines.

2.12.9 Released Varieties

S. No.	Variety	Origin	Distinct Features
1	*RMt-303*	RAU, Rajasthan	High seed yield; early; powdery mildew tolerant.
2	*AFg-3*	NRCSS Ajmer	High protein (24 %); dual-purpose.
3	*CO-1*	TNAU	Green-leaf type; aromatic.

2.12.10 Quality Evaluation

» **Diosgenin:** 1–3 % in seeds; extracted for steroid industry.

» **Protein:** 20–25 %; balanced amino acid profile.

» **Flavour:** Sotolone and trigonelline determine bitterness and aroma.

2.12.11 Molecular Advances

» SSR markers grouped germplasm into three clusters.

» Genes for sterol biosynthesis characterized.

» QTLs for powdery mildew and seed size identified.

2.12.12 Integrated Breeding Approaches

S. No.	Approach	Contribution
1	Selection + mutation	Yield and disease resistance.
2	Marker analysis	Diosgenin content tagging.
3	Tissue culture	Uniform plantlets for seed multiplication.

2.13 Ajwain (*Trachyspermum ammi* L.)

2.13.1 Introduction

Trachyspermum ammi L. (syn. *Carum copticum*), commonly known as **ajwain**, is a spice of the family **Apiaceae** renowned for its pungent aroma and distinctive flavour caused by the essential-oil compound **thymol**. Native to the eastern Mediterranean and West Asian region, it is widely grown in India, Iran, Egypt, and Pakistan. India is the largest producer, with major cultivation in **Rajasthan, Gujarat, and Madhya Pradesh**.

The crop is highly valued for its **seeds**, used as a spice and medicine, and for its **thymol-rich essential oil (2–4 %)** employed in pharmaceutical and cosmetic industries. Ajwain is a short-duration, drought-tolerant plant that thrives in semi-arid conditions, making it ideal for dryland farming systems.

2.13.2 Botanical Description and Cytology

» **Family:** Apiaceae

» **Genus:** *Trachyspermum*

» **Species:** *T. ammi* L.

» **Chromosome number:** 2n = 18

» **Habit:** Annual herb (30–90 cm) with erect, branched stems.

» **Leaves:** Finely divided, alternate, aromatic.

» **Inflorescence:** Compound umbel with 8–12 umbellules.

» **Flowers:** Small, white, bisexual, protandrous.

» **Fruit:** Ovoid schizocarp 2–3 mm long, grayish brown, ridged surface.

Self-pollinated with minor crossing via insects; diploid stable genome with moderate variability.

2.13.3 Economic Importance

1. **Culinary Uses:** Seeds impart pungent flavour to curries and baked goods.
2. **Medicinal Applications:** Thymol acts as antiseptic and carminative; ajwain oil used for cold and indigestion.
3. **Industrial Use:** Oil in toothpaste, mouthwash, perfumes.
4. **Agronomic Role:** Short (110–120 days) crop suited for intercropping and low input systems.

2.13.4 Breeding Objectives

1. High seed yield (> 1.2 t ha^{-1}).
2. Elevated thymol content (≥ 40 %).
3. Resistance to **blight** and **wilt**.
4. Tolerance to heat and water stress.
5. Compact plants for mechanical harvest.

2.13.5 Genetic Resources and Germplasm

S. No.	Institute / Location	Accessions Maintained	Remarks
1	NRCSS (Ajmer)	≈ 350	Core collection for seed spices.
2	NBPGR (New Delhi)	≈ 300	Global accessions with chemotype data.
3	Rajasthan SAUs	≈ 150	Local high-thymol types.

2.13.6 Floral Biology and Pollination

» Flowering 40–45 days after sowing.

» Protandry encourages partial out-crossing (15–20 %).

» Pollination via bees and flies.

» Seed set better at 20–25 °C under dry conditions.

2.13.7 Breeding and Hybridization Methods

a. Mass Selection

Field screening for bold seeds and high oil. Varieties like *AA-93* and *AA-1* developed by this method.

b. Hybridization

Used to combine yield and oil composition. Crosses made through manual pollination in bagged umbels.

c. Mutation Breeding

Gamma rays (100–300 Gy) produced early, compact, and wilt-tolerant mutants.

d. Molecular Approaches

» RAPD and ISSR markers used for chemotype differentiation.

» QTLs for thymol biosynthesis being mapped.

» Tissue culture protocols for micropropagation standardized.

2.13.8 Major Diseases and Resistance

S. No.	Disease / Pathogen	Symptoms	Resistant Sources
1	Leaf blight – *Alternaria alternata*	Brown spots, defoliation	*AA-93* resistant.
2	Wilt – *Fusarium oxysporum*	Vascular browning	*AA-1*, *RA-20*.
3	Powdery mildew	White coating on foliage	Moderate tolerance in *RAC-5*.

2.13.9 Released Varieties

S. No.	Variety	Origin / Institute	Salient Features
1	**AA-1**	NRCSS Ajmer	High thymol (42 %); blight tolerant.
2	**AA-93**	ARS Rajasthan	Early maturity (110 days); bold seeds.
3	**RA-20**	Rajasthan Agric. Univ.	Wilt resistant;
3	**RA-20**	Rajasthan Agric. Univ.	Wilt resistant; high yield (1.4 t ha^{-1}); strong aroma.
4	**RAC-5**	NRCSS Ajmer	High thymol (45 %); tolerant to powdery mildew.

2.13.10 Quality Evaluation and Biochemical Screening

» **Essential-oil content:** 2–4 %, obtained by steam distillation.

» **Oil composition:** Thymol (35–50 %), γ-terpinene (25 %), p-cymene (20 %), and β-pinene (5 %).

- **Aroma index:** Thymol / p-cymene ≥ 2 indicates premium flavour.
- **Oleoresin yield:** 7–9 % for industrial extraction.
- **Antimicrobial activity:** Strong inhibition against *E. coli* and *S. aureus*, directly correlated with thymol percentage.

2.13.11 Molecular and Genomic Advances

1. **Chemotypic diversity:** ISSR and RAPD markers distinguish high- and low-thymol populations.
2. **Candidate genes:** Thymol synthase and terpene-pathway genes characterized via transcriptome sequencing.
3. **Marker-assisted selection:** Early tagging of wilt-resistance loci.
4. **Molecular diagnostics:** PCR fingerprinting for varietal identification and purity testing.
5. **Metabolomics:** GC–MS profiling used to identify stable high-thymol chemotypes for export markets.

2.13.12 Propagation and Agronomic Standardization

- **Sowing:** Direct seeding at 6–8 kg ha^{-1}; spacing 30 × 20 cm.
- **Seed treatment:** Thiram 2 g kg^{-1} + *Trichoderma* bio-agent.
- **Fertilization:** 25 : 30 : 20 kg N : P : K ha^{-1} + farmyard manure.
- **Irrigation:** Light watering at branching, flowering, and seed-filling stages.
- **Harvesting:** At physiological maturity (seeds grayish-brown, aromatic).
- **Drying:** Seeds dried to < 9 % moisture for storage and oil extraction.

2.14 Dill (*Anethum graveolens* L.)

2.14.1 Introduction

Anethum graveolens L., commonly known as **dill**, is a delicate annual aromatic herb of the family **Apiaceae (Umbelliferae)**. It is cultivated for its **leaves (dill weed)** and **fruits (dill seeds)**, both of which are used extensively as flavouring agents and condiments. Dill imparts a characteristic **fresh, tangy, and slightly pungent aroma** derived from its essential oils, primarily **carvone** and **limonene**.

The crop is native to the **Mediterranean and Western Asian regions**, with evidence of its use dating back to ancient Egyptian and Roman times. It is now cultivated globally, with major production in **India, Egypt, Russia, Turkey, and parts of Europe**. In India, dill is popularly known as *Suva* or *Sowa*, and is primarily grown in **Rajasthan, Gujarat, and Madhya Pradesh** during the rabi season under irrigated and rainfed conditions.

Dill has gained increasing attention due to its **culinary versatility**, **nutraceutical value**, and **adaptability to dry, marginal soils**. Both leaves and seeds possess considerable **therapeutic properties**, being rich in **flavonoids, phenolic compounds, and essential oils**, which exhibit antioxidant, digestive, and antimicrobial activities. These attributes make dill a crop of dual significance, economic and medicinal, suitable for smallholder farming systems.

2.14.2 Botanical Description and Cytology

- **Family:** Apiaceae (Umbelliferae)
- **Genus:** *Anethum*
- **Species:** *A. graveolens* L.
- **Chromosome number:** 2n = 22
- **Growth habit:** Annual herb, 60–120 cm tall, aromatic and glabrous.
- **Root:** Taproot with fine laterals.
- **Stem:** Erect, hollow, finely grooved, bluish-green with waxy coating.
- **Leaves:** Finely dissected, pinnately compound, bluish-green; segments filiform.
- **Inflorescence:** Large compound umbel bearing 15–25 umbellules, each with 15–25 small yellow flowers.
- **Flowers:** Small, bisexual, actinomorphic, protandrous.
- **Fruit:** Elliptic, flattened schizocarp (3–5 mm long), aromatic; commonly referred to as dill seed.
- **Aroma:** Dominated by **carvone (35–60 %), limonene (30–40 %)**, and trace amounts of **anethofuran**.

Cytologically, dill is a **diploid (2n = 22)** species showing minimal chromosomal variability, but a wide range of **chemical polymorphism** among landraces for oil yield and composition.

2.14.3 Economic Importance

1. **Culinary Uses**
   - Tender leaves used as fresh herbs in salads, soups, and pickles.
   - Seeds used whole or ground as condiments in bread, curries, and preserved foods.
   - Essential oil applied as a natural flavour in beverages and confectionery.
2. **Medicinal and Nutraceutical Significance**
   - Seeds and leaves used as digestive, carminative, and lactation-promoting agents.
   - Exhibits antioxidant, antimicrobial, and cholesterol-lowering properties.
   - Dill water traditionally used to relieve colic and digestive distress in infants.
3. **Industrial Relevance**
   - Essential oil (2–4 %) rich in carvone used in flavour and perfumery industries.
   - Oleoresins used in pharmaceutical formulations.
4. **Agro-ecological Benefits**
   - Drought-tolerant, low-input crop suited for marginal lands.
   - Fits well in cereal-pulse-spice rotations under dryland conditions.

2.14.4 Breeding Objectives

1. **Enhanced Seed Yield** (≥ 1.5 t ha^{-1}) and stable performance under diverse environments.
2. **High Essential-Oil Content and Quality**, carvone ≥ 55 %, limonene ≤ 35 %.
3. **Resistance to Major Diseases and Pests**, powdery mildew, Alternaria blight, and aphids.
4. **Early and Uniform Maturity** for synchronous harvest.
5. **Improved Plant Type**, semi-dwarf, erect, with compact umbels and strong stems.
6. **Dual-Purpose Types** suitable for both leaf and seed harvest.

2.14.5 Genetic Resources and Germplasm

a. Centers of Diversity

Primary center – Mediterranean basin; secondary centers – India and Southwest Asia.

b. Germplasm Collections

S. No.	Institute / Location	Accessions Maintained	Remarks
1	ICAR–NRCSS (Ajmer)	> 350	Core collection of Indian and exotic lines.
2	NBPGR (New Delhi & Jodhpur)	≈ 300	Global accessions and wild relatives.
3	SAUs (Rajasthan, Gujarat, MP)	≈ 150	Local landraces and breeding lines.

2.14.6 Floral Biology and Pollination

» **Flowering:** Begins 50–55 days after sowing; continues for 15–25 days.

» **Inflorescence:** Large compound umbel; outer flowers open first (centrifugal blooming).

» **Pollination:** Predominantly cross-pollinated (60–70 %), mediated by honeybees, flies, and beetles.

» **Self-pollination:** Possible but limited due to protandry.

» **Seed Set:** Enhanced at 18–25 °C with low humidity.

Understanding pollination dynamics is crucial for hybridization and maintaining varietal purity in breeding programs.

2.14.7 Breeding and Hybridization Methods

a. Mass and Family Selection

Traditional method for improving locally adapted landraces. Superior plants selected for umbels per plant, seed weight, and oil yield; progeny tested over seasons for stability.

b. Hybridization

Used to combine high yield and disease resistance.

» Emasculation performed before anthesis.

» Controlled pollination with bagging of umbels.

» Segregants screened in F_2 for vigour, compactness, and oil profile.

c. Recurrent Selection

Applied to maintain heterogeneity while accumulating favorable alleles for polygenic traits such as yield and carvone content.

d. Mutation Breeding

Gamma irradiation (100–300 Gy) and EMS (0.2 %) induced mutants with early maturity, short stature, and high oil yield (3.2 %).

e. Biotechnological and Molecular Approaches

1. **Tissue Culture:** Regeneration from hypocotyl and leaf explants using MS + BAP (1 mg L^{-1}).
2. **Molecular Markers:** RAPD, ISSR, and SSR markers employed to estimate genetic diversity and identify duplicates.
3. **Marker-Trait Linkage:** Preliminary QTL mapping for carvone and limonene biosynthesis genes.
4. **Micropropagation:** Used to conserve elite chemotypes and produce disease-free planting material.

2.14.8 Major Diseases and Resistance Breeding

S. No.	Disease / Pathogen	Symptoms	Resistant / Tolerant Lines
1	Powdery mildew – *Erysiphe polygoni*	White coating on leaves, premature senescence	*AD-1*, *AJD-1* tolerant.
2	Alternaria blight – *A. alternata*	Brown spots on leaves and umbels	Moderate tolerance in *RUD-1*.
3	Root rot – *Fusarium oxysporum*	Yellowing, wilting	Biocontrol + tolerant lines (*GJSD-2*).
4	Aphid infestation – *Hyadaphis coriandri*	Leaf curling, stunted growth	Compact foliage lines less affected.

Integrated resistance management, combining tolerant varieties with organic fungicides and neem-based bioagents, ensures sustainable control.

2.14.9 Released Varieties and Performance

S. No.	Variety / Line	Origin / Type	Salient Features
1	**AD-1**	SDAU, Gujarat	High seed yield (1.8 t ha^{-1}); oil 2.8 %; mildew tolerant.
2	**AJD-1**	NRCSS, Ajmer	Drought tolerant; early maturity (120 days).
3	**RUD-1**	RAU, Rajasthan	High carvone (58 %); bold seeds.
4	**GJSD-2**	Gujarat	Compact umbels; uniform maturity.

2.14.10 Quality Evaluation and Biochemical Screening

» **Essential Oil Content:** 2–4 %, measured by steam distillation.

» **Oil Composition:** Carvone (40–60 %), limonene (30–40 %), and small amounts of α-phellandrene.

» **Aroma Quality:** Superior when carvone/limonene ratio ≥ 1.5.

» **Oleoresin Yield:** 8–10 %; key for export-grade flavouring.

» **Antioxidant and Antimicrobial Activities:** Strong radical-scavenging activity and inhibition of *E. coli* and *Candida albicans*.

Spice Crops

(Black Pepper | Cardamom | Ginger | Turmeric | Nutmeg | Cinnamon | Clove | Garcinia | Coriander | Cumin | Fennel | Fenugreek | Ajwain | Dill)

S. No.	Common Name	Botanical Name (Italicized)	Family	Key Constituents / Active Compounds	Major Breeding Objectives	Released Varieties / Highlights
1	Black Pepper	*Piper nigrum* L.	Piperaceae	Piperine, essential oil	High yield, Phytophthora resistance, vine vigour	*Panniyur 1–8, IISR Shakti, Karimunda*
2	Cardamom	*Elettaria cardamomum* (Mat.)	Zingiberaceae	1,8-Cineole, α-Terpinyl acetate	High capsule yield, quality aroma, resistance to thrips	*Njallani Green Gold, PV-1, IISR Avinash*
3	Ginger	*Zingiber officinale* Rosc.	Zingiberaceae	Gingerol, Shogaol, Zingerone	High rhizome yield, disease resistance, oleoresin quality	*Rio-de-Janeiro, Suprabha, IISR Varada*
4	Turmeric	*Curcuma longa* L.	Zingiberaceae	Curcumin, Demethoxycurcumin	High curcumin, rhizome yield, leaf spot resistance	*Pratibha, Sudarsana, Rajendra Sonia*
5	Nutmeg	*Myristica fragrans* Houtt.	Myristicaceae	Myristicin, Elemicin	High kernel oil, regular bearing, fruit-borer resistance	*Konkan Swad, IISR Keralashree*
6	Cinnamon	*Cinnamomum verum* J. Presl	Lauraceae	Cinnamaldehyde, Eugenol	High bark oil, coppicing ability, disease resistance	*Navashree, Nithyasree, CIN-1*
7	Clove	*Syzygium aromaticum* (L.) Merr. & Perry	Myrtaceae	Eugenol, β-Caryophyllene	High bud yield, oil content, wilt resistance	*Zanzibar, Kandy, CIM-Clove 1*
8	Garcinia (Kokum)	*Garcinia indica* Choisy	Clusiaceae	Hydroxycitric acid	High fruit yield, high HCA, seedless types	*Konkan Kokum-1, CIM-GA-2*
9	Coriander	*Coriandrum sativum* L.	Apiaceae	Linalool, Geraniol	High yield, early maturity, bold seed	*Rajendra Swarna, CO-4, RCR-41*

10	Cumin	*Cuminum cyminum* L.	Apiaceae	Cuminaldehyde, Terpenes	Wilt resistance, early maturity, bold seed	*GC-4, RZ-19, UC-198*
11	Fennel	*Foeniculum vulgare* Mill.	Apiaceae	Anethole, Fenchone	High oil yield, seed quality, bolting resistance	*GF-2, RF-205, Hisar Suraj*
12	Fenugreek	*Trigonella foenum-graecum* L.	Fabaceae	Trigonelline, Diosgenin	High seed yield, maturity synchrony, powdery mildew resistance	*CO-2, RMT-1, Pusa Early Bunching*
13	Ajwain	*Trachyspermum ammi* (L.) Sprague	Apiaceae	Thymol, Carvacrol	High thymol yield, wilt tolerance	*AA-93, RA-1, CIM-Ajwain 1*
14	Dill	*Anethum graveolens* L.	Apiaceae	Carvone, Limonene	High seed oil, disease resistance, early flowering	*GD-1, Rajendra Sowa 1, CIM-Dill 2*

Chapter - 3

Medicinal Crops

3.1 Senna (*Cassia angustifolia* Vahl.)

3.1.1 Introduction

Cassia angustifolia Vahl., commonly known as **Senna**, is an important **medicinal shrub** belonging to the family **Fabaceae (Leguminosae)**. Native to **Sudan and Arabia**, it is now widely cultivated in the **semi-arid tracts of India**, especially in **Tamil Nadu, Gujarat, and Rajasthan**, for its **purgative leaves and pods**.

Senna has immense pharmaceutical significance due to the presence of **anthraquinone glycosides**, primarily **sennosides A and B**, which act as natural laxatives. The dried leaflets and pods form the raw material for herbal formulations used globally in traditional and modern medicine.

3.1.2 Botanical Description and Cytology

- **Family:** Fabaceae
- **Genus:** *Cassia*
- **Species:** *C. angustifolia* Vahl.
- **Chromosome number:** 2n = 28
- **Habit:** Perennial, erect herb or undershrub, 50–100 cm tall.
- **Leaves:** Pinnately compound with 4–6 pairs of narrow leaflets.
- **Flowers:** Yellow, arranged in axillary or terminal racemes.
- **Pods:** Flat, oblong, 4–7 cm long, containing 5–8 brown seeds.
- **Root:** Deep taproot, suited to dry soils.

The plant shows **predominant self-pollination**, though cross-pollination by insects occurs occasionally.

DOI: 10.1201/9781042010332-4

3.1.3 Economic and Medicinal Importance

Pharmaceutical Use: Source of sennosides A & B used in laxative drugs.

Industrial Application: Extracts used in herbal teas and health products.

Export Significance: India is the largest producer and exporter of senna leaves and pods.

Agro-ecological Value: Drought tolerant, low-input crop suited for arid lands.

3.1.4 Breeding Objectives

Develop high-yielding genotypes with high **sennoside content** (> 2 %).

Improve **leaf size, leaflet density**, and **biomass yield.**

Enhance **drought and salinity tolerance.**

Incorporate **resistance** to leaf spot (*Alternaria cassiae*) and pod borers.

Short-duration varieties for multiple harvests annually.

3.1.5 Genetic Resources and Germplasm

S. No.	Institute / Location	Accessions Maintained	Remarks
1	CIMAP, Lucknow	> 250	Core germplasm and improved lines.
2	TNAU, Coimbatore	≈ 150	Drought-tolerant ecotypes.
3	NBPGR, New Delhi	120	Wild relatives and exotic types.

Notable lines include *Sona*, *KM-5*, and *ALFT-2*, which show superior leaf yield and glycoside concentration.

3.1.6 Floral Biology and Pollination

Flowering occurs 90–100 days after sowing.

» **Flower type:** Bisexual, bright yellow, zygomorphic.

» **Pollination:** Mostly autogamous; 10–15 % cross-pollination by bees.

» **Fertilization:** Occurs within 24 h of anthesis. Understanding floral morphology aids in controlled hybridization for trait improvement.

3.1.7 Breeding and Hybridization Methods

Mass Selection: Used to improve local landraces for leaf yield.

Pure-Line Selection: Based on sennoside content and pod set.

Hybridization: Limited due to small floral size; manual crossing done by emasculating immature buds and hand-pollinating with desired pollen.

Mutation Breeding: Gamma irradiation (200–400 Gy) used to induce early flowering and high sennoside mutants.

Biotechnological Tools: Callus culture for glycoside enhancement and in vitro mutagenesis.

3.1.8 Major Diseases and Resistance

S. No.	Disease	Causal Agent	Resistant Lines
1	Leaf spot	*Alternaria cassiae*	*Sona, KM-5.*
2	Wilt	*Fusarium oxysporum*	*ALFT-2.*
3	Pod borer	*Helicoverpa armigera*	Avoided by timely harvest.

3.1.9 Released Varieties

S. No.	Variety	Origin	Salient Features
1	*Sona*	CIMAP	High sennosides (2.5 %); short duration (100 days).
2	*KM-5*	TNAU	High leaf yield (2.8 t ha^{-1}); disease tolerant.
3	*ALFT-2*	CIMAP	Drought hardy; bold pods.

3.1.10 Phytochemical and Quality Evaluation

» **Sennosides:** Determined by HPLC; commercial grade ≥ 2 %.

» **Other Compounds:** Rhein, chrysophanol.

» **Quality Index:** Leaf to pod ratio and glycoside content determine market value.

3.1.11 Integration of Conventional and Modern Breeding

S. No.	Approach	Impact
1	Hybridization + Selection	Development of *Sona* and *KM-5.*
2	Mutation Breeding	Early and compact mutants.
3	Tissue Culture	Rapid clonal propagation.
4	Biochemical Screening	Enhanced sennoside content.

3.2 Periwinkle (*Catharanthus roseus* L.)

3.2.1 Introduction

Catharanthus roseus L. (syn. *Vinca rosea*) belongs to **Apocynaceae** and is a globally important **medicinal ornamental** known for its alkaloids **vincristine** and **vinblastine,** which are used as potent anticancer drugs. The plant is native to **Madagascar** but now cultivated in India, Africa, and South America.

India has become a key producer of periwinkle biomass for pharmaceutical extraction industries due to its adaptability and high alkaloid content.

3.2.2 Botanical Description and Cytology

» **Habit:** Perennial herb, 40–90 cm tall.

» **Stem:** Cylindrical, smooth, often branching.

- **Leaves:** Opposite, oblong-lanceolate, glossy, 3–5 cm.
- **Flowers:** Pink, white, or purple; axillary and solitary.
- **Fruit:** A pair of follicles 2–3 cm long containing small black seeds.
- **Chromosome number:** 2n = 16.

Periwinkle is primarily **self-pollinated** but cross-pollination occurs via insects.

3.2.3 Economic and Medicinal Importance

- Source of **vinblastine** and **vincristine**, used in chemotherapy for leukemia and Hodgkin's disease.
- Leaves and roots yield over **100 alkaloids**, including ajmalicine and serpentine.
- Used in Ayurveda as a hypotensive and antidiabetic herb.
- Significant export commodity for alkaloid extraction.

3.2.4 Breeding Objectives

Increase total alkaloid yield per hectare.

Develop varieties with high leaf biomass.

Breed early flowering and compact plant types.

Enhance drought and pest resistance.

3.2.5 Genetic Resources and Germplasm

S. No.	Institution	Accessions	Remarks
1	CIMAP, Lucknow	> 300	Alkaloid diversity collection.
2	TNAU, Coimbatore	≈ 200	Colour and chemotype diversity.
3	NBPGR	150	Global collection.

3.2.6 Floral Biology and Pollination

Flowering occurs 80–90 days after sowing; cross-pollination aided by bees.

Protandry flavours partial outcrossing; hence controlled pollination is required for hybridization.

3.2.7 Breeding and Hybridization Methods

- **Selection:** Based on leaf alkaloid content.
- **Hybridization:** Emasculation of pre-anthesis buds, manual pollination, bagging for purity.
- **Mutation Breeding:** EMS and irradiation to induce color and alkaloid mutants.
- **Biotechnological Tools:** Cell suspension cultures for enhanced vincristine yield.

3.2.8 Diseases and Resistance

S. No.	Disease	Causal Agent	Management / Resistant Lines
1	Leaf spot	*Alternaria alternata*	*Nirmal* tolerant.
2	Phytoplasma wilt	Phytoplasma spp.	*Dhawal* moderately tolerant.

3.2.9 Released Varieties

S. No.	Variety	Features
1	*Nirmal*	High vincristine (0.0006 %).
2	*Dhawal*	High alkaloid in leaves; drought tolerant.
3	*Arka Shubra*	White-flowered, early flowering.

3.2.10 Phytochemical Evaluation

» **Vincristine:** 0.0005–0.001 %.

» **Vinblastine:** 0.0008–0.0012 %.

» **Ajmalicine:** 0.15–0.25 % in roots.

3.2.11 Integration of Breeding and Biotechnology

S. No.	Approach	Impact
1	Hybridization + selection	High alkaloid recombinants.
2	In vitro culture	Enhanced secondary metabolites.
3	Mutation + molecular screening	Novel chemotypes.

3.3 Ashwagandha (*Withania somnifera* L.)

3.3.1 Introduction

Withania somnifera L. Dunal, commonly called **Ashwagandha** or **Indian Ginseng**, is a medicinal plant from the **Solanaceae** family. Its roots are used in Ayurvedic formulations for rejuvenation and stress relief. The chief bioactive constituents are **withanolides, withaferin A**, and **alkaloids** with adaptogenic and immunomodulatory properties.

India is the largest producer, with major cultivation in **Madhya Pradesh, Rajasthan, and Karnataka**.

3.3.2 Botanical Description and Cytology

» **Habit:** Perennial shrub, 30–100 cm tall.

» **Leaves:** Ovate, dull green, softly pubescent.

» **Flowers:** Small, greenish-yellow, axillary.

» **Fruit:** Orange-red berry with numerous seeds.

» **Chromosome number:** 2n = 48. The species is largely **self-pollinated,** but some cross-pollination occurs via insects.

3.3.3 Economic and Medicinal Importance

Roots are used as a tonic, antistress agent, and anti-inflammatory. Withanolides have shown anticancer and neuroprotective activities. It is a key export crop in Ayurvedic and nutraceutical industries.

3.3.4 Breeding Objectives

- Increase root yield and withanolide content (> 1.5 %).
- Develop short-duration and drought-tolerant lines.
- Improve disease resistance (*Alternaria* leaf spot).
- Breed uniform plant types for mechanized harvest.

3.3.5 Genetic Resources

S. No.	Institution	Accessions	Remarks
1	CIMAP, Lucknow	> 200	Diverse chemotypes.
2	MPUAT, Udaipur	150	Drought-tolerant types.
3	NBPGR	120	Global collection.

3.3.6 Breeding and Hybridization

- **Selection:** Based on root diameter and withanolide concentration.
- **Hybridization:** Manual crossing; bagging to prevent contamination.
- **Mutation Breeding:** Gamma rays for compact, early flowering lines.
- **Biotechnology:** Tissue culture for clonal propagation and somaclonal variation.

3.3.7 Released Varieties

S. No.	Variety	Features
1	*Poshita*	High root yield; withanolides 1.8 %.
2	*Jawahar Ashwagandha-20*	Early maturing; uniform root size.
3	*WS-134*	Resistant to leaf spot.

3.3.8 Phytochemical Evaluation

- **Withanolide content:** 1–2 %.
- **Active constituents:** Withaferin A, sitoindosides VII–X.
- **Root quality:** Determined by diameter, color, and texture.

3.3.9 Integration of Breeding Approaches

S. No.	Approach	Outcome
1	Selection	Stable yield and quality.
2	Mutation	Early maturing lines.
3	Marker-assisted screening	Identification of high-withanolide chemotypes.

3.4 Isabgol (*Plantago ovata* Forsk.)

3.4.1 Introduction

Plantago ovata Forsk., known as **Isabgol** or **Psyllium husk**, is a small annual medicinal crop of the **Plantaginaceae** family. It is cultivated for its **husk**, the outer seed coat rich in **mucilage** (psyllium), used as a natural laxative and cholesterol-lowering agent. India accounts for **> 80 % of global production**, centered in **Gujarat and Rajasthan.**

3.4.2 Botanical Description and Cytology

» **Habit:** Annual herb, 20–40 cm tall.

» **Leaves:** Narrow, lanceolate.

» **Inflorescence:** Cylindrical spikes on long stalks.

» **Flowers:** Small, white, sessile, bisexual.

» **Seed:** Boat-shaped, translucent; husk forms 25–30 % of seed weight.

» **Chromosome number:** 2n = 8. It is predominantly **self-pollinated.**

3.4.3 Medicinal Importance

Husk contains **mucilage (30 %)**, a polysaccharide complex with water absorption and gel-forming ability, useful in treating constipation and hyperlipidemia. Seeds also contain **albuminoids and fixed oil.**

3.4.4 Breeding Objectives

» High seed and husk yield (> 1.2 t ha^{-1}).

» Short-duration and lodging-resistant varieties.

» Resistance to downy mildew and aphids.

» Improved seed shattering resistance.

3.4.5 Genetic Resources

S. No.	Institute	Accessions	Remarks
1	NRCSS, Ajmer	400	Core germplasm.
2	NBPGR	250	Global collection.
3	SDAU, Gujarat	180	Regional lines.

3.4.6 Breeding and Hybridization

» **Selection:** Primary method, screen for yield and husk quality.

» **Hybridization:** Limited by small flowers; hand pollination required.

» **Mutation Breeding:** Created bold-seeded lines (*GI-2*).

» **Biotechnological Tools:** In vitro culture for germplasm conservation.

3.4.7 Major Diseases

S. No.	Disease	Causal Agent	Tolerant Varieties
1	Downy mildew	*Peronospora plantaginis*	*GI-2*, *Niharika*.
2	Fusarium wilt	*F. oxysporum*	Resistant types under selection.

3.4.8 Released Varieties

S. No.	Variety	Salient Features
1	*GI-2*	High yield (1.3 t ha^{-1}); mildew tolerant.
2	*Niharika*	Early maturing (110 days); bold seed.
3	*G-1*	High husk recovery (30 %).

3.4.9 Phytochemical and Quality Evaluation

- **Mucilage content:** 25–30 %.
- **Viscosity index:** 5,000–8,000 cps.
- **Color and swelling index:** Determine export quality.

3.4.10 Integration of Breeding Approaches

S. No.	Approach	Contribution
1	Selection	High-yielding stable genotypes.
2	Mutation breeding	Compact, lodging-resistant plants.
3	Marker-assisted selection	Quality-based early screening.

3.5 Sarpagandha (*Rauvolfia serpentina* Benth. ex Kurz.)

3.5.1 Introduction

Rauvolfia serpentina (family Apocynaceae) is one of India's most celebrated medicinal shrubs, commonly called **Sarpagandha** or **Indian Snakeroot**. The roots yield indole alkaloids, notably **reserpine, ajmaline, serpentine, and yohimbine**, that revolutionized modern psychopharmacology as the first effective **antihypertensive and tranquilizing drugs**. Traditionally, its powdered root was prescribed in Ayurveda for **insomnia, hypertension, hysteria, and snake-bite**. Because of over-exploitation from forests, it was brought under systematic cultivation across **Uttar Pradesh, Bihar, and Assam**.

3.5.2 Botanical Description and Cytology

- **Habit:** Perennial, woody, evergreen shrub 40–100 cm tall.
- **Root:** Tuberous, yellowish-brown, rich in latex and alkaloids.
- **Stem:** Cylindrical, glabrous, branched.
- **Leaves:** Simple, 7–12 cm long, arranged in whorls of 3–4.
- **Flowers:** Small, white-pink, borne in terminal cymes.

- **Fruit:** Globose drupe, purple-black when ripe, 2 seeds.
- **Chromosome number:** 2n = 22.

It is a **cross-pollinated** species; insect visitors (bees, thrips) ensure pollen movement.

3.5.3 Medicinal and Economic Importance

Pharmaceutical Use: Reserpine is used worldwide in antihypertensive formulations.

Industrial Demand: Ajmaline employed in anti-arrhythmic drugs.

Export Potential: India supplies > 80 % of global reserpine raw material.

Conservation Value: Listed under CITES Appendix II; genetic resources safeguarded through ex-situ collections.

3.5.4 Breeding Objectives

High root yield (≥ 2.5 t ha^{-1}) with **reserpine ≥ 0.12 %.**

Early maturity and synchronous flowering.

Resistance to root rot (*Fusarium oxysporum*).

Adaptation to light soils and rainfed conditions.

Conservation-linked improvement, domesticated high-alkaloid clones replacing wild harvests.

3.5.5 Genetic Resources and Germplasm

S. No.	Institute / Location	Accessions	Remarks
1	CIMAP Lucknow	> 200	Chemotypic collection (reserpine 0.06–0.18 %).
2	NBPGR New Delhi	≈ 150	Germplasm + wild relatives (*R. tetraphylla*).
3	TNAU Coimbatore	100	South-Indian ecotypes; early flowering.

3.5.6 Floral Biology and Pollination

- **Flowering:** 5–6 months after planting.
- **Pollination:** Predominantly entomophilous; out-crossing ≈ 60 %.
- **Fruit set:** Enhanced by bee activity.
- **Seed viability:** Short; fresh seed used for breeding.

3.5.7 Breeding and Hybridization Methods

Clonal Selection: Preferred since seed set is low; elite plants selected for high reserpine.

Hybridization: Crosses made between high-yield and high-alkaloid parents; hand pollination using fine brushes.

Mutation Breeding: Gamma rays (15–25 kR) created compact, early rooting mutants.

Polyploid Induction: Colchicine (0.1 %) used to enhance alkaloid biosynthesis.

Biotechnological Tools: Hairy-root cultures (using *Agrobacterium rhizogenes*) produce 5–8 × higher reserpine content.

3.5.8 Diseases and Resistance

S. No.	Disease / Causal Agent	Symptoms	Resistant Lines / Management
1	Root rot – *Fusarium oxysporum*	Wilting and blackened roots	Tolerant clone *CIM-RS-3*; Trichoderma seedling treatment.
2	Leaf spot – *Alternaria tenuissima*	Brown necrotic patches	Copper fungicides + resistant accessions.

3.5.9 Released Varieties

S. No.	Variety	Origin	Salient Features
1	*CIM-RS-1*	CIMAP	High reserpine (0.14 %); root yield 2.8 t ha^{-1}.
2	*CIM-RS-3*	CIMAP	Root-rot tolerant; early maturing.
3	*TNAU-1*	Coimbatore	Adapted to southern India.

3.5.10 Phytochemical Evaluation

» **Reserpine:** 0.08–0.16 % (dry root basis).

» **Ajmaline + Serpentine:** 0.1–0.2 %.

» **Quality Index:** Alkaloid sum ≥ 0.25 % and root diameter > 1.5 cm.

3.5.11 Integration of Approaches

S. No.	Approach	Contribution
1	Clonal selection + polyploidy	Higher reserpine and vigorous growth.
2	Mutation breeding	Compact plants suitable for dense cultivation.
3	Hairy-root culture	Sustainable alkaloid production ex vitro.

3.6 Opium Poppy (*Papaver somniferum* L.)

3.6.1 Introduction

Papaver somniferum L., the **Opium Poppy**, belongs to Papaveraceae and is cultivated for its latex that contains pharmaceutically vital alkaloids, **morphine, codeine, thebaine, papaverine, and narcotine**. India is the only country authorized by the UN to produce opium for medicinal purposes under strict licensing. The crop also provides poppy seeds ("khus-khus") rich in oil (40–45 %).

3.6.2 Botanical Description

» **Habit:** Annual herb, 60–150 cm tall.

» **Leaves:** Large, glaucous, lobed.

» **Flowers:** Showy, 4–10 cm across, white to purple.

- **Fruit:** Capsule (3–5 cm diameter) exuding milky latex when incised.
- **Chromosome number:** 2n = 22. It is **predominantly self-pollinated** but 10–20 % crossing occurs via insects.

3.6.3 Medicinal and Economic Value

- **Morphine** is a principal analgesic used in surgery and palliative care.
- **Codeine** serves as antitussive and mild analgesic.
- **Thebaine** is precursor for semisynthetic opioids like oxycodone.
- **Seeds and oil** used in bakery and nutraceutical products.

3.6.4 Breeding Objectives

High latex yield and morphine content (> 12 %).

Balanced alkaloid profile (low narcotine, high thebaine types).

Resistance to downy mildew and capsule borer.

Non-narcotic low-morphine types for oilseed use.

Uniform plant height for mechanized harvest.

3.6.5 Genetic Resources

S. No.	Institute	Accessions	Remarks
1	CIMAP Lucknow	> 400	Chemotypic variability for alkaloids.
2	DRDO Gwalior	≈ 250	High-thebaine lines.
3	NBPGR	200	Global accessions for diversity.

3.6.6 Breeding and Hybridization

- **Pure-line selection:** For morphine percentage and latex volume.
- **Hybridization:** Performed by emasculating flower buds before anthesis; hand-pollination next morning.
- **Mutation breeding:** EMS and gamma rays created low-morphine, high-thebaine lines.
- **Biotech:** Tissue culture for rapid multiplication and somaclonal variation screening.

3.6.7 Released Varieties

S. No.	Variety	Features
1	*Shweta*	Morphine 12 %; resistant to downy mildew.
2	*CIM-P-Op1*	High thebaine (3 %); low morphine.
3	*Jawahar Op-3*	High seed oil (44 %).

3.6.8 Quality Evaluation

- **Morphine:** 8–14 %.
- **Codeine:** 1–2 %.
- **Oil:** 40–45 %.
- Analyses by HPLC and GC–MS ensure export quality compliance.

3.6.9 Integrated Approach

Approach	Contribution
Selection + mutation	Diversified alkaloid profiles.
Molecular screening	Marker tags for high-thebaine alleles.
Tissue culture	Clean propagation material.

3.7 Glory Lily (*Gloriosa superba* L.)

3.7.1 Introduction

Gloriosa superba L. of the family *Colchicaceae* is a tuberous climber native to tropical Africa and India. It is renowned for its flaming red-yellow flowers and for its high alkaloid content, **colchicine and gloriosine,** used as anti-gout and plant mutagenic agents. Because colchicine is a precursor for chromosome-doubling agents, Glory Lily occupies a unique industrial niche among medicinal plants.

3.7.2 Botanical Description

- **Habit:** Perennial climber with tendrillar leaf tips.
- **Root:** V-shaped rhizomatous tubers.
- **Leaves:** Lanceolate with spiral tendrils.
- **Flowers:** Large, crinkled, reflexed petals, red with yellow margins.
- **Fruit:** Capsule with numerous round seeds.
- **Chromosome number:** 2n = 24.

Cross-pollinated; pollinators include butterflies and bees.

3.7.3 Medicinal Significance

- **Colchicine** (0.3–0.5 %) used in gout therapy and cytogenetics.
- **Gloriosine** has anti-inflammatory properties.
- Seeds and tubers serve as industrial raw materials for colchicine extraction export.

3.7.4 Breeding Objectives

- Increase tuber yield and colchicine content.
- Develop disease-free planting stock.
- Induce early maturity and high seed set.
- Enhance resistance to tuber rot and leaf spot.

3.7.5 Breeding and Propagation

- **Clonal selection** from wild populations.
- **Hybridization** rare due to low fertility; crosses attempted using controlled pollination.
- **Mutation breeding** and **in vitro mutagenesis** developed compact, high-colchicine lines.
- **Microtuber culture** and somatic embryogenesis** standardized for rapid propagation.

3.7.6 Released Lines

S. No.	Line / Variety	Features
1	*TNAU GS-1*	High colchicine (0.5 %); disease free.
2	*CIM-GS-2*	Compact vine; early flowering.

3.7.7 Integration of Approaches

Approach	Contribution
Micropropagation	Virus-free planting stock.
Mutation + selection	Enhanced colchicine yield.
Clonal	Clonal selection

3.8 Medicinal Coleus (*Coleus forskohlii* Briq.)

3.8.1 Introduction

Coleus forskohlii (syn. *Plectranthus barbatus*) of the **Lamiaceae** family is a semi-succulent perennial herb indigenous to India, Nepal, and Sri Lanka. It is cultivated for its tuberous roots that yield **forskolin**, a labdane diterpenoid that stimulates adenylate cyclase, increasing intracellular cyclic AMP. This mechanism forms the pharmacological base for **antihypertensive, bronchodilatory, and cardiotonic drugs**.

Because forskolin has both therapeutic and nutraceutical value, *C. forskohlii* has gained prominence as a **high-value medicinal crop** in India's dry tropical belts (notably Tamil Nadu, Karnataka, and Andhra Pradesh).

3.8.2 Botanical Description and Cytology

- **Habit:** Erect or decumbent herb, 30–80 cm tall.
- **Stem:** Quadrangular, succulent, often reddish.
- **Leaves:** Ovate, serrate, opposite; glandular hairs.
- **Flowers:** Bluish-purple, in terminal racemes.
- **Roots:** Thick, tuberous, aromatic, yellow-brown.
- **Chromosome number:** $2n = 30$.

Pollination is **entomophilous**, but plants are largely propagated vegetatively.

3.8.3 Economic and Medicinal Importance

- **Forskolin** is used in pharmaceuticals to treat heart failure, glaucoma, and asthma.
- **Root extract** forms a dietary supplement promoting weight management and lipid metabolism.
- **Essential oil** is valued in perfumery.
- **Agro-economic advantage:** Thrives under low rainfall and marginal soils.

3.8.4 Breeding Objectives

- High forskolin yield (> 0.4 % dry weight).
- Development of compact, high-root-mass genotypes.
- Drought and root-rot resistance.
- Elite clonal lines for uniform yield and active content.
- Genetic stability under vegetative propagation cycles.

3.8.5 Genetic Resources

S. No.	Institute / Location	Accessions	Remarks
1	CIMAP, Lucknow	220	High-forskolin chemotypes.
2	TNAU, Coimbatore	150	Drought-hardy ecotypes.
3	NBPGR	100	Global germplasm; African wild relatives.

3.8.6 Breeding and Propagation

- **Clonal Selection:** Main breeding route; elite plants identified for root yield and forskolin content.
- **Hybridization:** Rare, due to limited seed set; attempted to recombine traits like early maturity and root quality.
- **Mutation Breeding:** EMS treatments produced fast-growing, high-forskolin mutants.
- **In Vitro Techniques**
  - *Micropropagation* ensures large-scale multiplication of true-to-type, pathogen-free stock.
  - *In vitro selection* for drought tolerance under osmotic stress media.

3.8.7 Diseases and Management

S.No.	Disease / Pathogen	Symptoms	Control / Resistant Lines
1	Root rot – *Fusarium solani*	Browning of tubers	Resistant line *CIM-CF-3*.
2	Leaf spot – *Cercospora* sp.	Irregular brown lesions	Copper oxychloride + tolerant clones.

3.8.8 Released Varieties

S. No.	Variety / Clone	Features
1	*CIM-CF-1*	Forskolin 0.45 %; root yield 3.5 t ha^{-1}.
2	*CIM-CF-3*	Root-rot tolerant; compact canopy.
3	*Arka Sugandha*	Drought tolerant; good aroma profile.

3.8.9 Phytochemical and Quality Evaluation

- **Forskolin range:** 0.2–0.5 % (dry root).
- **Extraction method:** Soxhlet extraction or supercritical CO_2.
- **Standard marker:** HPLC quantification of forskolin retention at 6.8 min.

3.8.10 Integration of Breeding Approaches

S. No.	Approach	Contribution
1	Clonal selection	Enhanced yield and forskolin recovery.
2	Mutation breeding	High-forskolin compact lines.
3	In vitro propagation	Uniform elite planting stock.
4	Chemoprofiling	Direct selection for high diterpenoid chemotypes.

3.9 *Mucuna Pruriens* (L.) DC.

3.9.1 Introduction

Mucuna pruriens (L.) DC., also known as **Velvet Bean** or **Kaunch**, belongs to the **Fabaceae** family and is an ancient medicinal legume highly valued for its seeds, which contain the neurotransmitter precursor **L-DOPA (L-3,4-dihydroxyphenylalanine)**, the primary natural source used in the treatment of **Parkinson's disease**.

In Ayurveda, *Mucuna* is used as a **tonic, aphrodisiac, and nervine stimulant**, while in modern phytomedicine, it has gained prominence for **neuroprotective, antioxidant, and antidiabetic properties**. The crop also enriches soil fertility through nitrogen fixation, making it valuable for intercropping and sustainable systems.

3.9.2 Botanical Description and Cytology

- **Habit:** Climbing annual or perennial legume, 3–15 m in length.
- **Leaves:** Trifoliate with ovate-lanceolate leaflets, densely pubescent.
- **Flowers:** Axillary racemes; pale purple to white; papilionaceous.
- **Pods:** Oblong, covered with stinging hairs, containing 3–6 seeds.
- **Seeds:** Ovoid, mottled brown or black.
- **Chromosome number:** 2n = 22.

Mucuna is **self-pollinated**, though 5–10 % cross-pollination may occur by insects.

3.9.3 Economic and Medicinal Importance

- **Pharmaceutical:** Seeds contain 3–6 % L-DOPA; used in Parkinson's therapy.
- **Nutritional:** High-protein (25–30 %) pulse after detoxification of anti-nutrients.
- **Agronomic:** Nitrogen-fixing cover crop improving soil organic carbon.
- **Industrial:** L-DOPA extraction industries in India, Thailand, and Brazil.

3.9.4 Breeding Objectives

- Increase **L-DOPA content** (> 5 % dry seed).
- Develop **non-itchy, smooth-podded** cultivars.
- Enhance **seed yield and uniform maturity**.
- Improve **pest and disease tolerance** (anthracnose, powdery mildew).
- Adaptation to varied agro-climates and mechanized harvest systems.

3.9.5 Genetic Resources

S. No.	Institute / Center	Accessions Maintained	Remarks
1	ICAR–CIMAP, Lucknow	200	Diverse Indian ecotypes
2	NBPGR, New Delhi	180	Global and wild *Mucuna* species
3	TNAU, Coimbatore	120	South Indian smooth-podded lines.

3.9.6 Floral Biology and Pollination

- **Flowering:** 90–120 days after sowing.
- **Anthesis:** Early morning; anthers dehisce before flower opening (cleistogamy).
- **Pollination:** Mostly self-pollinated.
- **Fertilization:** Occurs within 8–10 hours after anthesis.

Controlled hybridization is feasible through **manual tripping** of keel petals and dusting pollen.

3.9.7 Breeding and Hybridization

- **Mass Selection:** For high-yield and smooth pod texture.
- **Pure-Line Selection:** Based on seed size, color, and L-DOPA level.
- **Hybridization:** Crossing between *M. pruriens* and *M. deeringiana* for disease resistance.
- **Mutation Breeding:** EMS and gamma-ray mutants with 6.5 % L-DOPA reported.
- **Biotechnological Approaches:**
  - In vitro cell suspension cultures for enhanced L-DOPA synthesis.
  - Genetic transformation targeting tyrosinase and DOPA decarboxylase genes.

3.9.8 Diseases and Resistance

S. No.	Disease	Causal Agent	Resistant Source / Strategy
1	Anthracnose	*Colletotrichum truncatum*	Resistant line *MP-20*.
2	Powdery mildew	*Erysiphe polygoni*	*MP-42* tolerant.
3	Pod borer	*Helicoverpa armigera*	Avoid late sowing; neem oil spray.

3.9.9 Released Varieties

S. No.	Variety	Origin	Salient Features
1	*Arka Dhanvantari*	IIHR, Bengaluru	Smooth pods; 4.8 % L-DOPA.
2	*CIM-MP-1*	CIMAP	High yield (1.8 t ha^{-1}); 5 % L-DOPA.
3	*TNAU MP-2*	TNAU	Early maturing (140 days); 4.5 % L-DOPA.

3.9.10 Quality Evaluation

- **L-DOPA quantification:** HPLC or spectrophotometry (λmax 280 nm).
- **Target index:** ≥ 4.5 % L-DOPA with seed weight 15–20 g per 100 seeds.
- **Seed detoxifiction:** Roasting or soaking reduces trypsin inhibitors.

3.9.11 Integration of Approaches

S. No.	Approach	Contribution
1	Hybridization + mutation	High L-DOPA and smooth pods.
2	In vitro enhancement	10× L-DOPA yield via tissue culture.
3	Marker-assisted screening	Early selection for tyrosinase gene.

3.10 *Ocimum Sanctum* L. (TULSI)

3.10.1 Introduction

Ocimum sanctum L. (syn. *O. tenuiflorum*), known as **Tulsi** or **Holy Basil**, belongs to the **Lamiaceae** family. It holds immense spiritual, medicinal, and aromatic significance across South Asia. The leaves and essential oil contain **eugenol, ursolic acid, and rosmarinic acid**, which exhibit antimicrobial, adaptogenic, and anti-inflammatory properties.

Tulsi is central to Ayurvedic formulations for respiratory disorders, stress, and immunity enhancement, and its essential oil is widely used in perfumery and aromatherapy industries.

3.10.2 Botanical Description

- **Habit:** Erect, aromatic perennial herb, 30–80 cm tall.
- **Leaves:** Opposite, ovate, with glandular trichomes.
- **Flowers:** Small, purple or white, in terminal spikes.

» **Fruit:** Nutlets with mucilaginous coating.

» **Chromosome number:** 2n = 32.

Species include *O. sanctum* (Krishna Tulsi) and *O. gratissimum* (Ram Tulsi).

3.10.3 Economic and Medicinal Importance

» **Essential oil:** 0.5–1.5 %, rich in eugenol (40–70 %).

» **Pharmaceutical:** Immunomodulatory, antipyretic, antidiabetic.

» **Aromatic industry:** Used in mouthwash, incense, and flavoring.

» **Export:** Dried leaves and oil exported globally.

3.10.4 Breeding Objectives

» High **essential-oil yield** (> 1.5 %).

» Enhanced **eugenol** or **methyl eugenol** content depending on end-use.

» Drought and pest tolerance.

» Compact, high-branching plant type for mechanical harvesting.

3.10.5 Germplasm Resources

S. No.	Institute	Accessions	Remarks
1	CIMAP Lucknow	250	High-oil chemotypes.
2	TNAU Coimbatore	120	Drought-tolerant lines.
3	NBPGR	150	Global species diversity.

3.10.6 Breeding and Hybridization

» **Mass Selection:** Based on oil content and biomass.

» **Intervarietal Crosses:** Between *O. sanctum* × *O. gratissimum* for vigor.

» **Mutation Breeding:** Gamma rays produced early, compact types.

» **Polyploidy:** Induced tetraploids show 20–30 % higher oil yield.

» **Molecular Breeding:** RAPD and ISSR markers used for chemotype differentiation.

3.10.7 Released Varieties

S. No.	Variety	Features
1	*CIM-Ayu*	High eugenol (70 %); oil 1.8 %.
2	*CIM-Saumya*	High biomass; drought hardy.
3	*Krishna Tulsi*	Rich purple leaves; aromatic.

3.10.8 Quality Evaluation

» **Eugenol content:** 60–70 %.

- **Oil yield:** 1.5–2.0 % on fresh weight basis.
- **GC–MS Profiling:** Confirms chemotype composition.

3.10.9 Integration of Approaches

S. No.	Approach	Contribution
1	Hybridization	Vigor and high-oil genotypes.
2	Mutation + selection	Early flowering compact types.
3	Chemoprofiling	Identification of high-eugenol chemotypes.
4	Molecular markers	Parentage verification and variety protection.

3.11 *CENTELLA ASIATICA* (L.) URB.

3.11.1 Introduction

Centella asiatica (L.) Urb., commonly called **Gotu Kola** or **Mandukaparni,** is a creeping perennial herb of the family **Apiaceae** (Umbelliferae). It is used extensively in **Ayurveda, Chinese, and African traditional medicine** for improving memory, wound healing, and skin health. The primary active compounds are **triterpenoid saponins, asiaticoside, madecassoside, and asiatic acid.**

3.11.2 Botanical Description

- **Habit:** Creeping herb with slender stolons.
- **Leaves:** Reniform, petiolate, glabrous.
- **Flowers:** Small, pinkish, in axillary umbels.
- **Fruits:** Cremocarps with mericarps.
- **Chromosome number:** 2n = 18.

It propagates both sexually (seeds) and vegetatively (stolons).

3.11.3 Medicinal and Economic Importance

- **Phytochemicals:** Asiaticoside (0.5–1.5 %), madecassoside (0.2–0.8 %).
- **Pharmacological uses:** Memory enhancer, anti-ulcer, wound healing, and anti-aging.
- **Export:** Dried herb and extracts used in nutraceuticals and cosmetics.

3.11.4 Breeding Objectives

- High triterpenoid yield per unit biomass.
- Compact, fast-spreading, high-leaf-mass types.
- Drought and shade adaptability.
- Resistance to leaf spot and root rot.

3.11.5 Genetic Resources

S. No.	Institute	Accessions	Remarks
1	CIMAP Lucknow	180	Asiaticoside variability 0.4–1.8 %.
2	Kerala Agri. Univ.	100	Moisture-loving ecotypes.
3	NBPGR	80	Regional accessions.

3.11.6 Breeding and Propagation

» **Clonal Selection:** Main method; high-asiaticoside lines multiplied via stolons.

» **Mutation Breeding:** EMS (0.2 %) produced compact high-yield mutants.

» **Tissue Culture:** Nodal and shoot-tip cultures for mass propagation.

» **Somaclonal Variation:** Used to select stress-tolerant clones.

3.11.7 Diseases and Management

S. No.	Disease	Causal Agent	Control / Resistance
1	Leaf spot	*Cercospora centellae*	Resistant line *CA-5*.
2	Root rot	*Rhizoctonia solani*	Raised-bed cultivation.

3.11.8 Quality Evaluation

» **Asiaticoside:** ≥ 1.2 % (HPLC standard).

» **Total triterpenes:** 1.5–2 %.

» **Bioassay:** Wound-healing index and collagen synthesis in vitro.

3.11.9 Integration of Approaches

S. No.	Approach	Contribution
1	Clonal selection	High saponin yield and stability.
2	Mutation + tissue culture	Compact and uniform clones.
3	Molecular screening	Identification of elite asiaticoside lines.

3.12 *BACOPA MONNIERI* (L.) WETTST.

3.12.1 Introduction

Bacopa monnieri (L.) Wettst., known as **Brahmi**, is a creeping aquatic herb of **Plantaginaceae** (formerly Scrophulariaceae). It is a revered memory enhancer in Ayurveda and is used as a **nootropic, antioxidant, and neuroprotective** herb. The principal active compounds are **bacosides A and B**, triterpenoid saponins responsible for cognitive improvement.

3.12.2 Botanical Description

» **Habit:** Creeping herb with succulent stems and opposite leaves.

- **Flowers:** Pale blue or white, solitary, axillary.
- **Fruit:** Ovoid capsule with minute seeds.
- **Chromosome number:** 2n = 64. The plant reproduces mainly by **stem cuttings**.

3.12.3 Economic and Medicinal Importance

- Used in **memory and learning enhancement** formulations.
- Extracts possess **antioxidant, adaptogenic, and hepatoprotective** properties.
- Commercial demand in **pharmaceutical and nutraceutical industries**.

3.12.4 Breeding Objectives

- High **bacoside A + B** content (> 2.5 %).
- Rapid growth and high stolon yield.
- Tolerance to waterlogging and salinity.
- Compact architecture for dense cultivation.

3.12.5 Genetic Resources

S. No.	Institute	Accessions	Remarks
1	CIMAP Lucknow	250	Bacoside variability 1.0–3.0 %.
2	KAU Kerala	150	Aquatic ecotypes.
3	NBPGR	80	Germplasm from eastern India.

3.12.6 Breeding and Propagation

- **Clonal selection:** Elite plants selected for bacoside content.
- **Mutation breeding:** EMS (0.1–0.3 %) developed early, high-yield mutants.
- **In vitro culture:** Node and leaf explants regenerated via organogenesis.
- **Somaclonal variation:** Screened for drought and salinity tolerance.

3.12.7 Quality Evaluation

- **Bacoside A:** 1.5–2.5 %.
- **Bacoside B:** 0.8–1.2 %.
- **Standardization:** HPLC and TLC fingerprinting for active compound quantification.

3.12.8 Integration of Breeding Approaches

S. No.	Approach	Outcome
1	Clonal selection	Stable, high-bacoside clones.
2	Mutation + tissue culture	Enhanced metabolite accumulation.
3	Marker-assisted screening	Distinction of high- and low-bacoside genotypes.

3.13 *ANDROGRAPHIS PANICULATA* (BURM.F.) WALL. EX NEES

3.13.1 Introduction

Andrographis paniculata, commonly known as **Kalmegh** or "King of Bitters," belongs to the family **Acanthaceae.** This herbaceous annual is one of India's most important medicinal crops due to its broad therapeutic value. Its chief bioactive compound, **andrographolide,** possesses **anti-inflammatory, antiviral, hepatoprotective, antimalarial, and immunomodulatory** properties.

Owing to its efficacy against liver disorders and infectious diseases, *A. paniculata* has been recognized by the World Health Organization as a potential global herbal resource. The crop thrives well in tropical and subtropical regions with minimal agronomic inputs, making it ideal for medicinal-farming systems.

3.13.2 Botanical Description

- **Habit:** Erect, branched annual herb, 30–110 cm tall.
- **Stem:** Quadrangular, green, slightly pubescent.
- **Leaves:** Opposite, lanceolate, 5–8 cm long, smooth margins.
- **Flowers:** Small, white with purple spots, in axillary or terminal racemes.
- **Fruit:** Linear capsule, 1.5–2.0 cm long, containing 10–15 seeds.
- **Chromosome number:** 2n = 50. Pollination is mostly **self-pollinated,** though insects may assist minor outcrossing.

3.13.3 Economic and Medicinal Importance

- **Andrographolide** (0.5–2.5 %) is the principal diterpenoid lactone responsible for therapeutic value.
- Extracts show **antiviral and immunostimulant** activity, supporting their use in modern formulations.
- Large-scale industrial demand for **herbal hepatoprotective and antipyretic preparations.**
- High export potential from India and Southeast Asia for standardized extracts.

3.13.4 Breeding Objectives

- Enhance **andrographolide content** (> 2 %).
- Develop **early, high-biomass, and disease-tolerant** cultivars.
- Improve **seed set and viability** (often poor in cultivated lines).
- Develop varieties suited for **mechanized harvesting and uniform maturity.**

3.13.5 Genetic Resources

S. No.	Institute / Centre	Accessions	Remarks
1	CIMAP, Lucknow	> 200	High-andrographolide chemotypes.
2	TNAU, Coimbatore	120	South Indian germplasm; drought tolerance.
3	NBPGR	150	Global diversity collection.

3.13.6 Floral Biology and Pollination

» **Flowering:** 60–75 days after planting.

» **Pollination type:** Predominantly self-pollinated; anther dehiscence precedes stigma receptivity slightly.

» **Hybridization:** Manual emasculation and dusting possible; flower size allows fine-scale crossing with forceps.

3.13.7 Breeding and Hybridization

» **Mass Selection:** For plant height, leaf area, and andrographolide yield.

» **Hybridization:** Recombination between high-yield × high-andrographolide parents.

» **Mutation Breeding:** Gamma and EMS mutagenesis induced early flowering and compact forms.

» **Polyploidy Induction:** Colchicine (0.2 %) doubled chromosome number, increasing andrographolide yield 30 %.

Biotechnological Tools:

» *Callus culture* and *elicitation* using jasmonic acid enhance secondary metabolite synthesis.

» *Marker-assisted selection* identifies high-producing chemotypes via RAPD/ISSR markers.

3.13.8 Diseases and Tolerance

S. No.	Disease	Causal Agent	Resistant / Management Practice
1	Leaf spot	*Cercospora andrographidis*	Resistant line *AP-5*; avoid water stagnation.
2	Root rot	*Rhizoctonia solani*	Well-drained soil + Trichoderma.

3.13.9 Released Varieties

S. No.	Variety	Origin	Salient Traits
1	*CIM-Megha*	CIMAP	Andrographolide 2.4 %; early flowering.
2	*KAL-1*	KAU	Compact canopy; drought tolerant.
3	*CIM-AP-3*	CIMAP	High andrographolide + biomass.

3.13.10 Quality Evaluation

» **Andrographolide quantification:** HPLC or TLC fingerprinting; standard retention 5.4 min.

» **Acceptable range:** ≥ 2 % in dry leaf tissue.

» **Bioassay correlation:** High andrographolide lines exhibit > 60 % antioxidant index in vitro.

3.13.11 Integration of Breeding Approaches

S. No.	Approach	Outcome
1	Hybridization + mutation	Early, high-metabolite types.
2	Polyploid induction	Enhanced andrographolide yield.
3	Tissue culture + elicitation	Scalable metabolite production.

3.14 ALOE VERA (L.) BURM. F.

3.14.1 Introduction

Aloe vera (L.) Burm. f., family **Asphodelaceae,** is one of the oldest and most widely used medicinal succulents. The fleshy leaves yield **aloin (anthraquinone glycoside)** and **acemannan (a polysaccharide),** responsible for **laxative, anti-inflammatory, and wound-healing** properties. The gel and latex are commercialized in cosmetic, pharmaceutical, and nutraceutical industries globally.

India is among the leading producers, with large-scale cultivation in **Rajasthan, Gujarat, and Tamil Nadu** under arid and semi-arid conditions.

3.14.2 Botanical Description

» **Habit:** Perennial, stemless succulent, 30–60 cm tall.

» **Leaves:** Thick, fleshy, lanceolate, serrated margins, arranged in rosettes.

» **Flowers:** Yellowish-orange, tubular, borne on tall spikes.

» **Chromosome number:** 2n = 14. *Aloe* reproduces mainly **vegetatively** via suckers or tissue culture.

3.14.3 Medicinal and Economic Significance

» **Gel:** Contains acemannan, enzymes, and amino acids, used in cosmetics and dermatology.

» **Latex (yellow sap):** Source of aloin, a purgative compound.

» **Global market:** Major export commodity for the herbal and cosmetic industries.

» **Agro-economics:** Drought-tolerant perennial suitable for marginal soils.

3.14.4 Breeding Objectives

» Increase **gel yield** and **acemannan concentration.**

» Select for **low anthraquinone (aloin)** content in edible gel types.

» Develop **salt- and drought-tolerant** clones.

» Ensure **disease-free, pathogen-indexed** planting material.

3.14.5 Genetic Resources

S. No.	Institute	Accessions	Remarks
1	CIMAP, Lucknow	100	Core germplasm for gel yield and acemannan.
2	NBPGR	80	Exotic *Aloe* species collection.
3	KAU	50	Southern ecotypes with large leaves.

3.14.6 Breeding and Propagation

» **Clonal Selection:** Primary method; elite plants identified for high gel yield.

» **Hybridization:** Limited by sterility and low seed set; somatic hybridization explored experimentally.

» **Mutation Breeding:** Gamma rays (10–20 Gy) induced early sucker formation and thicker leaves.

» **Tissue Culture:** Micropropagation through shoot-tip and meristem cultures ensures disease-free stock.

3.14.7 Diseases and Resistance

S. No.	Disease	Causal Agent	Management
1	Leaf spot	*Alternaria alternata*	Resistant line *AV-3*.
2	Root rot	*Pythium sp.*	Raised-bed cultivation, Trichoderma.

3.14.8 Released Varieties

S. No.	Variety	Features
1	*CIM-Sheetal*	High gel recovery (65 %); low aloin.
2	*AL-1*	Thick leaves; drought tolerant.
3	*Arka Aloe*	Early sucker formation; compact growth.

3.14.9 Quality Evaluation

» **Acemannan content:** 0.2–0.4 % (dry basis).

» **Aloin level:** < 20 ppm (for edible grade).

» **Gel recovery:** 55–70 % of fresh-leaf weight.

3.14.10 Integrated Approaches

Approach	Outcome
Tissue culture + clonal selection	Large-scale propagation.
Mutation breeding	Early, thick-leaved clones.
Somatic hybridization	Experimental recombinants.

3.15 *PHYLLANTHUS AMARUS SCHUM.* & THONN.

3.15.1 Introduction

Phyllanthus amarus is a small herb of the **Phyllanthaceae** family, commonly known as **Bhui Amla** or **Stone Breaker**. It is valued for hepatoprotective and antiviral activities, particularly against **Hepatitis B virus**. The active constituents include **phyllanthin, hypophyllanthin, and lignans** such as niranthin and corilagin.

3.15.2 Botanical Description

- **Habit:** Annual, erect herb, 30–60 cm tall.
- **Leaves:** Small, alternate, resembling tamarind leaflets.
- **Flowers:** Small, greenish-yellow, axillary.
- **Fruit:** Globose capsule, dehiscent.
- **Chromosome number:** 2n = 52.

Pollination is mostly autogamous, though small insects contribute to minor cross-pollination.

3.15.3 Medicinal Importance

- **Phyllanthin and hypophyllanthin** act as hepatoprotective and antioxidant agents.
- Used in traditional medicine for **jaundice, kidney stones, and diabetes.**
- Exported in large quantities for herbal formulations and nutraceuticals.

3.15.4 Breeding Objectives

- High **phyllanthin + hypophyllanthin content** (> 0.8 %).
- Early and synchronous maturity.
- Tolerance to drought and leaf spot.
- Compact plant type suitable for dense planting.

3.15.5 Breeding Methods

- **Mass Selection:** For biomass and phytochemical yield.
- **Hybridization:** Limited; manual crossing possible between divergent types.
- **Mutation Breeding:** EMS-induced high-lignan mutants identified (*PA-20*).
- **Tissue Culture:** Shoot-tip regeneration for uniform planting material.

3.15.6 Released Varieties

S. No.	Variety	Origin	Features
1	*CIM-Bhui Amla*	CIMAP	Phyllanthin 0.45 %; high biomass.
2	*PA-20*	Mutation line	Compact, early maturing.
3	*TNAU-Phyll-3*	TNAU	Drought tolerant; uniform canopy.

3.15.7 Quality Evaluation

» **Phyllanthin content:** 0.3–0.5 %.

» **HPLC quantification:** λmax 280 nm.

» **Total lignan yield:** 0.6–1.0 % (dry basis).

3.15.8 Integrated Breeding Approaches

Approach	Contribution
Mutation breeding + selection	High-lignan mutants.
Tissue culture propagation	Clonal fidelity and disease-free stock.
Biochemical screening	Standardized metabolite evaluation.

3.16 EUCALYPTUS SPP.

3.16.1 Introduction

Eucalyptus, belonging to the **Myrtaceae** family, comprises over 700 species native to Australia, cultivated globally for timber, pulpwood, and essential oil. Medicinally, its **leaves and oil** (rich in **cineole, pinene, and limonene**) possess **antiseptic, expectorant, and anti-inflammatory** properties.

Plantations in India span millions of hectares, serving both **industrial forestry** and **phytopharmaceutical** industries.

3.16.2 Economic and Medicinal Importance

1. **Eucalyptus oil (cineole 70–85 %)** used in cough syrups, balms, and aromatherapy.
2. **Leaves** employed in traditional inhalants and antiseptic formulations.
3. **Wood** provides raw material for paper and charcoal.
4. **Agroforestry benefit:** Excellent carbon sequestration and windbreak species.

3.16.3 Breeding Objectives

1. High **oil yield and cineole content.**
2. Fast-growing, straight-boled trees.
3. Resistance to **rust and bacterial wilt.**
4. Hybrid vigor and clonal uniformity for plantation use.

3.16.4 Breeding and Hybridization

» **Inter-specific hybridization:** e.g., *E. tereticornis* × *E. camaldulensis* for growth and drought tolerance.

» **Clonal selection:** Superior trees cloned via coppice shoots or tissue culture.

» **Controlled pollination:** Achieved by bagging flower buds before anthesis.

» **Mutation and Polyploid Breeding:** Induced tetraploids show thicker leaves and higher oil yield.

3.16.5 Released Clones

S. No.	Clone	Species / Hybrid	Features
1	*IFGTB-Et-2*	*E. tereticornis*	High growth, cineole 82 %.
2	*CIM-Euca-1*	*E. globulus*	High oil yield (2.5 %).
3	*TNAU-Hyb-1*	*E. camaldulensis* × *E. tereticornis*	Drought hardy.

3.16.6 Integration of Approaches

Approach	Contribution
Hybridization + cloning	Enhanced vigor and oil yield.
Tissue culture	Rapid clonal multiplication.
Chemoprofiling	Identification of cineole chemotypes.

3.17 *AEGLE MARMELOS* (L.) CORR. (BAEL)

3.17.1 Introduction

Aegle marmelos (L.) Corr., known as **Bael** or **Bilva**, belongs to **Rutaceae** and is a sacred medicinal tree of India. The fruits, leaves, and roots are used in **Ayurvedic formulations** for gastrointestinal, diabetic, and cardiac ailments. The major bioactives include **marmelosin, aegeline, and coumarins**.

3.17.2 Botanical Description

» **Habit:** Medium-sized deciduous tree, 8–12 m tall.

» **Leaves:** Trifoliate, aromatic.

» **Flowers:** Greenish-white, sweet-scented.

» **Fruit:** Globose, hard-shelled, containing aromatic pulp.

» **Chromosome number:** 2n = 18.

Cross-pollinated species, primarily insect-pollinated (bees).

3.17.3 Breeding Objectives

» High pulp yield and **marmelosin** content.

» Early fruiting and regular bearing.

» Resistance to fruit fly and powdery mildew.

» Compact, dwarf tree forms for high-density orchards.

3.17.4 Breeding and Propagation

» **Clonal selection** from high-yielding wild trees.

» **Hybridization:** Manual crossing between elite parents for pulp quality.

» **Mutation Breeding:** Gamma-irradiated seedlings showed early bearing.

- **Micropropagation:** Nodal cultures for uniform orchard establishment.

S. No.	Variety / Clone	Features
1	*CIM-Bael-1*	Early bearing, high pulp recovery (65 %).
2	*NBPGR-B-4*	Thick rind; fruit weight 1.5 kg; good aroma.
3	*TNAU-BL-3*	Compact, high-yielding tree for dense orchards.

3.17.5 Quality Evaluation

- **Marmelosin content:** 0.4–0.7 %.
- **Pulp yield:** 60–70 %.
- **Sugar–acid ratio:** 10–12, ideal for processing.
- **Vitamin C:** 30–40 mg per 100 g pulp.

3.17.6 Integration of Breeding Approaches

Approach	Contribution
Clonal selection	True-to-type high-quality lines.
Hybridization	Genetic recombination for pulp traits.
Mutation breeding	Early bearing and dwarf types.
Micropropagation	Mass propagation of elite trees.

3.18 *Hyoscyamus Niger L.* (HENBANE)

3.18.1 Introduction

Hyoscyamus niger L., known as **Henbane**, is a biennial or annual herb of the **Solanaceae** family, native to Europe and Asia. The plant is valued for **tropane alkaloids, hyoscyamine, scopolamine, and atropine,** used as antispasmodic and sedative drugs.

Historically regarded as a psychoactive plant, it is now strictly cultivated under controlled medicinal licensing for pharmaceutical extraction.

3.18.2 Botanical Description

- **Habit:** Erect, viscid, glandular-pubescent herb, 30–100 cm tall.
- **Leaves:** Alternate, lobed, covered with sticky hairs.
- **Flowers:** Yellowish with purple veins, funnel-shaped.
- **Fruit:** Capsule enclosed in a persistent calyx.
- **Chromosome number:** 2n = 28. Predominantly self-pollinated with some entomophily.

3.18.3 Medicinal and Economic Importance

- **Hyoscyamine and scopolamine** used in ophthalmology and as anticholinergic agents.
- Raw material for atropine synthesis.

- Export commodity for the pharmaceutical industry in Europe and India.

3.18.4 Breeding Objectives

- High **alkaloid content** (≥ 0.5 %).
- Uniform maturity and synchronized flowering.
- Resistance to root rot and aphids.
- Compact, high-leaf-yield plant type.

3.18.5 Germplasm and Genetic Resources

S. No.	Institute / Centre	Accessions	Remarks
1	CIMAP, Lucknow	100	High-scopolamine chemotypes.
2	CSIR–NBRI, Lucknow	80	Tissue-culture-maintained accessions.
3	NBPGR	60	Exotic lines from Europe.

3.18.6 Breeding and Hybridization

- **Selection:** Based on leaf alkaloid yield and morphology.
- **Hybridization:** Manual pollination; interspecific crosses with *H. muticus* explored.
- **Polyploidy Breeding:** Tetraploids show 40 % higher alkaloid content.
- **Mutation Breeding:** EMS-induced compact, high-alkaloid mutants.
- **Tissue Culture:** Hairy-root cultures (using *Agrobacterium rhizogenes*) for continuous scopolamine production.

3.18.7 Released Lines

S. No.	Variety / Clone	Features
1	*CIM-HN-1*	Hyoscyamine 0.6 %; uniform leaf yield.
2	*CIM-HN-3*	Compact habit; scopolamine 0.4 %.
3	*HN-Var-5*	Drought tolerant; high biomass.

3.18.8 Quality Evaluation

- **Alkaloid content:** Total tropanes 0.5–0.8 %.
- **HPLC standards:** Hyoscyamine (Rt 9.4 min), scopolamine (Rt 11.2 min).
- **Dry-leaf yield:** 1.2–1.8 t ha^{-1} under irrigated conditions.

3.18.9 Integration of Breeding Approaches

Approach	Contribution
Selection + mutation	Enhanced alkaloid and compact plant type.
Polyploidy	Vigorous, high-biomass variants.
Hairy-root culture	Continuous alkaloid production system.

Table: Summary of Medicinal Crops

S. No.	Common Name	Botanical Name (Italicized)	Family	Key Bioactive Compounds	Major Breeding Objectives	Released Varieties / Highlights
1	Senna	*Cassia angustifolia* Vahl.	Fabaceae	Sennosides A & B	High sennoside yield, drought tolerance, disease resistance	*Sona, KM-5, ALFT-2*
2	Periwinkle	*Catharanthus roseus* (L.) G. Don	Apocynaceae	Vincristine, Vinblastine, Ajmalicine	High alkaloid yield, early flowering, pest resistance	*Nirmal, Dhawal, Arka Shubra*
3	Ashwagandha	*Withania somnifera* (L.) Dunal	Solanaceae	Withanolides, Withaferin A	High root yield, high withanolide, drought tolerance	*Poshita, Jawahar Ashwagandha-20, WS-134*
4	Isabgol (Psyllium)	*Plantago ovata* Forsk.	Plantaginaceae	Mucilage (Psyllium)	High seed & husk yield, disease resistance, lodging resistance	*GI-2, Niharika, G-1*
5	Sarpagandha	*Rauvolfia serpentina* Benth. ex Kurz	Apocynaceae	Reserpine, Ajmaline, Serpentine	High root yield, high reserpine, disease resistance	*CIM-RS-1, CIM-RS-3, TNAU-1*
6	Opium Poppy	*Papaver somniferum* L.	Papaveraceae	Morphine, Codeine, Thebaine	High latex & morphine yield, low narcotine, disease resistance	*Shweta, CIM-P-Op1, Jawahar Op-3*
7	Glory Lily	*Gloriosa superba* L.	Colchicaceae	Colchicine, Gloriosine	High colchicine yield, tuber disease resistance	*TNAU GS-1, CIM-GS-2*
8	Medicinal Coleus	*Coleus forskohlii* Briq.	Lamiaceae	Forskolin	High forskolin content, drought tolerance, root-rot resistance	*CIM-CF-1, CIM-CF-3, Arka Sugandha*
9	Velvet Bean (Kaunch)	*Mucuna pruriens* (L.) DC.	Fabaceae	L-DOPA	High L-DOPA, non-itchy pods, yield stability	*Arka Dhanvantari, CIM-MP-1, TNAU MP-2*
10	Holy Basil (Tulsi)	*Ocimum sanctum* L.	Lamiaceae	Eugenol, Ursolic acid, Rosmarinic acid	High essential oil, high eugenol, drought tolerance	*CIM-Ayu, CIM-Saumya, Krishna Tulsi*
11	Gotu Kola	*Centella asiatica* (L.) Urb.	Apiaceae	Asiaticoside, Madecassoside	High saponin yield, drought & shade tolerance	*CA-5, CIM-CA-3* (elite clones)

12	Brahmi	*Bacopa monnieri* (L.) Wettst.	Plantaginaceae	Bacosides A & B	High bacoside content, salinity tolerance, high stolon yield	*CIM-Brahmi, BM-5*
13	Kalmegh	*Andrographis paniculata* (Burm.f.) Wall. ex Nees	Acanthaceae	Andrographolide	High andrographolide yield, early maturity, compact canopy	*CIM-Megha, KAL-1, CIM-AP-3*
14	Aloe	*Aloe vera* (L.) Burm.f.	Asphodelaceae	Aloin, Acemannan	High gel recovery, low aloin, disease-free clones	*CIM-Sheetal, AL-1, Arka Aloe*
15	Bhui Amla	*Phyllanthus amarus* Schum. & Thonn.	Phyllanthaceae	Phyllanthin, Hypophyllanthin	High lignan yield, drought tolerance	*CIM-Bhui Amla, PA-20, TNAU-Phyll-3*
16	Eucalyptus	*Eucalyptus* spp.	Myrtaceae	Cineole, Pinene, Limonene	High cineole, fast growth, rust resistance	*IFGTB-Et-2, CIM-Euca-1, TNAU-Hyb-1*
17	Bael	*Aegle marmelos* (L.) Corr.	Rutaceae	Marmelosin, Coumarins, Aegeline	High pulp & marmelosin, early fruiting, dwarf types	*CIM-Bael-1, NBPGR-B-4, TNAU-BL-3*
18	Henbane	*Hyoscyamus niger* L.	Solanaceae	Hyoscyamine, Scopolamine, Atropine	High alkaloid yield, uniform flowering, disease resistance	*CIM-HN-1, CIM-HN-3, HN-Var-5*

Chapter - 4

Aromatic Crops

4.1 Lemongrass (*Cymbopogon flexuosus* (Steud.) Wats. & *C. citratus* (DC.) Stapf)

4.1.1 Introduction

Lemongrass represents one of the most economically important aromatic grasses of the genus *Cymbopogon* in the **Poaceae** family. It is cultivated extensively across tropical Asia, Africa, and South America for its essential oil rich in **citral**, a compound used as a precursor for vitamin A synthesis, perfumery, and flavouring industries. Two major commercial species dominate global cultivation: *Cymbopogon flexuosus* (East Indian lemongrass) and *C. citratus* (West Indian lemongrass). India remains a global leader, contributing over 80 % of the world's lemongrass oil supply, primarily from Kerala, Karnataka, Uttar Pradesh, and Assam.

4.1.2 Botanical Description and Cytology

- **Habit:** Perennial tufted grass, 1.0–1.5 m tall, with slender leaves and erect culms.
- **Leaves:** Linear-lanceolate, 50–90 cm long, glaucous green, with sharp margins.
- **Inflorescence:** Loose panicles with numerous racemes.
- **Flowers:** Small, bisexual, arranged in pairs of sessile and pedicelled spikelets.
- **Chromosome number:** 2n = 60 (polyploid).

Cytological variation (2n = 40–80) across species accounts for wide morphological diversity and adaptability.

4.1.3 Economic and Industrial Significance

1. **Essential oil yield:** 0.4–0.8 % of fresh herbage; world demand ~1,200 tonnes/year.
2. **Key constituents:** Citral (70–85 %), geraniol, myrcene, and limonene.
3. **Industrial uses:** Fragrance, flavouring, soap, detergent, insect repellents, and vitamin A intermediates.

DOI: 10.1201/9781042010332-5

4. **By-products:** Distillation residue used as organic mulch and biofuel feedstock.
5. **Agro-ecological benefit:** Soil-binding perennial grass suitable for erosion control on slopes.

4.1.4 Major Breeding Objectives

1. High herbage and oil yield per hectare.
2. Elevated **citral content (> 80 %)** for industrial use.
3. Resistance to leaf blight (*Curvularia* spp.) and rust (*Puccinia nakanishikii*).
4. Drought and cold tolerance for year-round production.
5. Rapid tillering and long ratoon life (> 3 years).

4.1.5 Genetic Resources and Germplasm

S. No.	Institute / Centre	No. of Accessions	Remarks
1	CSIR–CIMAP, Lucknow	> 250	Major germplasm centre for *Cymbopogon.*
2	NBPGR, New Delhi	150	Core collection with cytogenetic diversity.
3	TNAU, Coimbatore	80	Southern ecotypes for high citral.

Wild relatives such as *C. pendulus* and *C. jwarancusa* contribute drought and disease resistance genes.

4.1.6 Floral Biology and Reproductive System

Lemongrass flowers are **highly cross-pollinated** and exhibit **protandry,** limiting self-fertilization. Seed set is often poor, so **vegetative propagation** by slips is common. Controlled hybridization requires careful synchronization of flowering panicles and isolation.

4.1.7 Breeding and Hybridization

1. **Clonal Selection:** Primary breeding route; superior clumps identified for oil yield and citral content.
2. **Hybridization:** Intervarietal crosses (*C. flexuosus* × *C. pendulus*) developed high-citral hybrids.
3. **Mutation Breeding:** Gamma radiation (10–15 kr) created compact, high-oil mutants (e.g., *Krishna*).
4. **Polyploidy Induction:** Colchicine (0.2 %) doubled chromosome sets, improving oil biosynthesis.
5. **Molecular Tools:** ISSR and RAPD markers used for diversity and purity analysis.

4.1.8 Major Diseases and Pests

S. No.	Disease / Pest	Pathogen / Insect	Remarks
1	Leaf blight	*Curvularia lunata*	Severe under humid conditions.
2	Rust	*Puccinia nakanishikii*	Controlled by tolerant line *CIM-AP-1.*
3	Shoot borer	*Chilo infuscatellus*	Managed by clonal resistance.

4.1.9 Released Varieties

S. No.	Variety	Features
1	*CIM-Shikha*	Citral 82 %, yield 120 kg oil/ha.
2	*Krishna*	Gamma mutant; robust tillering.
3	*Pragati*	Early harvest, 80 % citral.
4	*Neema*	Resistant to leaf blight.

4.1.10 Essential Oil Composition and Quality

Compound	Range (%)
Citral (Geranial + Neral)	70–85
Myrcene	5–7
Limonene	3–5
Geraniol	1–3
β-Caryophyllene	0.5–1.0

Citral is the industrial determinant of oil grade, quantified by GC or HPLC fingerprinting.

4.1.11 Integration of Breeding Approaches

Approach	Impact
Clonal and mutation selection	Foundation of all modern cultivars.
Cytogenetic hybridization	Broadened gene base and yield potential.
Molecular marker analysis	Assisted varietal identification and chemotype purity.
Tissue-culture propagation	Mass multiplication of elite clones.

4.2 Palmarosa (*Cymbopogon martinii* (Roxb.) Wats. var. *motia*)

4.2.1 Introduction

Palmarosa, a fragrant grass of *Cymbopogon martinii* var. *motia*, is cultivated for its essential oil rich in **geraniol**, a primary ingredient in perfumery, soaps, and cosmetics. The name "Rosha grass" derives from its characteristic rose-like aroma. India is a global leader, exporting over 100 tonnes of oil annually, with Karnataka, Madhya Pradesh, and Gujarat as primary production zones.

4.2.2 Botanical Description

- **Habit:** Perennial tufted grass, 1.2–1.8 m tall.
- **Leaves:** Linear, green with pinkish tinge at base.
- **Inflorescence:** Loose, purple panicles bearing spikelets in pairs.
- **Chromosome number:** 2n = 60. Highly cross-pollinated; fertile seed production limited, hence vegetative propagation predominates.

4.2.3 Economic and Industrial Importance

1. **Geraniol content:** 70–90 % of total oil.
2. **Applications:** Perfume, soap, cosmetic formulations, and aroma therapy.
3. **Pharmacological use:** Antimicrobial, antioxidant, anti-inflammatory.
4. **Agro-economic value:** Thrives in semi-arid zones; offers 4–5 years of profitable ratoons.

4.2.4 Major Breeding Objectives

1. High **herbage and oil yield (> 100 kg/ha)**.
2. Elevated **geraniol content (> 85 %)**.
3. Resistance to rust and leaf spot.
4. Drought and heat tolerance.
5. Uniform growth and synchronous flowering for industrial harvests.

4.2.5 Germplasm and Genetic Resources

S. No.	Centre	Accessions	Traits
1	CIMAP, Lucknow	150	Chemotype diversity for geraniol.
2	NBPGR	80	Wild relatives of *C. martinii.*
3	TNAU	60	Drought-tolerant clones.

4.2.6 Breeding and Hybridization

» **Clonal Selection:** Most varieties are selections from natural populations.

» **Hybridization:** Limited success due to poor seed set; crosses with *C. jwarancusa* improved drought tolerance.

» **Mutation Breeding:** EMS and gamma rays induced compact, high-oil lines.

» **Polyploidy:** Induced tetraploids show thick leaves and higher biomass.

» **Tissue Culture:** Nodal cultures produce pathogen-free elite clones.

4.2.7 Major Diseases and Resistance

S. No.	Disease	Causal Agent	Resistance Source / Remarks
1	Rust	*Puccinia nakanishikii*	*PRC-1* line resistant.
2	Leaf blight	*Curvularia sp.*	Tolerant clones identified at CIMAP.

4.2.8 Released Varieties

S. No.	Variety	Features
1	*CIM-Pawan*	Geraniol 88 %, high ratoon vigour.
2	*RRL-82*	Early maturing; 85 % geraniol.
3	*Trishna*	Compact, drought hardy.

4.2.9 Essential Oil Composition

Compound	Range (%)
Geraniol	70–90
Geranyl acetate	3–8
Linalool	2–5
Citronellol	1–2
Myrcene	1–2

Quality assessment depends on the **geraniol:acetate ratio**, with ≥10 considered premium.

4.2.10 Integration of Breeding Approaches

Approach	Impact
Clonal selection	Foundation for varietal improvement.
Mutation breeding	Compact, early maturing clones.
In vitro propagation	Rapid, disease-free multiplication.
Chemotype screening	Direct selection for geraniol-rich lines.

4.3 Citronella (*Cymbopogon winterianus* Jowitt ex Bor)

4.3.1 Introduction

Citronella is one of the most widely cultivated aromatic grasses, valued for its **citronellal- and geraniol-rich oil** used in perfumery, mosquito repellents, and household disinfectants. Two main commercial types exist: *C. winterianus* (Java type) and *C. nardus* (Ceylon type). The Java type predominates industrially due to its higher citronellal content and stronger lemon-like aroma.

4.3.2 Botanical Description

» **Habit:** Perennial, tufted grass reaching 1.5–2.0 m.

» **Leaves:** Broad, glaucous, strongly aromatic.

» **Inflorescence:** Large panicles with purple tinge.

» **Chromosome number:** 2n = 40–60. Reproduction occurs mainly **vegetatively** as most commercial clones are sterile.

4.3.3 Economic and Industrial Significance

1. **Oil yield:** 1.0–1.2 % of fresh biomass (~100–120 kg oil/ha).
2. **Constituents:** Citronellal (35–45 %), geraniol (20–25 %), citronellol (10–15 %).
3. **Applications:** Perfume, soap, insect repellents, floor cleaners.
4. **By-products:** Residue used in agarbatti (incense) production and compost.

4.3.4 Major Breeding Objectives

1. High herbage and oil yield.
2. High **citronellal (> 40 %)** and balanced **geraniol content**.
3. Disease resistance (leaf blight and rust).
4. Drought tolerance and prolonged ratoon life.
5. Compact growth suitable for dense planting.

4.3.5 Germplasm Resources

S. No.	Centre	Accessions	Traits
1	CIMAP, Lucknow	200	Major diversity collection.
2	IIHR, Bengaluru	80	Drought and pest tolerance.
3	NBPGR	100	Wild *Cymbopogon* species.

4.3.6 Breeding and Hybridization

- **Clonal selection:** Core breeding approach.
- **Mutation breeding:** Gamma rays (15–25 kr) produced high citronellal mutants (*Manjusha*).
- **Interspecific hybridization:** Attempts between *C. winterianus* × *C. nardus* to combine aroma and yield.
- **Polyploidy induction:** Tetraploids had higher leaf biomass but variable oil composition.
- **Tissue culture:** Used for rapid multiplication and virus-free stock.

4.3.7 Diseases and Resistance

S. No.	Disease	Pathogen	Resistance Source / Control
1	Leaf blight	*Curvularia lunata*	Resistant line *Java-II.*
2	Rust	*Puccinia* sp.	Tolerant clone *CIMAP-04.*

4.3.8 Released Varieties

S. No.	Variety	Features
1	*Manjusha*	Gamma mutant; high citronellal (45 %).
2	*Mandakini*	High oil (1.2 %); rust tolerant.
3	*CIMAP Bio-13*	Elite line for industrial distillation.
4	*Java-II*	Improved tillering, long ratoon life.

4.3.9 Essential Oil Composition

Compound	Range (%)
Citronellal	35–45
Geraniol	20–25
Citronellol	10–15
Limonene	4–6
Eugenol & others	<2

4.3.10 Integration of Breeding Approaches

Approach	Contribution
Mutation breeding	Key source of high citronellal lines.
Clonal selection	Enhanced yield and ratooning.
Tissue-culture propagation	Maintained uniformity and disease freedom.
Chemoprofiling	Used for oil-quality standardization.

4.4 Vetiver (*Chrysopogon zizanioides* (L.) Roberty)

4.4.1 Introduction

Vetiver, also known as Khus, is a perennial aromatic grass of the family Poaceae, cultivated for its fragrant roots and essential oil used in perfumery and cosmetics. India is the genetic homeland of *Chrysopogon zizanioides*, though commercial production is centered in Tamil Nadu, Uttar Pradesh, Rajasthan, and Madhya Pradesh.

The roots yield a highly viscous amber oil, **vetiver oil**, renowned as "the oil of tranquillity," which forms the base note in numerous luxury perfumes. Additionally, vetiver is employed in **soil conservation, erosion control, carbon sequestration, and wastewater remediation,** making it both an economic and ecological crop.

4.4.2 Botanical Description and Cytology

- **Habit:** Deep-rooted perennial grass forming dense clumps up to 1.5 m tall.
- **Leaves:** Long, narrow, rough-margined blades; 50–100 cm long.
- **Roots:** Strongly aromatic, fibrous, penetrating up to 3–4 m deep.
- **Inflorescence:** Panicle 15–30 cm long with paired spikelets.
- **Chromosome number:** 2n = 20, 40, 60 (varies among ecotypes).

Commercial types are **sterile and propagated vegetatively**, ensuring genetic uniformity.

4.4.3 Economic and Industrial Importance

- **Essential oil yield:** 0.4–0.6 % of fresh root weight.
- **Constituents:** Vetiverol, khusimol, α-vetivone, β-vetivone.

» **Uses:** Perfume fixative, aromatherapy, incense, and soaps.

» **Industrial demand:** ~250 tonnes/year globally; India contributes 60 %.

» **Environmental services:** Root network stabilizes soil and recharges groundwater.

4.4.4 Major Breeding Objectives

1. High root and oil yield.
2. High vetiverol and vetivone content (> 60 %).
3. Drought and heat tolerance.
4. Resistance to root rot and termites.
5. Compact growth with strong root systems for agro-ecological applications.

4.4.5 Genetic Resources

S. No.	Centre	Accessions	Remarks
1	CSIR–CIMAP, Lucknow	180	Core germplasm of wild and cultivated types.
2	NBPGR, New Delhi	80	Indigenous and exotic lines.
3	TNAU, Coimbatore	60	Drought-tolerant accessions.

4.4.6 Breeding and Hybridization

1. **Clonal Selection:** Key approach; superior clumps selected for oil yield and composition.
2. **Mutation Breeding:** Gamma irradiation created high-root-oil mutant *KS-1*.
3. **Interspecific Hybridization:** Limited due to sterility; experimental crosses with *C. nigritanus*.
4. **Polyploidy Induction:** Colchicine (0.2 %) improved root girth and oil percentage.
5. **Tissue Culture:** Micropropagation ensures rapid multiplication of disease-free elite clones.

4.4.7 Diseases and Resistance

S. No.	Disease	Pathogen / Pest	Management / Resistance
1	Root rot	*Fusarium oxysporum*	Resistant clone *KS-3*.
2	Termite attack	*Odontotermes* sp.	Controlled by tolerant ecotypes from arid zones.

4.4.8 Released Varieties

S. No.	Variety	Features
1	*KS-1*	High root yield; oil 0.6 %.
2	*CIM-Khus-2*	Vetiverol 45 %, khusimol 20 %.
3	*Sugandha*	Fast ratooning; strong roots.

4.4.9 Essential Oil Composition

Compound	Range (%)
Vetiverol	20–30
Khusimol	15–25
α- and β-Vetivone	20–35
Terpenes	5–10

4.5 Mint (*Mentha* spp.)

4.5.1 Introduction

Mints (*Mentha* spp.) constitute one of the most important aromatic crop groups cultivated for **menthol and essential oils**. The genus comprises 25–30 species, with *Mentha arvensis*, *M. piperita*, and *M. spicata* being most significant commercially. India is the world's largest producer of mint oil, accounting for over **75 % of global menthol supply**.

Mint oils are central to **pharmaceuticals, toothpastes, confectionery, cosmetics, and aromatherapy** industries, valued for their cooling sensation and antibacterial activity.

4.5.2 Botanical Description

» **Habit:** Perennial, herbaceous plants with stoloniferous spreading habit.

» **Leaves:** Simple, opposite, serrated, gland-dotted.

» **Flowers:** Terminal spikes, small, violet or pink.

» **Chromosome number:** 2n = 96 in *M. arvensis* (polyploid complex).

4.5.3 Economic and Industrial Significance

1. **Global production:** > 30,000 tonnes of mint oil annually.
2. **Major constituents:** Menthol, menthone, and isomenthone.
3. **Applications:** Toothpaste, chewing gum, cold rubs, analgesic balms, and perfumery.
4. **Agro-industrial value:** Fit for intercropping; generates high returns per hectare.

4.5.4 Major Breeding Objectives

1. High herbage and oil yield.
2. High **menthol content (> 75 %)**.
3. Resistance to leaf spot and rust.
4. Heat and waterlogging tolerance.
5. Dwarf, uniform plants for mechanized harvesting.

4.5.5 Genetic Resources

S. No.	Centre	Accessions	Remarks
1	CIMAP, Lucknow	350	Global Mentha collection.
2	NBPGR, New Delhi	200	Germplasm and wild relatives.
3	IIIM, Jammu	80	High-menthol chemotypes.

4.5.6 Breeding and Hybridization

1. **Inter-specific Hybridization:** Crosses such as *M. arvensis* × *M. spicata* produced *Menthol mint (Kosi type).*
2. **Mutation Breeding:** EMS-induced compact, early-maturing types.
3. **Polyploidy Manipulation:** Colchicine tetraploids yield higher biomass.
4. **Molecular Tools:** ISSR markers differentiate chemotypes.
5. **In Vitro Propagation:** Nodal explants used for rapid multiplication.

4.5.7 Diseases and Resistance

S. No.	Disease	Pathogen	Control / Resistant Line
1	Leaf spot	*Alternaria alternata*	Line *CIM-Mint-3* tolerant.
2	Rust	*Puccinia menthae*	Managed through resistant clones.
3	Root rot	*Rhizoctonia solani*	Avoided by raised-bed cultivation.

4.5.8 Released Varieties

S. No.	Variety	Species / Origin	Key Features
1	*Kosi*	*M. arvensis* hybrid	Menthol 78 %, high oil yield.
2	*Gomti*	*M. arvensis*	Rust tolerant; high biomass.
3	*Himalaya*	*M. piperita* type	Peppermint aroma; menthol 72 %.
4	*CIM-Mint-3*	Mutation line	Compact, early flowering.

4.5.9 Essential Oil Composition

Compound	Range (%)
Menthol	70–80
Menthone	10–15
Isomenthone	5–8
Menthyl acetate	2–5

4.6 Sweet Basil (*Ocimum basilicum* L.)

4.6.1 Introduction

Sweet basil (*Ocimum basilicum*), belonging to **Lamiaceae,** is an annual aromatic herb prized for its essential oil used in perfumery, culinary, and pharmaceutical industries. Known as **Tulsi or Babui Tulsi**, it is distinct from *O. sanctum* and valued for its high **linalool** and **methyl chavicol** contents.

Cultivation occurs across **India, Egypt, Italy, and the USA**, primarily for extraction of essential oil and as a culinary herb in international markets.

4.6.2 Botanical Description

- » **Habit:** Erect herb, 30–90 cm tall.
- » **Leaves:** Broad-ovate, aromatic, green or purple, gland-dotted.
- » **Flowers:** Small, white or purple, arranged in terminal racemes.
- » **Chromosome number:** 2n = 48.

Pollination is largely **entomophilous**, with partial cross-pollination.

4.6.3 Economic Importance

1. **Essential oil yield:** 0.3–1.0 % of herbage.
2. **Principal compounds:** Linalool (40–60 %), methyl chavicol (20–40 %).
3. **Industrial uses:** Perfumes, flavouring, pharmaceuticals, aromatherapy.
4. **Export demand:** Major component of international spice–aromatic trade.

4.6.4 Major Breeding Objectives

1. High herbage and oil yield.
2. Desirable oil chemotype (linalool-rich).
3. Resistance to downy mildew and leaf spot.
4. Uniform canopy and short growth cycle.

4.6.5 Germplasm Resources

S. No.	Institute	Accessions	Remarks
1	CIMAP, Lucknow	300	High-linalool and methyl chavicol types.
2	NBPGR	120	Wild *Ocimum* species.
3	TNAU	80	Disease-resistant germplasm.

4.6.6 Breeding and Hybridization

- » **Mass and Clonal Selection:** Main method for elite types.
- » **Hybridization:** Intervarietal crosses to combine linalool and yield traits.

- **Mutation Breeding:** EMS and gamma rays yielded compact high-linalool lines.
- **Polyploidy:** Tetraploids with enhanced oil yield but late flowering.
- **Biotechnological Tools:** RAPD markers used for chemotype fingerprinting.

4.6.7 Diseases and Resistance

S. No.	Disease	Pathogen	Resistant Source / Variety
1	Downy mildew	*Peronospora belbahrii*	Resistant *CIM-Surabhi*.
2	Leaf spot	*Cercospora* spp.	Controlled by tolerant lines.

4.6.8 Released Varieties

S. No.	Variety	Key Traits
1	*CIM-Surabhi*	Linalool 60 %, downy mildew tolerant.
2	*CIM-Saumya*	Methyl chavicol 40 %, high herbage.
3	*Arka Tulasi*	Early maturity, compact growth.

4.6.9 Essential Oil Composition

Compound	Range (%)
Linalool	40–60
Methyl chavicol	20–40
Eugenol	5–10
1,8-Cineole	1–3

4.7 Lavender (*Lavandula angustifolia* Mill.)

4.7.1 Introduction

Lavender (*Lavandula angustifolia*), of the **Lamiaceae** family, is one of the most valued aromatic plants cultivated worldwide for its **highly fragrant essential oil** used in perfumery, aromatherapy, and pharmaceutical products. Native to the Mediterranean region, it is now extensively cultivated in **France, Bulgaria, India, and China**.

In India, lavender is primarily grown in the **temperate Himalayas (Jammu & Kashmir, Himachal Pradesh, Uttarakhand)**, promoted under the **Aroma Mission** due to its export potential and high returns for mountain farmers.

4.7.2 Botanical Description and Cytology

- **Habit:** Perennial sub-shrub, 30–80 cm tall.
- **Leaves:** Linear, opposite, grey-green, aromatic due to glandular trichomes.
- **Flowers:** Small, purple-blue in terminal spikes; highly scented.
- **Fruit:** Schizocarp with four nutlets.
- **Chromosome number:** 2n = 48.

Pollination is mainly by **bees (entomophilous)**, and cross-pollination is common, favouring genetic diversity.

4.7.3 Economic and Industrial Importance

1. **Essential oil yield:** 0.8–1.5 % of fresh flowers.
2. **Key constituents:** Linalool (25–35 %), linalyl acetate (30–45 %).
3. **Applications:** Fine perfumery, aromatherapy, cosmetics, antiseptics, and stress-relief formulations.
4. **Market value:** Global lavender oil trade exceeds 1500 tonnes annually.
5. **Agro-ecological benefit:** Thrives in temperate, low-input hill regions, enhancing rural incomes.

4.7.4 Major Breeding Objectives

1. High oil yield and high **linalool + linalyl acetate** content.
2. Early flowering and prolonged blooming period.
3. Cold and frost tolerance for temperate Indian zones.
4. Resistance to root rot and Fusarium wilt.
5. Compact bush form suitable for mechanical harvesting.

4.7.5 Genetic Resources

S. No.	Institute / Location	Accessions / Lines	Remarks
1	IIIM, Jammu	100+	Core germplasm collection.
2	CIMAP, Lucknow (field station Pantnagar)	60	Temperate-adapted accessions.
3	NBPGR	40	Wild *Lavandula* species for hybridization.

4.7.6 Breeding and Hybridization

- **Clonal Selection:** From natural variability in oil yield and composition.
- **Hybridization:** *L. angustifolia* × *L. latifolia* produces **Lavandin hybrids** with superior oil yield.
- **Mutation Breeding:** EMS-induced compact mutants (IIIM-Lav-3).

- **Polyploidy:** Tetraploids showed higher biomass but slightly reduced oil quality.
- **Tissue Culture:** Nodal and shoot-tip cultures for rapid propagation and germplasm preservation.
- **Molecular Tools:** RAPD and ISSR markers used for chemotype identification.

4.7.7 Diseases and Resistance

S. No.	Disease	Causal Agent	Management / Resistance Source
1	Root rot	*Fusarium oxysporum*	Resistant *IIIM-Lav-2.*
2	Leaf blight	*Alternaria alternata*	Controlled by tolerant genotypes.

4.7.8 Released Varieties

S. No.	Variety	Key Features
1	*IIIM-Lav-1*	High oil yield (1.4 %), 40 % linalyl acetate.
2	*IIIM-Lav-3*	Compact mutant, early flowering.
3	*Him-Lavender*	Temperate hardy, uniform inflorescence.

4.7.9 Essential Oil Composition

Compound	Range (%)
Linalool	25–35
Linalyl acetate	30–45
Camphor	< 2
β-Caryophyllene	1–3

4.7.10 Integration of Breeding Approaches

Approach	Impact
Hybridization (Lavandin)	Enhanced yield and adaptability.
Mutation breeding	Compact growth for mechanization.
Tissue culture	Uniform plantlets and clonal fidelity.

4.8 Geranium (*Pelargonium graveolens*)

4.8.1 Introduction

Geranium (*Pelargonium graveolens*), also called **Rose-scented Geranium,** is a semi-perennial aromatic plant of the **Geraniaceae** family. It is cultivated for its essential oil used as a substitute for rose oil in perfumery, cosmetics, and pharmaceuticals.

Native to South Africa, it is now widely grown in **India (Nilgiris, Himachal Pradesh, and Uttar Pradesh)** due to its adaptability to subtropical and temperate climates.

4.8.2 Botanical Description

- **Habit:** Erect, branched herb, 60–100 cm tall.
- **Leaves:** Deeply lobed, velvety, glandular-hairy, strongly aromatic.
- **Flowers:** Pink or lilac, borne in umbels.
- **Chromosome number:** 2n = 88.

Cross-pollination is limited due to floral structure; **vegetative propagation** is most common.

4.8.3 Economic and Industrial Importance

1. **Essential oil yield:** 0.2–0.3 % (herbage basis).
2. **Principal constituents:** Citronellol (35–40 %), geraniol (20–25 %), linalool (10–15 %).
3. **Applications:** High-grade perfumery, soap, and deodorant industries.
4. **Agro-industrial value:** Multi-cut crop; ratoons harvested 3–4 times/year.

4.8.4 Major Breeding Objectives

1. High oil yield and citronellol + geraniol ratio (~1:1).
2. Resistance to root-knot nematodes and leaf spot.
3. Drought and cold tolerance.
4. Compact canopy and long ratoon life.

4.8.5 Germplasm and Genetic Resources

S. No.	Centre	Accessions	Remarks
1	CIMAP, Lucknow	250	High citronellol types.
2	IIHR, Bengaluru	100	South Indian germplasm.
3	NBPGR	70	Exotic *Pelargonium* species.

4.8.6 Breeding and Hybridization

- **Clonal Selection:** Primary improvement method.
- **Hybridization:** *P. graveolens* × *P. radens* yields vigourous F_1 hybrids.
- **Mutation Breeding:** Compact dwarf line *CIM-Gerani-2*.
- **Polyploidy:** Tetraploids produce larger leaves and thicker oil glands.
- **In Vitro Propagation:** Microshoot cultures standardised for large-scale multiplication.

4.8.7 Diseases and Resistance

S. No.	Disease / Pest	Causal Agent	Resistance / Control
1	Root-knot nematode	*Meloidogyne incognita*	Resistant line *GP-3*.
2	Leaf spot	*Alternaria* sp.	Controlled by resistant clone *CIM-Gerani-1*.

4.8.8 Released Varieties

S. No.	Variety	Features
1	*CIM-Gerani-1*	High oil yield (0.35 %), nematode tolerant.
2	*CIM-Gerani-2*	Dwarf mutant; rich in citronellol.
3	*Hemanti*	Vigorous, drought-tolerant.

4.8.9 Essential Oil Composition

Compound	Range (%)
Citronellol	35–40
Geraniol	20–25
Linalool	10–15
Citronellyl formate	5–8

4.9 Patchouli (*Pogostemon cablin* (Blanco) Benth.)

4.9.1 Introduction

Patchouli, a member of **Lamiaceae**, is cultivated for its strongly scented leaves, yielding **patchouli oil**, one of the most stable and long-lasting natural fixatives in perfumery. Native to the Philippines and Indonesia, it is now widely grown in India, particularly in **Andhra Pradesh, Karnataka, and Assam**.

The oil is valued for its unique earthy fragrance, used in **perfumes, incense, soaps, and cosmetics**.

4.9.2 Botanical Description

- **Habit:** Perennial, bushy herb, 60–120 cm tall.
- **Leaves:** Broad, ovate, densely glandular-hairy, strongly aromatic.
- **Flowers:** Pale purple in terminal spikes.
- **Chromosome number:** 2n = 32. Reproduction is mainly **vegetative** through cuttings.

4.9.3 Economic and Industrial Importance

1. **Oil yield:** 0.3–0.5 % of fresh leaves (~40–60 kg/ha).
2. **Constituents:** Patchouli alcohol (35–40 %), α-bulnesene, α-guaiene, pogostol.
3. **Applications:** Perfume base, soap, incense, and flavouring.

4. **By-products:** Distillation residue used for compost and vermiculture.

4.9.4 Major Breeding Objectives

1. High leaf and oil yield.
2. High **patchouli alcohol content (> 40 %)**.
3. Resistance to leaf blight and root rot.
4. Drought and pest tolerance.
5. Compact, high-branching canopy.

4.9.5 Germplasm Resources

S. No.	Centre	Accessions	Traits
1	CIMAP, Lucknow	150	High oil chemotypes.
2	IIHR, Bengaluru	70	Compact growth, drought-tolerant clones.
3	NBPGR	50	Indonesian and Malaysian accessions.

4.9.6 Breeding and Hybridization

- **Clonal Selection:** Main approach for uniform oil quality.
- **Mutation Breeding:** EMS treatments produced *CIM-Patch-3* with 45 % patchouli alcohol.
- **In Vitro Selection:** Callus culture with PEG simulated drought-tolerant variants.
- **Polyploidy:** Tetraploids showed increased leaf thickness and oil glands.
- **Molecular Tools:** RAPD profiles differentiate Indonesian vs Indian chemotypes.

4.9.7 Diseases and Resistance

S. No.	Disease	Pathogen	Resistant Line / Management
1	Leaf blight	*Cercospora* spp.	Resistant line *CIM-Patch-2*.
2	Root rot	*Rhizoctonia solani*	Managed by raised beds + tolerant clones.

4.9.8 Released Varieties

S. No.	Variety	Features
1	*CIM-Patch-1*	Patchouli alcohol 38 %; early harvest.
2	*CIM-Patch-3*	Mutation line; 45 % patchouli alcohol.
3	*Sugantha*	Drought-tolerant, long ratoon life.

4.9.9 Essential Oil Composition

Compound	Range (%)
Patchouli alcohol	35–45
α-Bulnesene	10–15
α-Guaiene	8–10
Pogostol	5–8

4.10 Artemisia (*Artemisia annua* L. and related species)

4.10.1 Introduction

The genus *Artemisia* (family **Asteraceae**) comprises more than 400 species distributed across temperate and subtropical regions of Asia, Europe, and North America. Among them, *Artemisia annua*, commonly known as **Sweet Wormwood** or **Qinghao**, holds exceptional global importance as the **natural source of artemisinin**, an endoperoxide sesquiterpene lactone with potent anti-malarial properties.

In addition to artemisinin, the plant also yields **aromatic essential oils** rich in camphor, 1,8-cineole, and artemisia ketone, which are used in perfumery and herbal medicine. India cultivates *A. annua*, *A. pallens*, and *A. vulgaris* across **Himachal Pradesh, Jammu & Kashmir, Uttarakhand, and parts of South India**, mainly for pharmaceutical and aromatic purposes.

4.10.2 Botanical Description

» **Habit:** Annual herb, 0.5–2.5 m tall, highly branched, with fine feathery leaves.

» **Leaves:** Alternate, bipinnate, deep green with glandular trichomes.

» **Flowers:** Small, yellow, arranged in panicles; heterogamous capitula.

» **Fruit:** Achene, light brown, tiny (≈1 mm).

» **Chromosome number:** 2n = 18.

4.10.3 Economic and Industrial Importance

1. **Artemisinin yield:** 0.5–1.5 % of dry leaf weight in improved lines.
2. **Essential oil yield:** 0.3–0.8 % of herbage; used in aromatherapy.
3. **Applications:**
   » Artemisinin – anti-malarial, antiviral, anticancer drug precursor.
   » Essential oils – perfumery, insect repellents, flavouring agents.
4. **Global demand:** Driven by pharmaceutical industry; China, Vietnam, and India are key producers.
5. **Agro-industrial value:** Short-duration (6–7 months) aromatic–pharmaceutical crop adaptable to diverse soils.

4.10.4 Major Breeding Objectives

1. High artemisinin yield per hectare.
2. Enhanced essential oil content and improved aroma.
3. Drought, cold, and lodging tolerance.
4. Uniform maturity and high biomass.
5. Resistance to powdery mildew and Alternaria leaf spot.

4.10.5 Germplasm and Genetic Resources

S. No.	Centre	Accessions	Notable Features
1	CSIR–CIMAP, Lucknow	>300	Global and indigenous accessions.
2	NBPGR, New Delhi	120	Wild and cultivated species.
3	IIIM, Jammu	100	Himalayan ecotypes high in artemisinin.

4.10.6 Breeding and Hybridization

» **Mass Selection:** Early method for high-artemisinin genotypes.

» **Hybridization:** Intraspecific crosses between early and high-yielding types.

» **Mutation Breeding:** EMS mutants (*CIM-Arogya*) with 1.5 % artemisinin.

» **Polyploidy:** Colchicine-induced tetraploids show higher biomass and essential oil yield.

» **Molecular Breeding:** Marker-assisted selection using *amorpha-4,11-diene synthase* (ADS) and *CYP71AV1* gene markers.

» **Tissue Culture:** Somatic embryogenesis and in vitro regeneration to maintain elite lines.

4.10.7 Diseases and Resistance

S. No.	Disease	Causal Agent	Control / Resistant Source
1	Powdery mildew	*Erysiphe cichoracearum*	Resistant line *CIM-Aarti*.
2	Leaf spot	*Alternaria alternata*	Managed by tolerant *A. pallens* types.

4.10.8 Released Varieties

S. No.	Variety	Key Traits
1	*CIM-Arogya*	1.5 % artemisinin; high oil yield.
2	*Jeevan Raksha*	High biomass and uniform growth.
3	*CIM-Aarti*	Powdery mildew resistant.

4.10.9 Phytochemical Composition

Compound	Range (%)
Artemisinin	0.5–1.5
Artemisia ketone	20–25
1,8-Cineole	10–15
Camphor	8–10

4.11 Rosemary (*Rosmarinus officinalis* L.)

4.11.1 Introduction

Rosemary (*Rosmarinus officinalis*), a woody perennial shrub belonging to the **Lamiaceae** family, is native to the Mediterranean basin but now cultivated widely in **Europe, Africa, and India**. It is renowned for its essential oil, rich in **1,8-cineole, camphor, borneol, and verbenone**, which find applications in perfumery, culinary flavouring, and pharmaceuticals.

In India, cultivation is expanding in **Himachal Pradesh, Tamil Nadu, Karnataka, and Uttarakhand** under herbal and aromatic crop diversification initiatives.

4.11.2 Botanical Description

» **Habit:** Evergreen shrub, 0.5–1.5 m tall.

» **Leaves:** Opposite, narrow, leathery, aromatic with revolute margins.

» **Flowers:** Pale blue or lilac, borne in short racemes.

» **Chromosome number:** 2n = 50.

» **Pollination:** Entomophilous (bees); natural cross-pollination 10–15 %.

4.11.3 Economic and Industrial Importance

1. **Essential oil yield:** 0.8–1.2 % (fresh biomass basis).
2. **Main constituents:** 1,8-Cineole (20–40 %), camphor (10–20 %), borneol (5–10 %).
3. **Applications:** Aromatherapy, cosmetics, meat preservatives, antioxidant extracts.
4. **Medicinal uses:** Memory enhancement, anti-inflammatory, hepatoprotective effects.
5. **Agro-climatic adaptability:** Tolerates mild drought and limestone soils.

4.11.4 Major Breeding Objectives

1. High oil yield and cineole–borneol balance.
2. Enhanced antioxidant and phenolic content.
3. Drought and cold tolerance.
4. Compact canopy and high branching for dense plantations.

5. Disease resistance, particularly to powdery mildew.

4.11.5 Germplasm and Genetic Resources

S. No.	Centre	Accessions	Remarks
1	CIMAP, Lucknow	100	Indian and exotic collections.
2	IIIM, Jammu	60	Temperate adapted ecotypes.
3	NBPGR	40	Mediterranean genotypes.

4.11.6 Breeding and Hybridization

- **Clonal Selection:** Main method, selecting genotypes for cineole and yield.
- **Hybridization:** Limited intervarietal crossing (low fertility).
- **Mutation Breeding:** EMS-induced compact mutants (*CIM-Roshni*).
- **Polyploidy:** Induced tetraploids exhibit thicker leaves and higher oil yield.
- **Molecular Profiling:** RAPD and SSR markers identify chemotype differences.
- **Tissue Culture:** Microcuttings for large-scale multiplication.

4.11.7 Diseases and Resistance

S. No.	Disease	Causal Agent	Resistance / Control
1	Powdery mildew	*Oidium* sp.	Resistant *CIM-Roshni*.
2	Root rot	*Fusarium oxysporum*	Managed via raised-bed planting.

4.11.8 Released Varieties

S. No.	Variety	Key Features
1	*CIM-Roshni*	Compact mutant, high cineole (40 %).
2	*CIM-Hans*	Drought-tolerant, high antioxidant content.

4.11.9 Essential Oil Composition

Compound	Range (%)
1,8-Cineole	20–40
Camphor	10–20
Borneol	5–10
Verbenone	3–6

4.12 Thyme (*Thymus vulgaris* L.)

4.12.1 Introduction

Thyme (*Thymus vulgaris*), another member of the **Lamiaceae** family, is a low-growing perennial herb native to Southern Europe. It is cultivated for its **thymol-rich essential oil** used in pharmaceuticals, flavouring, and perfumery. The crop also possesses significant antimicrobial and antioxidant properties, making it a valuable addition to herbal industries.

In India, thyme is now being promoted in **temperate regions of Himachal Pradesh, Jammu, and Uttarakhand**, often in integrated herbal farms.

4.12.2 Botanical Description

- **Habit:** Perennial, semi-shrubby plant (20–40 cm tall).
- **Leaves:** Small, linear-ovate, strongly aromatic.
- **Flowers:** Pinkish-white, arranged in dense clusters.
- **Chromosome number:** 2n = 30.
- **Pollination:** Entomophilous, moderate cross-pollination (20–25 %).

4.12.3 Economic and Industrial Importance

1. **Oil yield:** 0.5–1.0 % of herbage.
2. **Main constituents:** Thymol (35–55 %), carvacrol (10–15 %), p-cymene (5–10 %).
3. **Uses:** Food flavouring, antiseptics, mouthwashes, aromatherapy, natural preservatives.
4. **Pharmacological activity:** Antimicrobial, expectorant, and antioxidant properties.
5. **Agro-ecological importance:** Thrives in poor soils and semi-arid climates.

4.12.4 Major Breeding Objectives

1. High thymol and carvacrol content.
2. Compact and erect growth habit.
3. Resistance to leaf spot and powdery mildew.
4. Cold and drought tolerance.
5. High herbage and multi-cut potential.

4.12.5 Germplasm and Genetic Resources

S. No.	Centre	Accessions	Remarks
1	CIMAP, Lucknow	80	Global chemotypes (thymol, carvacrol types).
2	NBPGR	60	Wild *Thymus* species.
3	IHBT, Palampur	40	Temperate-adapted accessions.

4.12.6 Breeding and Hybridization

» **Clonal and Mass Selection:** To isolate thymol-rich lines.

» **Hybridization:** Intraspecific crosses for balanced oil profiles.

» **Mutation Breeding:** EMS mutants (*CIM-Thy-3*) with 55 % thymol.

» **Polyploidy:** Tetraploids exhibited larger leaves but reduced oil strength.

» **Molecular Tools:** ISSR and SSR markers to characterize chemotypes.

4.12.7 Diseases and Resistance

S. No.	Disease	Causal Agent	Resistant Line / Control
1	Leaf spot	*Alternaria alternata*	Resistant *CIM-Thy-1*.
2	Powdery mildew	*Erysiphe* sp.	Controlled by tolerant varieties.

4.12.8 Released Varieties

S. No.	Variety	Features
1	*CIM-Thy-1*	Thymol 50 %, high yield.
2	*CIM-Thy-3*	EMS mutant; compact growth.
3	*Him-Thyme*	Cold-hardy, high carvacrol.

4.12.9 Essential Oil Composition

Compound	Range (%)
Thymol	35–55
Carvacrol	10–15
p-Cymene	5–10
γ-Terpinene	2–5

4.13 Sage (*Salvia officinalis* L.)

4.13.1 Introduction

Sage (*Salvia officinalis*), belonging to the family **Lamiaceae**, is a semi-shrubby perennial aromatic herb native to the Mediterranean basin and cultivated globally for its essential oil, culinary herb use, and medicinal value. The name "sage" derives from the Latin *salvare*, meaning "to heal," reflecting its long history in traditional medicine.

In India, sage cultivation has expanded in **Himachal Pradesh, Uttarakhand, Jammu & Kashmir, and parts of South India**, supported by herbal industry demand for **camphor-, cineole-, and thujone-rich oils.**

Sage oil is used in **cosmetics, oral care, perfumery, and pharmaceuticals**, and its extracts exhibit **antioxidant, antimicrobial, and neuroprotective properties**.

4.13.2 Botanical Description and Cytology

» **Habit:** Evergreen perennial herb, 40–80 cm tall, with woody base and greyish pubescent leaves.

» **Leaves:** Opposite, oblong, wrinkled, gland-dotted, aromatic.

» **Flowers:** Violet-blue, bilabiate, borne in terminal spikes.

» **Chromosome number:** 2n = 14, 28, or 60 (depending on species/polyploidy).

» **Pollination:** Entomophilous; cross-pollination 40–60 %.

4.13.3 Economic and Industrial Importance

1. **Essential oil yield:** 0.5–2.0 % of fresh leaves.
2. **Main constituents:** α-thujone, β-thujone, camphor, and 1,8-cineole.
3. **Applications:** Flavouring agent, perfumery, oral hygiene formulations, and antioxidant extracts.
4. **Medicinal properties:** Cognitive enhancer, antimicrobial, and anti-inflammatory agent.
5. **Agro-ecological importance:** Drought-tolerant, suited for poor soils in semi-temperate zones.

4.13.4 Major Breeding Objectives

1. High herbage and oil yield.
2. Balanced thujone–camphor content to meet safety standards.
3. Resistance to powdery mildew and rust.
4. Cold and drought tolerance.
5. Compact canopy for high-density planting.

4.13.5 Germplasm and Genetic Resources

S. No.	Centre / Institute	Accessions	Remarks
1	CSIR–CIMAP, Lucknow	120	Core collection for Indian breeding.
2	NBPGR, New Delhi	60	Wild *Salvia* species.
3	IHBT, Palampur	80	Temperate-adapted accessions.

4.13.6 Breeding and Hybridization

» **Mass Selection:** Identified chemotypes rich in cineole or thujone.

» **Intraspecific Hybridization:** Used to combine oil yield and quality traits.

» **Mutation Breeding:** EMS and gamma rays produced *CIM-Sage-3* (low thujone).

» **Polyploidy:** Tetraploids show higher biomass but delayed flowering.

» **Molecular Breeding:** RAPD and SSR markers used for chemotype profiling.

» **In Vitro Techniques:** Micropropagation and somatic embryogenesis standardized.

4.13.7 Diseases and Resistance

S. No.	Disease	Causal Agent	Resistance / Control
1	Powdery mildew	*Erysiphe cichoracearum*	Resistant *CIM-Sage-1*.
2	Rust	*Puccinia menthae*	Controlled by tolerant varieties.

4.13.8 Released Varieties

S. No.	Variety	Key Traits
1	*CIM-Sage-1*	High oil yield (1.6 %); cineole 35 %.
2	*CIM-Sage-3*	Low thujone (<10 %); safe for medicinal use.
3	*Him-Salvia*	Cold hardy, compact growth.

4.13.9 Essential Oil Composition

Compound	Range (%)
α-Thujone	25–40
β-Thujone	5–10
Camphor	15–25
1,8-Cineole	10–20

4.13.10 Integration of Breeding Approaches

Approach	Impact
Hybridization	Combined yield and aroma quality.
Mutation breeding	Reduced thujone content.
Molecular markers	Chemotype authentication.
In vitro propagation	Rapid elite stock production.

4.14 Marjoram (*Origanum majorana* L.)

4.14.1 Introduction

Marjoram, known as sweet marjoram, is an aromatic perennial herb of the family Lamiaceae and is closely related to oregano (*Origanum vulgare*). It is cultivated for its essential oil and leaves used in flavouring, perfumery, and pharmaceuticals.

The plant produces an essential oil rich in **terpinen-4-ol, linalool, sabinene, and carvacrol,** widely used in spice and food industries. India cultivates marjoram in **Karnataka, Tamil Nadu, Gujarat, and Uttarakhand** under aromatic crop diversification programs.

4.14.2 Botanical Description

- **Habit:** Perennial herb, 30–60 cm tall, branched, with soft pubescent stems.
- **Leaves:** Opposite, oval, light green, aromatic.
- **Flowers:** Small, white or pink, borne in compact clusters.
- **Chromosome number:** 2n = 30.
- **Pollination:** Cross-pollinated (up to 40 %); entomophilous.

4.14.3 Economic and Industrial Importance

1. **Oil yield:** 0.3–0.7 % of herbage.
2. **Principal constituents:** Terpinen-4-ol (20–30 %), linalool (15–20 %), sabinene (10–15 %), carvacrol (5–10 %).
3. **Applications:** Food flavouring, perfumery, aromatherapy.
4. **Medicinal properties:** Antiseptic, expectorant, and digestive stimulant.
5. **Agro-industrial relevance:** Multi-cut aromatic herb adaptable to light soils.

4.14.4 Major Breeding Objectives

1. High herbage and oil yield.
2. High terpinen-4-ol and linalool content.
3. Resistance to downy mildew and leaf spot.
4. Compact and uniform growth.
5. Tolerance to heat and semi-arid climates.

4.14.5 Germplasm and Genetic Resources

S. No.	Centre	Accessions	Traits
1	CSIR–CIMAP, Lucknow	100	High-terpinen chemotypes.
2	NBPGR, New Delhi	60	Wild *Origanum* species.
3	IIHR, Bengaluru	50	Drought-tolerant accessions.

4.14.6 Breeding and Hybridization

- **Mass and Clonal Selection:** Basis for varietal development.
- **Hybridization:** Crosses among chemotypes to combine oil yield and aroma.
- **Mutation Breeding:** Gamma ray (10 kr) induced compact mutant *CIM-Major-2*.
- **Polyploidy:** Tetraploids with increased oil glands and thicker leaves.
- **Molecular Tools:** ISSR profiling used to identify chemotype diversity.
- **Tissue Culture:** Microshoot cultures for large-scale clonal propagation.

4.14.7 Diseases and Resistance

S. No.	Disease	Causal Agent	Resistant Source / Variety
1	Downy mildew	*Peronospora* sp.	Resistant *CIM-Major-1*.
2	Leaf spot	*Cercospora* sp.	Controlled by tolerant clones.

4.14.8 Released Varieties

S. No.	Variety	Key Traits
1	*CIM-Major-1*	Terpinen-4-ol 25 %, high oil yield.
2	*CIM-Major-2*	Mutation line; compact, heat-tolerant.
3	*Surabhi-Major*	High linalool content (20 %).

4.14.9 Essential Oil Composition

Compound	Range (%)
Terpinen-4-ol	20–30
Linalool	15–20
Sabinene	10–15
Carvacrol	5–10

4.14.10 Integration of Breeding Approaches

Approach	Impact
Hybridization	Combined aroma and yield traits.
Mutation breeding	Compact, drought-hardy plants.
Molecular profiling	Chemotype differentiation.
Tissue culture	Rapid clonal propagation.

4.15 Feverfew (*Tanacetum parthenium* (L.) Sch. Bip.)

4.15.1 Introduction

Feverfew (*Tanacetum parthenium*), a perennial herb of the **Asteraceae** family, is an aromatic and medicinal plant cultivated for its **parthenolide-rich essential oil**, which exhibits anti-migraine, anti-inflammatory, and insect-repellent properties. The crop is native to Southeastern Europe but now cultivated globally, including India, as a dual-purpose **aromatic–pharmaceutical** species.

4.15.2 Botanical Description

» **Habit:** Erect, herbaceous perennial, 40–80 cm tall.

» **Leaves:** Alternate, pinnately divided, aromatic.
» **Flowers:** Daisy-like, white ray florets with yellow disc florets.
» **Fruit:** Achene, small, ribbed.
» **Chromosome number:** 2n = 18.
» **Pollination:** Cross-pollinated by insects.

4.15.3 Economic and Industrial Importance

1. **Oil yield:** 0.2–0.4 % of herbage.
2. **Principal constituents:** Parthenolide (1–2 %), camphor (20–30 %), bornyl acetate (10–15 %).
3. **Applications:** Pharmaceutical formulations (anti-migraine), perfumery, insect repellents.
4. **Agro-industrial value:** Multi-cut herb suited to temperate climates.
5. **Export potential:** Increasing due to global demand for parthenolide-based drugs.

4.15.4 Major Breeding Objectives

1. High parthenolide content in leaves and oil.
2. High herbage yield and uniform flowering.
3. Resistance to leaf spot and root rot.
4. Drought and frost tolerance.
5. Compact growth for mechanical harvest.

4.15.5 Germplasm and Genetic Resources

S. No.	Centre	Accessions	Remarks
1	CIMAP, Lucknow	120	High-parthenolide lines.
2	IHBT, Palampur	70	Temperate germplasm.
3	NBPGR	60	Wild *Tanacetum* species.

4.15.6 Breeding and Hybridization

» **Clonal Selection:** Identified high-parthenolide clones (*CIM-Fee-1*).
» **Hybridization:** *T. parthenium* × *T. vulgare* improved vigor and oil yield.
» **Mutation Breeding:** Gamma rays induced compact mutant (*CIM-Fee-3*).
» **Polyploidy:** Tetraploids exhibited higher parthenolide concentration.
» **Molecular Tools:** Gene-based markers for parthenolide synthase pathways identified.

4.15.7 Diseases and Resistance

S. No.	Disease	Causal Agent	Resistant Source / Control
1	Leaf spot	*Alternaria alternata*	Resistant *CIM-Fee-1*.
2	Root rot	*Rhizoctonia solani*	Managed by raised-bed cultivation.

4.15.8 Released Varieties

S. No.	Variety	Features
1	*CIM-Fee-1*	High parthenolide (1.8 %).
2	*CIM-Fee-3*	Compact, early flowering mutant.
3	*Him-Fee*	Cold-tolerant, uniform flowering.

4.15.9 Essential Oil Composition

Compound	Range (%)
Parthenolide	1–2
Camphor	20–30
Bornyl acetate	10–15
β-Caryophyllene	3–5

Aromatic Crops

S. No.	Crop	Botanical Name	Major Constituents	Key Released Varieties	Distinctive Traits / Remarks
1	Lemongrass	*Cymbopogon flexuosus / C. citratus*	Citral (70–85 %), Myrcene, Geraniol	*Krishna, Pragati, Neema, CIM-Shikha*	High citral, rust resistance, 3-year ratoon life
2	Palmarosa	*Cymbopogon martinii* var. *motia*	Geraniol (70–90 %), Geranyl acetate	*CIM-Pawan, RRL-82, Trishna*	Rose-like aroma, drought-tolerant
3	Citronella	*Cymbopogon winterianus*	Citronellal (35–45 %), Geraniol (20–25 %)	*Manjusha, Mandakini, CIMAP Bio-13, Java-II*	High citronellal mutant lines
4	Vetiver (Khus)	*Chrysopogon zizanioides*	Vetiverol, Khusimol, α/β-Vetivone	*KS-1, CIM-Khus-2, Sugandha*	Deep-rooted, high vetiverol, anti-erosion
5	Mint (Menthol Mint)	*Mentha arvensis / M. piperita*	Menthol (70–80 %), Menthone, Isomenthone	*Kosi, Gomti, Himalaya, CIM-Mint-3*	High menthol yield, rust tolerance
6	Sweet Basil	*Ocimum basilicum*	Linalool (40–60 %), Methyl chavicol (20–40 %)	*CIM-Surabhi, CIM-Saumya, Arka Tulasi*	Linalool-rich, downy mildew resistant

7	Lavender	*Lavandula angustifolia*	Linalool (25–35 %), Linalyl acetate (30–45 %)	*IIIM-Lav-1, IIIM-Lav-3, Him-Lavender*	Temperate hardy, perfumery-grade oil
8	Geranium (Rose-scented)	*Pelargonium graveolens*	Citronellol (35–40 %), Geraniol (20–25 %)	*CIM-Gerani-1, CIM-Gerani-2, Hemanti*	Substitute for rose oil, nematode tolerant
9	Patchouli	*Pogostemon cablin*	Patchouli alcohol (35–45 %), α-Bulnesene	*CIM-Patch-1, CIM-Patch-3, Sugantha*	High fixative value, compact canopy
10	Artemisia (Sweet Wormwood)	*Artemisia annua*	Artemisinin (0.5–1.5 %), Artemisia ketone	*CIM-Arogya, Jeevan Raksha, CIM-Aarti*	Anti-malarial source, high artemisinin mutant
11	Rosemary	*Rosmarinus officinalis*	1,8-Cineole (20–40 %), Camphor (10–20 %), Borneol	*CIM-Roshni, CIM-Hans*	Antioxidant-rich, drought-tolerant
12	Thyme	*Thymus vulgaris*	Thymol (35–55 %), Carvacrol (10–15 %)	*CIM-Thy-1, CIM-Thy-3, Him-Thyme*	Thymol chemotype, cold hardy
13	Sage	*Salvia officinalis*	α-Thujone (25–40 %), Camphor (15–25 %), 1,8-Cineole	*CIM-Sage-1, CIM-Sage-3, Him-Salvia*	Medicinal and aromatic; low-thujone safe mutant
14	Marjoram (Sweet Marjoram)	*Origanum majorana*	Terpinen-4-ol (20–30 %), Linalool (15–20 %), Sabinene	*CIM-Major-1, CIM-Major-2, Surabhi-Major*	Culinary spice and perfumery herb
15	Feverfew	*Tanacetum parthenium*	Parthenolide (1–2 %), Camphor (20–30 %), Bornyl acetate	*CIM-Fee-1, CIM-Fee-3, Him-Fee*	Anti-migraine, compact, cold-hardy type

Chapter - 5

Genetic Resources

1. Introduction

1.1 Concept and Scope of Plantation Crop Genetic Resources

Plantation crops comprise a specialized group of long-lived perennials cultivated mainly in the humid tropics and subtropics for industrial raw materials such as oil, beverages, fibre, latex, and nuts. Major representatives include **coconut (*Cocos nucifera L.*)**, **arecanut (*Areca catechu L.*)**, **oil palm (*Elaeis guineensis Jacq.*)**, **rubber (*Hevea brasiliensis Muell. Arg.*)**, **tea (*Camellia sinensis (L.) O. Kuntze*)**, **coffee (*Coffea arabica L.* and *C. canephora Pierre ex Froehner*)**, **cocoa (*Theobroma cacao L.*)**, and **cashew (*Anacardium occidentale L.*)**. Together, they contribute billions of dollars annually to global trade, sustain rural livelihoods, and occupy over 30 million ha of the world's agricultural land.

The genetic resources of these crops encompass every form of heritable material – wild relatives, landraces, improved varieties, clones, hybrids, and breeding lines – that can be used to create new variability. They embody the evolutionary memory of each species and constitute the foundation for **crop improvement, climate resilience, and biotechnological innovation**. Because most plantation crops are **perennial, cross-pollinated, and clonally propagated**, their gene pools display complex evolutionary patterns distinct from annual field crops.

1.2 Importance of Genetic Resources in Perennial Systems

Unlike annual crops that can be rapidly re-selected every season, plantation crops are exposed to long-term climatic fluctuations over decades. A wide genetic base is therefore essential to buffer environmental stresses and emerging pathogens. Genetic resources enable:

» **Yield stability** through heterozygosity and polyclonality.

» **Adaptation** to marginal environments such as saline coasts (*Cocos nucifera*) or lateritic uplands (*Anacardium occidentale*).

» **Improved quality traits** – for instance, fatty-acid composition in *Elaeis guineensis*, caffeine and catechin balance in *Camellia sinensis*, or aromatic volatiles in *Coffea arabica*.

DOI: 10.1201/9781042010332-6

» **Resistance genes** against major diseases such as coffee leaf rust (*Hemileia vastatrix*), bud rot of coconut, or South American leaf blight of rubber.

Conservation and characterization of these genetic assets underpin every modern breeding, genomics, and biotechnology programme. Without them, the long-term sustainability of plantation agriculture is jeopardized.

1.3 Historical Evolution and Geographic Origin

Most plantation crops trace their origins to ancient centres of biodiversity:

Region	Representative Crops and Centers of Origin
Southeast Asia	*Cocos nucifera, Areca catechu, Camellia sinensis var. assamica*
West and Central Africa	*Elaeis guineensis, Coffea canephora*
South America (Amazon Basin)	*Hevea brasiliensis, Theobroma cacao, Anacardium occidentale*
East Africa (Ethiopian Highlands)	*Coffea arabica*

Dispersal followed natural routes (sea currents for coconuts, animal or human migration for coffee and cocoa) and later colonial trade, leading to domestication far from their original gene pools. Such **secondary centres of diversity** – e.g., Sri Lanka for tea, Kerala for coconut, Bahia for cocoa – became focal points of human selection and adaptive divergence.

1.4 Stages of Genetic Resource Development

1. **Exploration and Collection:** Discovery of wild and cultivated forms through expeditions (e.g., FAO/IBPGR missions for coconut and oil palm).
2. **Characterization:** Morphological, biochemical, and molecular profiling to identify unique alleles.
3. **Evaluation:** Screening for agronomic and stress-tolerance traits.
4. **Conservation:** *Ex situ* (field genebanks, cryopreservation) and *in situ* (wild reserves, on-farm) methods.
5. **Utilization:** Incorporation into breeding pipelines via crossing, cloning, or biotechnological transfer.

1.5 Objectives of Studying Genetic Resources

» To decipher **evolutionary processes** governing diversity.

» To delineate **phylogenetic relationships** and taxonomic boundaries.

» To provide raw material for **hybridization, mutagenesis, and molecular breeding**.

» To support **conservation policies** ensuring equitable benefit-sharing of tropical biodiversity.

The subsequent sections examine the evolutionary mechanisms shaping these genetic resources.

2. Evolutionary Mechanisms in Plantation Crops

2.1 Overview of Evolutionary Forces

Plant genomes evolve under five principal forces – **mutation, recombination, gene flow, genetic drift, and natural selection** – with additional contributions from **polyploidy, hybridization, and human-mediated selection.** Their relative intensity varies with reproductive biology: cross-pollinated species like *Hevea* or *Elaeis* maintain high heterozygosity, whereas predominantly self-pollinated forms like *Coffea arabica* accumulate fixed alleles and evolve more slowly.

2.2 Mutation – The Primary Source of Novelty

2.2.1 Spontaneous Mutations

Spontaneous point or structural mutations arise during DNA replication or through transposable elements. In perennials, somatic mutations can be perpetuated vegetatively, leading to **clonal sports**:

» *Camellia sinensis* has produced numerous "bud sports" differing in leaf morphology and theaflavin content.

» Dwarf coconuts, such as *Chowghat Orange Dwarf* originated from spontaneous germ-line mutations conferring precocity.

» In *Hevea brasiliensis*, mutations influencing latex vessel density underlie elite clones like RRIM 600 and PB 235.

Because of their stability in clonally propagated crops, these variants constitute invaluable germplasm for micropropagation and molecular mapping.

2.2.2 Induced Mutagenesis

Artificial mutagenesis (gamma rays, EMS, sodium azide) supplements natural variation. The **FAO/IAEA mutation-breeding programmes** demonstrated that irradiated seedlings of *Areca catechu* and *Coffea arabica* exhibited desirable alterations in fruit size, caffeine level, and disease tolerance. In vegetatively propagated species, chimeric mutations can be stabilized by somatic embryogenesis. Mutagenesis thus remains a vital adjunct where sexual recombination is restricted.

2.3 Recombination and Genetic Shuffling

During meiosis, homologous chromosomes exchange segments, generating new allelic combinations. In heterozygous tree crops, recombination contributes vastly more variability than mutation. Controlled hybridization programmes exploit this natural process:

» In *Elaeis guineensis*, reciprocal crosses between **Dura** (thick-shelled) and **Pisifera** (shell-less) palms yield **Tenera** hybrids with superior oil yield – a classic recombinational outcome.

» Crosses between *Coffea arabica* (high cup quality) and *C. canephora* (rust resistance) produced hybrids such as "Catimor" and "Sarchimor."

» In *Cocos nucifera*, recombination between tall and dwarf palms has generated *Tall × Dwarf hybrids* combining high yield with earliness.

The rate of recombination is influenced by chromosomal architecture, presence of transposons, and epigenetic modifications; hence, cytogenetic studies are integral to understanding trait inheritance.

2.4 Genetic Drift and Population Bottlenecks

Genetic drift entails random fluctuations of allele frequencies, particularly severe in small founder populations typical of early introductions:

» The global rubber industry descends from merely 22 seeds collected in 1876 from the Amazon – an extreme **founder bottleneck.**

» Commercial cocoa clones largely derive from narrow **Criollo** and **Forastero** gene pools.

» Limited seed exchange in colonial coconut plantations caused localized genetic uniformity, predisposing stands to pest epidemics.

Drift reduces effective population size (*Ne*) and heterozygosity (H), increasing inbreeding depression. Modern breeding therefore emphasizes **broad-based parental populations** and **recurrent selection** to restore variability.

2.5 Gene Flow and Migration

Gene flow (pollen or seed migration) maintains connectivity between populations, counteracting drift. In perennial plantation landscapes, it occurs through:

1. **Natural Agents** – wind, insects, birds, and ocean currents. For instance, coconut fruits float for weeks, facilitating trans-oceanic gene exchange.
2. **Human Agency** – movement of planting material across colonies introduced new alleles but also exotic diseases (e.g., coffee leaf rust from Sri Lanka to Brazil).

2.6 Natural Selection and Adaptive Differentiation

Natural selection acts on phenotypic variation to enhance survival under specific ecological regimes:

» **Abiotic adaptation:** Salt-tolerant coconut ecotypes thrive along India's coastal fringes; drought-hardy cashew landraces prosper in arid tracts.

» **Biotic pressure:** Coffee populations exposed to *Hemileia vastatrix* evolved hypersensitive response genes (e.g., *SH1–SH9* series).

» **Co-adaptation:** Rubber clones selected under tapping stress exhibit improved wound-healing latex metabolism.

2.7 Hybridization and Introgression

When distinct yet compatible species interbreed, hybridization creates heterotic combinations and novel allele assemblages:

» *Hevea brasiliensis* × *H. benthamiana* hybrids show increased tolerance to leaf blight and wind.

» *Coffea arabica* × *C. congensis* hybrids exhibit improved fertility and cup flavour.

» *Cocos nucifera (Tall* × *Dwarf)* crosses yield precocious high-yielding palms exploited in India, Sri Lanka, and the Philippines.

Subsequent backcrossing to the recurrent parent allows **introgression** of specific genes (e.g., disease resistance) while retaining the desired phenotype. Molecular markers (SSR, SNP) now track introgression blocks with precision, accelerating perennial breeding cycles.

2.8 Polyploidy and Chromosomal Evolution

Polyploidy – whole-genome duplication – is widespread in angiosperms and underlies many evolutionary breakthroughs.

- *Coffea arabica* (2n = 4x = 44) is an **allotetraploid** derived from spontaneous hybridization between diploid *C. eugenioides* and *C. canephora*, combining quality and robustness.
- *Camellia sinensis* populations exhibit diploid (2n = 30), triploid, and tetraploid cytotypes influencing leaf size and secondary-metabolite content.
- Polyploid *Hevea* mutants often display enhanced latex yield and thicker bark.

Polyploidization buffers deleterious mutations, increases allelic dosage, and fosters **gene neofunctionalization**, enabling adaptation to diverse ecological niches.

2.9 Coevolution and Symbiotic Associations

Perennial plantations sustain intricate ecological relationships:

- **Pollinators:** Cashew and cocoa depend on specific insect guilds (Thrips, Ceratopogonidae) – coevolution of floral morphology ensures effective pollination.
- **Symbionts:** Mycorrhizal fungi in coffee and tea enhance nutrient uptake; endophytic bacteria modulate stress tolerance.
- **Defense synergy:** Polyphenol biosynthesis in tea coevolved as deterrence against herbivory, now prized for antioxidant value.

Such coevolutionary dynamics continuously shape gene frequencies in both partners, linking crop evolution to ecosystem processes.

2.10 Human-Mediated Selection and Domestication Pressures

Human preference constitutes a potent selective force that transforms wild perennials into domesticated plantations:

- Selection for **larger nuts** and **higher oil content** in coconut and oil palm.
- Preference for **cup quality** in *Coffea arabica* and **aroma intensity** in *Camellia sinensis.*
- Clonal propagation of **high-latex Hevea** and **high-kernel cashew** genotypes ensuring uniformity.

Over millennia, such **anthropogenic selection** led to domestication syndromes – reduced seed dispersal, increased yield, and dependency on human cultivation. Modern genomics reveals domestication-related genes (e.g., *Sh4*, *QTLs* for fruit size) analogous to annual crops, though fixation occurs more slowly in long-generation trees.

Table 1 – Principal Evolutionary Mechanisms and Representative Examples in Plantation Crops

S. No.	Mechanism	Process Description	Representative Crops / Examples	Evolutionary Significance
1	Mutation	Spontaneous or induced DNA change	Dwarf coconut; high-latex Hevea clones	Primary source of new alleles
2	Recombination	Exchange of chromosomal segments	*Elaeis guineensis* (Dura × Pisifera → Tenera)	Generates novel gene combinations
3	Genetic Drift	Random allele loss in small populations	Narrow rubber base; Criollo cocoa	Reduces heterozygosity
4	Gene Flow	Pollen/seed migration between populations	Oceanic dispersal of coconut	Maintains connectivity, homogenizes allele frequencies
5	Natural Selection	Differential survival of adaptive phenotypes	Drought-tolerant cashew; salt-tolerant coconut	Local adaptation and speciation
6	Hybridization & Introgression	Crossing and gene transfer between taxa	*C. arabica* × *C. canephora* hybrids	Combines desirable traits
7	Polyploidy	Whole-genome duplication	Tetraploid *Coffea arabica*	Genetic buffering and vigour
8	Human Selection	Artificial choice for desired traits	High-flavour tea, high-oil palm	Domestication and improvement

3. Adaptation and Domestication

3.1 Adaptation

Adaptation is the evolutionary adjustment of an organism to its environment, resulting from the cumulative effect of natural selection acting on heritable variation. In plantation crops, perennial species with life spans often exceeding 30 years, adaptation is expressed through physiological, morphological, and biochemical traits that ensure survival, reproduction, and economic yield in diverse, fluctuating tropical environments.

Key adaptive mechanisms include:

1. **Morphological plasticity**, such as variable leaf orientation, stomatal density, and canopy structure (e.g., *Cocos nucifera* palms adjust leaf angles to optimize light interception in coastal versus inland sites).
2. **Physiological tolerance** to abiotic stresses – salt, drought, shade, and temperature extremes.
3. **Biochemical responses**, including synthesis of osmolytes, antioxidants, and secondary metabolites conferring resistance to pests and pathogens.

4. **Reproductive flexibility**, allowing adjustment of flowering, fruiting, or pollination timing to seasonal cues.

3.2 Definition and Scope of Domestication

Domestication refers to the evolutionary process by which wild species are transformed into forms dependent on humans for propagation and survival. It involves **genetic, morphological, and behavioural changes** induced by selection and cultivation. In plantation crops, domestication is particularly complex because many were semi-domesticated before systematic breeding began. Unlike cereals or pulses domesticated within the last 10000 years, several plantation crops remained in **pre-domesticated or incipient stages** until the colonial era. Their domestication trajectories often span thousands of years of human–plant interaction, local selection, and later, scientific improvement.

4. Adaptive Mechanisms in Major Plantation Crops

4.1 Coconut (*Cocos nucifera L.*)

Coconut exemplifies adaptation to **coastal saline ecosystems.**

- **Morphological traits**: fibrous husk aiding buoyancy, thick cuticle, and root pneumatophores promoting oxygen exchange.
- **Physiological tolerance**: capacity to exclude Na^+ and Cl^- ions via selective root uptake and accumulation of compatible solutes (proline, sugars).
- **Reproductive strategy**: monoecious inflorescences with protandry reduce self-pollination, enhancing genetic diversity for local adaptation.
- **Ecotypic variation**: coastal tall varieties (e.g., West Coast Tall) contrast with inland dwarfs (e.g., Chowghat Green Dwarf) in stature, precocity, and nut composition, reflecting climatic and edaphic adaptation.

4.2 Oil Palm (*Elaeis guineensis Jacq.*)

Originating in humid West Africa, oil palm displays strong adaptation to **high-rainfall equatorial zones.**

- **Canopy architecture** with self-shading fronds prevents photoinhibition under intense irradiance.
- **Photosynthetic resilience** enables productivity even at 35–38 °C.
- **Root system** with abundant pneumatodes allows survival in water-logged soils.
- **Phenotypic plasticity** underlies its successful introduction to Southeast Asia, where nutrient-use efficiency and response to management improved through selection.

4.3 Arecanut (*Areca catechu L.*)

Native to the Indo-Malayan region, arecanut exhibits **shade tolerance** and **moisture dependence.**

- Adapted to mid-elevation, high-humidity belts (700–1 200 m asl).

- **Physiological dormancy** during dry spells and **synchronous flowering** with monsoon onset indicate fine climatic adaptation.
- Genetic variability in carotenoid content and alkaloid biosynthesis correlates with micro-climatic gradients across India, Myanmar, and Indonesia.

4.4 Rubber (*Hevea brasiliensis Muell. Arg.*)

Rubber trees evolved in the **Amazon Basin**, characterized by alternating wet and dry seasons.

- Possess a deep tap-root system and high leaf-area index supporting sustained latex metabolism.
- **Leaf shedding** during drought minimizes transpiration, while **photosynthetic recovery** is rapid upon refoliation.
- **Latex flow** physiology adapted to periodic wounding (tapping); clones differ in turgor maintenance and laticifer regeneration, traits under genetic control shaped by human selection.
- When introduced to Asia, adaptive evolution through selection for wind tolerance, disease resistance (against *Microcyclus ulei*), and phenological adjustment allowed naturalization beyond the Amazon.

4.5 Tea (*Camellia sinensis (L.) O. Kuntze*)

Tea is adapted to **humid subtropical and tropical highlands**.

- Exhibits **evergreen habit**, enabling year-round plucking cycles.
- Rooting depth and leaf pubescence vary along altitudinal gradients, reflecting micro-adaptation to temperature and moisture.
- **Shade tolerance** arises from efficient chloroplast acclimation and high chlorophyll b content.
- **Biochemical adaptation** involves the synthesis of catechins and theaflavins that deter herbivores and oxidative stress – coincidentally defining beverage quality.

4.6 Coffee (*Coffea arabica L.* and *C. canephora Pierre ex Froehner*)

- *C. arabica* evolved in **Ethiopian highlands** (cool, shaded montane forests), whereas *C. canephora* originated in **lowland equatorial Africa** (hot, humid habitats).
- **Arabica** exhibits self-compatibility, a narrow genetic base, and sensitivity to heat – adapted to 18–22 °C.
- **Robusta** (*C. canephora*) tolerates 30 °C and higher rainfall, with open-pollenated flowers promoting variability.
- Adaptive divergence between species underpins breeding of inter-specific hybrids ("Catimor", "Sarchimor"), combining stress tolerance with beverage quality.

4.7 Cocoa (*Theobroma cacao L.*)

Native to the Amazon rainforest, cocoa thrives in **low-light understory conditions**.

- **Shade adaptation**: high chlorophyll b/a ratio, efficient photon capture at low intensity.

- **Root architecture**: fine feeder roots with high mycorrhizal colonization enabling nutrient uptake in acidic soils.
- **Disease adaptation**: coevolution with fungal pathogens (*Phytophthora palmivora*) produced genotypes with constitutive defense enzymes (peroxidase, chitinase).
- Domesticated "Criollo", "Forastero", and "Trinitario" groups represent adaptive complexes shaped by human preference for flavour, yield, and stress resistance.

4.8 Cashew (*Anacardium occidentale L.*)

Cashew originated in northeastern Brazil's dry tropics and demonstrates **exceptional drought tolerance**.

- **Xeromorphic leaves** with thick cuticles, sunken stomata, and high water-use efficiency.
- **Taproot dominance** ensures moisture access in semi-arid soils.
- **Phenological synchronization** with dry-season flowering prevents fungal infection of inflorescences.
- Genetic divergence between Indian and African germplasm reflects independent adaptive trajectories to local pests and rainfall regimes.

5. Domestication Pathways and Processes

5.1 Stages of Domestication

1. **Initial association** – humans begin utilizing wild stands for food or material (e.g., wild *Hevea* latex).
2. **Protection and management** – selective retention or transplanting of superior phenotypes near settlements.
3. **Cultivation and propagation** – intentional planting from seed or vegetative material.
4. **Selective breeding** – conscious selection for yield, quality, or ease of harvest.
5. **Scientific improvement** – application of genetics, tissue culture, and molecular tools.

5.2 Domestication Centers and Secondary Diversification

Each plantation crop has a distinct **primary domestication center** (natural origin) and one or more **secondary diversification centers** (regions where human selection generated new diversity).

S. No.	Crop	Primary Center of Origin	Secondary Centers of Diversification
1	*Cocos nucifera*	Indo-Pacific islands	South India, Sri Lanka, Philippines
2	*Elaeis guineensis*	West Africa	Malaysia, Indonesia
3	*Areca catechu*	Malaya Archipelago	Southern India, Sri Lanka
4	*Hevea brasiliensis*	Amazon Basin	Southeast Asia
5	*Camellia sinensis*	Southern China / Assam	Kenya, Sri Lanka

6	*Coffea arabica*	Ethiopia	Latin America
7	*Theobroma cacao*	Upper Amazon	West Africa
8	*Anacardium occidentale*	Northeast Brazil	India, Mozambique

5.3 Genetic Consequences of Domestication

Domestication leads to recognizable genomic signatures:

» **Reduced heterozygosity** due to selective sweeps (e.g., uniformity in *Hevea* clones).

» **Fixation of domestication traits** – larger fruit, higher oil or sugar content, altered branching pattern.

» **Loss of seed dormancy and dispersal ability**, increasing dependence on humans.

» **Introgression** from wild relatives occasionally restores diversity, as in *Coffea arabica* × *C. canephora* hybrids.

Molecular analyses using SSR, AFLP, and chloroplast DNA markers confirm domestication bottlenecks and reveal cryptic diversity useful for broadening breeding populations.

5.4 Role of Human Selection in Domestication Syndrome

Over centuries, farmers and breeders unconsciously selected phenotypes offering better utility:

» **Yield traits** – large nuts in coconut, thicker shell and high kernel in cashew, high latex in rubber.

» **Quality traits** – low acidity in coffee, high catechin in tea, desirable fatty-acid profile in palm oil.

» **Management traits** – compact canopy or dwarfism facilitating harvest.

» **Non-shattering or parthenocarpy** in some cocoa and arecanut lines.

This artificial selection reinforced specific alleles, producing distinct **domestication syndromes** recognizable in today's cultivars.

6. Adaptive Genetics and Domestication Syndromes

6.1 Morphological Changes

Domesticated plantation crops display consistent morphological modifications:

1. **Gigantism** – increased fruit or nut size (coconut, cashew).
2. **Loss of defense structures** – reduced spines or secondary metabolites in cultivated *Hevea*.
3. **Canopy reconfiguration** – shorter internodes in dwarf coconut or compact canopy in tea.
4. **Synchrony in flowering and fruiting** aiding uniform harvest.

6.2 Physiological and Biochemical Changes

Selection for metabolic efficiency has altered physiological pathways:

» Enhanced **photosynthetic rate per unit leaf area** in improved *Elaeis guineensis*.

» Elevated **sucrose and lipid biosynthesis enzymes** in high-yield clones.

» Modified **secondary-metabolite flux** determining flavour (tea polyphenols, coffee alkaloids, cocoa flavanols).

6.3 Reproductive Changes

Domestication often modifies mating systems:

» Transition from **self-incompatibility to partial self-compatibility** in *Coffea arabica*, allowing stable seed production.

» **Shift toward parthenocarpy or vegetative propagation** (coconut tissue culture, Hevea budding).

» Reduced seed dispersal mechanisms (loss of abscission zones).

7. Environmental and Socio-Cultural Drivers of Domestication

7.1 Climatic Pressures

Changing Pleistocene climates shaped primary adaptation – species surviving glacial fluctuations acquired broad thermal tolerance, facilitating later domestication in different continents.

7.2 Cultural Practices and Human Dispersal

Human migration and trade catalyzed **biocultural co-evolution**:

» Polynesians carried coconuts across the Pacific.

» Arab traders spread coffee and dates.

» European colonials established cocoa and rubber in new continents.

Each introduction was accompanied by **founder selection**, resulting in region-specific gene pools and novel adaptive trajectories.

8. Modern Perspectives on Adaptation and Domestication

8.1 Molecular Insights

Genomic studies reveal candidate genes controlling domestication traits:

» *SH4*, *QTLs* for kernel size, *LIM* transcription factors for lignin biosynthesis in *Hevea*, *WRKY* regulators of secondary metabolism in tea. These insights enable **genome-assisted domestication**, accelerating the traditionally slow improvement of perennials.

8.2 Climate Change and Future Adaptation

Projected temperature increases threaten current production zones of *C. arabica* and *Camellia sinensis*. Harnessing wild relatives from hotter or drier habitats and exploiting epigenetic plasticity (DNA methylation, histone modification) become vital for adaptation.

8.3 Integrating Traditional Knowledge

Local farmers maintain diverse landraces adapted to micro-climates – for example, coastal "Kalparasa" coconut ecotypes or drought-resistant cashew of Goa. Incorporating such **indigenous adaptive knowledge** into formal breeding enhances sustainability.

Table 2. Summary of Adaptive Traits and Domestication Syndromes in Major Plantation Crops

S. No.	Crop	Key Adaptive Traits	Domestication Changes	Principal Selection Goal
1	*Cocos nucifera*	Salt tolerance, floating fruit, fibrous husk	Dwarfism, large nut	Early bearing, copra yield
2	*Elaeis guineensis*	High photosynthetic efficiency, water-logging tolerance	Tenera hybrid, mesocarp oil enrichment	Oil yield and composition
3	*Areca catechu*	Shade tolerance, deep roots	Compact crown, uniform fruiting	Alkaloid content, nut size
4	*Hevea brasiliensis*	Drought deciduousness, latex regeneration	Clonal uniformity, thick bark	Latex productivity
5	*Camellia sinensis*	Shade tolerance, pest-resistant polyphenols	Bush architecture, reduced seed set	Leaf quality, pluckability
6	*Coffea arabica / C. canephora*	Temperature specialization	Hybrid derivatives	Flavour + disease resistance
7	*Theobroma cacao*	Shade adaptation, disease resistance	Cultivar groups (Criollo, Forastero)	Flavour, yield
8	*Anacardium occidentale*	Drought tolerance, pest resistance	Tree compactness, larger nut	Kernel recovery

10. Concept of Genetic Resources

10.1 Definition and Scope

Genetic resources refer to all heritable materials, genes, genotypes, populations, and species, that possess actual or potential economic or scientific value. In plantation crops, they include *wild relatives*, *landraces*, *improved cultivars*, *elite clones*, and *experimental hybrids*. Because these perennials are propagated for decades without frequent replacement, the continuity of their genetic base is crucial for maintaining productivity and resilience.

Plantation genetic resources therefore, encompass:

» **Intraspecific diversity** – variation within cultivated species (e.g., tall vs. dwarf coconuts).

- **Interspecific diversity** – related wild taxa providing novel alleles (e.g., *Hevea benthamiana*, *Coffea eugenioides*).
- **Ecosystem-level diversity** – the landscapes and microbial symbionts supporting crop health.

10.2 Categories of Genetic Resources

1. **Primary gene pool (GP I):** cultivated and easily cross-compatible species; crossing produces fertile offspring (e.g., *Coffea arabica* × *C. canephora*).
2. **Secondary gene pool (GP II):** related species producing partially fertile hybrids (e.g., *Hevea brasiliensis* × *H. benthamiana*).
3. **Tertiary gene pool (GP III):** distant relatives requiring advanced techniques such as embryo rescue or somatic hybridization (e.g., *Theobroma grandiflorum* × *T. cacao*).

Understanding and exploiting all three pools broadens breeding possibilities, especially under climate stress.

11. Exploration and Collection of Genetic Resources

11.1 Historical Explorations

Early colonial botanical expeditions were the first organized efforts to collect plantation germplasm:

- **19th-century rubber transfers** from Brazil to Kew Gardens and later to Ceylon and Malaya (1876) marked a milestone in global germplasm movement.
- **Coffee and tea expeditions** by British and French explorers established the foundations of current plantations.
- **Coconut and arecanut surveys** by FAO, IPGRI, and national institutes in the 20th century systematically documented diversity across Asia and the Pacific.

11.2 Modern Collection Strategies

Present-day exploration emphasizes both **geographical coverage** and **ecological representativeness**:

1. **Target areas:** wild habitats, farmers' fields, old plantations, and remote islands.
2. **Sampling designs:** random, stratified, or core-subset approaches capturing maximum allelic diversity.
3. **Documentation:** passport data (origin, altitude, soil type), herbarium vouchers, and digital geo-referencing.
4. **Bio-safety protocols:** phytosanitary inspection, quarantine, and DNA fingerprinting to avoid duplicate introductions.

12. Characterization and Evaluation of Genetic Resources

12.1 Morphological Characterization

Each crop has standardized **descriptors** (IPGRI/IBPGR guidelines) covering vegetative, floral, and fruit characters, for example:

» *Cocos nucifera:* stem girth, nut shape, husk thickness, copra content.

» *Hevea brasiliensis:* leaflet size, latex colour, bark anatomy.

» *Camellia sinensis:* leaf pubescence, internode length, bud colour.

» *Coffea arabica:* berry size, bean shape, and branching pattern.

Morphological characterization provides preliminary grouping, though phenotypes can be influenced by environment.

12.2 Biochemical Characterization

Analysis of **isozymes, alkaloids, oils, and polyphenols** reveals intra-population diversity:

» **Fatty-acid profiles** in *Elaeis guineensis* differentiate Dura, Pisifera, and Tenera types.

» **Caffeine and chlorogenic-acid spectra** partition *Coffea* accessions.

» **Catechin ratios** distinguish Assam and China types of tea. Such biochemical fingerprints link genotype to product quality and geographic origin.

12.3 Molecular Characterization

Modern DNA-based tools revolutionized germplasm evaluation:

Marker Type	Utility in Plantation Crops
RAPD / ISSR	Rapid detection of polymorphism in *Areca catechu*, *Anacardium occidentale*.
SSR / Microsatellites	Genetic distance estimation in *Hevea* and *Elaeis* core collections.
AFLP / SRAP	Assessment of hybrid purity in *Coffea* and *Cocos*.
SNP arrays & Genotyping-by-Sequencing	Genome-wide diversity scans; marker–trait associations.

Molecular data permit **cluster analysis**, **phylogenetic reconstruction**, and **core-collection development**, critical for managing large field genebanks.

13. Conservation of Genetic Resources

13.1 In Situ Conservation

In situ conservation maintains evolutionary processes in natural habitats or traditional farms.

1. **Wild reserves:** Protected populations of *Hevea spruceana* in Brazil or *Coffea arabica* forests in Ethiopia safeguard allelic richness.
2. **On-farm conservation:** farmers' coconut groves or mixed agroforestry systems preserve local landraces.
3. **Community gene sanctuaries:** in India and Sri Lanka, participatory models allow villagers to manage diverse palms, integrating conservation with livelihoods.

13.2 Ex Situ Conservation

Because many plantation species are **recalcitrant seeders** (seeds cannot be dried or frozen), conventional seed banks are inadequate. Alternatives include:

- **Field genebanks** – living collections such as the Central Plantation Crops Research Institute (CPCRI) coconut germplasm, Malaysian Palm Oil Board's oil-palm repository, and the Tocklai tea collection.
- **In vitro and cryogenic storage** – tissue cultures of embryogenic callus or somatic embryos maintained at –196 °C; successful for *Coffea*, *Theobroma cacao*, and *Hevea*.
- **Pollen banks** – long-term storage for controlled hybridization (tea, cocoa).

13.3 Core and Mini-Core Collections

To manage thousands of accessions efficiently, representative subsets capturing maximum diversity are defined statistically. For example:

- Coconut core collection ≈ 650 accessions representing global diversity.
- Coffee core collection ≈ 300 entries covering Ethiopia, Congo, and Madagascar. Such cores enable cost-effective evaluation and reduce redundancy.

14. Utilization of Genetic Resources

14.1 Conventional Breeding Use

- **Parental selection** – identification of complementary lines for hybrid programs (e.g., Dura × Pisifera in oil palm).
- **Recurrent selection** – recycling of superior clones in *Hevea* to accumulate favourable alleles.
- **Inter-specific hybridization** – incorporation of resistance genes from wild relatives.

14.2 Biotechnological Applications

1. **Somaclonal variation:** exploited for traits such as dwarfism in coconut or disease resistance in coffee.
2. **Molecular marker-assisted selection:** SNP markers linked to oil yield or caffeine genes enable early screening.
3. **Transgenic and genome-editing approaches:** introduction of *Hevea* antifungal chitinases, CRISPR editing of *Coffea* caffeine synthase for low-caffeine varieties.
4. **Metabolomic profiling:** correlating biochemical diversity with genotype for nutraceutical product development.

14.3 Pre-Breeding and Introgression

Wild species harbor stress-resistance genes absent from cultivated forms. Pre-breeding bridges this gap:

- *Hevea benthamiana* contributed leaf-blight tolerance.
- *Coffea racemosa* provided drought tolerance to *C. arabica*.

- *Theobroma grandiflorum* donated disease-resistance loci to cocoa.

Marker-aided backcrossing stabilizes such introgressed traits while retaining commercial quality.

15. Assessment of Genetic Diversity and Divergence

15.1 Quantitative Assessment

Mahalanobis D^2 statistics, Euclidean distance, and cluster analysis quantify morphological diversity in coconut, tea, and cashew. Principal-component analysis identifies trait combinations contributing most to divergence (e.g., nut weight, leaf area, oil content).

15.2 Molecular Assessment

Molecular diversity indices, polymorphic-information content (PIC), heterozygosity (H), Shannon index, are computed to prioritize unique accessions. High genetic divergence ensures heterosis and broad adaptation in hybrid progeny.

15.3 Phylogeographic Studies

Chloroplast and mitochondrial haplotypes trace evolutionary lineages and migration routes:

- **Coconut:** Two major haplogroups, Pacific and Indo-Atlantic, correspond to independent domestication events.
- **Coffee:** Ethiopian highland populations distinct from lowland *C. canephora.*
- **Cocoa:** Ten recognized genetic clusters reflecting Amazonian river basins.

Such information guides **germplasm exchange** while avoiding genetic erosion.

16. Documentation and Information Management

Digitization has transformed genetic-resource management:

- **GRIN-Global**, **Genesys**, and **ICAR-NAGS** databases store passport, characterization, and evaluation data.
- **GIS integration** enables mapping of diversity hot-spots and climate-matching for new introductions.
- **Bar-coding and DNA IDs** provide traceability in international exchange.Transparent documentation ensures compliance with **Convention on Biological Diversity (CBD)** and **Nagoya Protocol** on Access and Benefit-Sharing.

17. Challenges and Prospects in Genetic-Resource Management

17.1 Genetic Erosion

Monoclonal cultivation, common in *Hevea*, *Elaeis*, and *Camellia*, has narrowed variability, heightening vulnerability to pests and climate shocks. Rejuvenating plantations with diverse clones is imperative.

17.2 Climate Change

Rising temperature, erratic rainfall, and salinity threaten existing germplasm. **Predictive modelling** identifies future-suitable ecotypes; **assisted migration** of tolerant genotypes is being tested.

17.3 Policy and Legal Issues

Benefit-sharing and intellectual-property rights under the CBD demand equitable frameworks so source countries of germplasm gain from commercialization.

17.4 Integration of Traditional Knowledge

Farmer custodians of coconut, arecanut, and cashew landraces possess empirical selection knowledge; participatory programmes linking them with research institutions strengthen on-farm conservation.

Table 3 – Global Status of Ex Situ Collections of Major Plantation Crops

S. No.	Crop	Principal Repositories / Institutions	Approx. No. of Accessions	Remarks
1	*Cocos nucifera*	CPCRI (India), Philippine Coconut Authority	> 2 000	Field collections of tall and dwarf types world-wide
2	*Elaeis guineensis*	Malaysian Palm Oil Board, NIFOR (Nigeria)	1 500	Core and breeding collections
3	*Hevea brasiliensis*	RRIM (Malaysia), IRRDB network	1 200	Clonal repository with wild relatives
4	*Camellia sinensis*	Tocklai (India), TRI (Sri Lanka)	2 800	Covers Assam, China, Cambodia types
5	*Coffea arabica / C. canephora*	CATIE (Costa Rica), ICO collections	> 5 000	Most comprehensive coffee germplasm
6	*Theobroma cacao*	ICQC (Trinidad), CRIG (Ghana)	3 200	Safety-duplicate cryopreserved samples
7	*Anacardium occidentale*	NRC Cashew (India), EMBRAPA (Brazil)	800	Diverse landraces and hybrids

19. Concept and Significance of Genetic Divergence

19.1 Definition

Genetic divergence refers to the degree of genetic differentiation among individuals, populations, or taxa within a species. It represents accumulated variation caused by mutation, recombination, selection, drift, and isolation. In plantation crops, which are perennial and often cross-pollinated, divergence occurs both **within populations** (intrapopulation) and **between populations or regions** (inter-population or inter-regional).

19.2 Importance in Breeding and Conservation

Genetic divergence analysis is fundamental to:

1. **Parental selection** – Highly divergent parents are preferred for hybridization to exploit heterosis.
2. **Core-collection development** – Ensures representation of all genetic groups.
3. **Evolutionary inference** – Reveals phylogeographic relationships and domestication pathways.
4. **Conservation prioritization** – Identifies unique or rare gene pools requiring urgent protection.

For long-lived tree crops, divergence studies also guide clonal exchange, germplasm quarantine, and region-specific breeding.

20. Sources and Causes of Genetic Divergence

20.1 Mutation and Recombination

Continuous accumulation of point and structural mutations creates allelic variants. Recombination reshuffles these alleles during meiosis. Over centuries of cross-pollination, this process generates **micro-evolutionary differentiation**, for example, numerous coconut ecotypes adapted to coastal versus inland habitats.

20.2 Geographic and Reproductive Isolation

20.2.1 Geographic Isolation

Island, mountain, or continental barriers restrict gene flow, leading to **allopatric divergence**. *Example:* Pacific and Indo-Atlantic coconut populations show distinct chloroplast haplotypes due to oceanic separation. Similarly, African and Asian oil-palm populations represent independent gene pools.

20.2.2 Reproductive Isolation

Differences in flowering time, floral morphology, or incompatibility systems impede interbreeding (e.g., *Coffea arabica* vs. *C. canephora*). These barriers gradually lead to reproductive isolation and the formation of subspecies.

20.3 Selection and Adaptation

Natural and human selection drive divergence through local adaptation:

» **Environmental gradients** (temperature, rainfall, soil) shape physiological and morphological variation.

» **Cultural selection**, farmers favour particular nut or cup quality, accelerating phenotypic differentiation.

» **Directional breeding** for yield or quality may narrow diversity within breeding lines but increase divergence among regions.

20.4 Genetic Drift and Founder Effects

Repeated bottlenecks, such as the rubber seed transfer from the Amazon to Asia, eliminate alleles randomly, fixing unique combinations and establishing divergent clonal lineages. Small population size intensifies drift, as in early cocoa and coffee introductions.

20.5 Polyploidy and Hybridization

Genome duplication or hybridization introduces dosage effects and novel gene interactions. The allotetraploid *Coffea arabica* differs substantially from its diploid progenitors, representing a classic example of **instantaneous divergence**.

21. Assessment of Genetic Divergence

21.1 Morphological and Agronomic Approaches

Traditional breeders quantify divergence using multivariate statistics applied to morphological and yield traits.

21.1.1 Mahalanobis D^2 Statistics

Widely used in plantation crops to cluster genotypes based on multiple quantitative traits (nut weight, leaf area, yield components). *Example:* D^2 analysis in *Areca catechu* grouped 60 accessions into five clusters, revealing that inter-cluster crosses produced superior heterotic progenies.

21.1.2 Principal Component Analysis (PCA)

PCA reduces complex trait data to principal axes representing maximum variance. In coconut, the first three PCs often explain >70 % of total variation, primarily linked to fruit size and husk thickness.

21.1.3 Cluster and Canonical Analysis

Hierarchical or non-hierarchical clustering (UPGMA, Tocher's method) delineates genetic groups. Canonical variate analysis further discriminates populations and identifies key differentiating traits.

21.2 Biochemical Markers

Isozyme polymorphism reflects allelic variation at enzyme-coding loci:

- **Esterase and peroxidase patterns** distinguish tall and dwarf coconuts.
- **Polyphenol oxidase profiles** separate Assam and China teas.
- **Latex-protein electrophoresis** differentiates *Hevea* clones.

These biochemical markers are inexpensive and useful for rapid screening where molecular laboratories are limited.

21.3 Molecular Marker Techniques

Molecular data provide the most precise estimates of genetic distances.

Marker System	Application in Plantation Crops	Advantages
RAPD / ISSR	Assess inter-varietal variation in coconut and arecanut	Quick, low cost
SSR (Microsatellites)	Diversity mapping in tea, coffee, rubber	Co-dominant, reproducible
AFLP / SRAP	Germplasm fingerprinting in oil palm, cocoa	Genome-wide coverage
SNP Genotyping	High-resolution population structure in *Hevea*, *Coffea*	High throughput, quantifiable

Genetic distance (Nei's, Jaccard's, or Dice coefficients) and AMOVA (Analysis of Molecular Variance) quantify intra- and inter-population differentiation. Bayesian clustering (STRUCTURE, DAPC) reveals admixture levels and hidden lineages.

22. Patterns of Genetic Divergence in Major Plantation Crops

22.1 Coconut (*Cocos nucifera L.*)

Two principal genetic lineages exist:

1. **Pacific Group:** predominantly dwarf, self-pollinated types; higher oil but smaller nuts.
2. **Indo-Atlantic Group:** tall, cross-pollinated types; larger nuts and greater heterozygosity.

Molecular data suggest independent domestication events from ancestral coastal palms. Gene flow via ocean currents maintained partial connectivity, while localized selection for copra and toddy yield accentuated regional divergence.

22.2 Oil Palm (*Elaeis guineensis Jacq.*)

Divergence between **Guinean, Congolese**, and **Zairean** populations corresponds to African rainforest refugia. Malaysian and Indonesian breeding materials form a separate Asiatic lineage derived from a few African introductions, explaining limited allelic diversity in Asian plantations. Breeding programs now introgress wild *Elaeis oleifera* (South American oil palm) to broaden genetic base and incorporate disease resistance.

22.3 Arecanut (*Areca catechu L.*)

Microsatellite analyses reveal high intra-population variability but low inter-regional divergence across South and Southeast Asia, indicating substantial gene flow through human exchange of seed nuts. Distinct morphotypes (tall vs. semi-tall) reflect adaptive differentiation to local humidity and soil fertility rather than deep genetic separation.

22.4 Rubber (*Hevea brasiliensis Muell. Arg.*)

Because Asian plantations originated from limited Amazonian germplasm, overall diversity is narrow. However, within the Amazon basin, wild populations show profound divergence linked to river-basin isolation (Tapajós, Madeira, Purus). These wild relatives harbor alleles for disease resistance, frost tolerance, and latex quality. Current breeding integrates Amazonian wild clones (e.g., FX 3864, RO 38) into Asian breeding pools to restore variability.

22.5 Tea (*Camellia sinensis (L.) O. Kuntze*)

Three major gene pools:

1. **China Type (*C. sinensis var. sinensis*)** – small-leaved, frost-hardy, high-quality beverage.
2. **Assam Type (*C. sinensis var. assamica*)** – large-leaved, heat-tolerant, high yield.
3. **Cambod Type (*C. sinensis var. assamica subsp. lasiocalyx*)** – intermediate morphology.

Molecular clustering confirms strong differentiation among them, yet extensive hybridization in India, Sri Lanka, and Kenya has produced admixture populations combining desirable attributes.

22.6 Coffee (*Coffea arabica L., C. canephora Pierre ex Froehner*)

Arabica, a self-fertile allotetraploid, shows low diversity and high population structure. *Robusta*, an outcrossing diploid, exhibits broad genetic variation across the Congo Basin. Divergence between them (~1 Myr) predates human domestication. Breeding hybrids such as Catimor and Sarchimor re-combines their genomes, parially restoring diversity in cultivated coffee.

22.7 Cocoa (*Theobroma cacao L.*)

Ten major genetic clusters correspond to Amazonian river systems (Maranon, Nanay, Purus, etc.). Two cultivated groups, **Criollo** and **Forastero**, represent divergent domestication streams; the hybrid **Trinitario** bridges both. SSR and SNP markers confirm high differentiation (FST ≈ 0.35) between Criollo and Forastero populations, reflecting long-term isolation and human selection for flavour versus vigour.

22.8 Cashew (*Anacardium occidentale L.*)

African and Indian germplasm pools diverge significantly. Brazilian wild accessions possess unique alleles for pest resistance and dwarf stature, while Indian accessions show broader phenotypic diversity due to cross-breeding. Genetic distance (Jaccard 0.42–0.67) indicates moderate divergence, adequate for heterotic hybrid development.

23. Quantitative Indices of Genetic Divergence

23.1 Genetic Distance and Similarity Coefficients

Commonly applied indices:

- **Nei's Genetic Distance (D)** – measures accumulated allele-frequency differences.
- **Jaccard's Similarity Index** – used for presence/absence marker data.
- **Rogers' Distance** – useful for quantitative traits. Interpretation: $D < 0.05$ = very close; 0.05–0.25 = moderately divergent; > 0.25 = highly divergent.

23.2 Analysis of Molecular Variance (AMOVA)

Partitions total genetic variance into **within-population** and **among-population** components.

Example: In global coffee collections, AMOVA revealed 65 % of diversity within populations and 35 % among populations, indicating that conservation should emphasize multiple individuals per site rather than numerous sites.

23.3 Fixation Index (FST)

Quantifies population differentiation: FST = (Va)/(Va + Vw). Values near 0 → panmixia; near 1 → complete isolation. Typical FST: *Hevea* 0.12 (moderate), *Cocoa* 0.35 (high), *Coconut* 0.18 (intermediate).

24. Factors Influencing Maintenance or Erosion of Divergence

24.1 Reproductive Biology

Self-compatible species (e.g., *Coffea arabica*) maintain distinct lineages, whereas cross-pollinated ones (e.g., *Cocos nucifera*, *Hevea brasiliensis*) experience continual gene flow reducing differentiation.

24.2 Human Intervention

Monoclonal plantations and global germplasm exchange homogenize alleles, while *localized farmer selection* preserves unique ecotypes. Balancing improvement with diversity conservation remains a challenge.

24.3 Environmental Heterogeneity

Climatic mosaics across tropical landscapes sustain divergent selection pressures, e.g., salinity tolerance in coastal coconuts vs. drought tolerance in inland cashew, thus maintaining regional divergence despite gene flow.

25. Practical Applications of Genetic-Divergence Studies

1. **Hybridization Planning:** Crossing genetically distant parents maximizes heterosis (oil palm Dura × Pisifera; coconut Tall × Dwarf).
2. **Germplasm Management:** Cluster-based sampling avoids redundancy and captures total variation.
3. **Evolutionary Breeding:** Combines controlled gene flow with natural selection to generate dynamic, diverse populations.
4. **Molecular Breeding and Genomics:** Genome-wide association studies (GWAS) use divergent panels to identify alleles controlling yield, oil quality, or stress tolerance.

Table 4 – Representative Estimates of Genetic Divergence in Major Plantation Crops

S. No.	Crop	Marker Type / Approach	Mean Genetic Distance (D or FST)	Remarks / Interpretation
1	*Cocos nucifera*	SSR markers	0.18	Moderate divergence between Pacific and Indo-Atlantic groups
2	*Elaeis guineensis*	AFLP	0.22	Distinct African vs. Asian lineages
3	*Hevea brasiliensis*	SSR + SNP	0.12	Low diversity in Asian clones; high in Amazon wilds
4	*Camellia sinensis*	ISSR	0.27	Strong differentiation among China, Assam, Cambod types
5	*Coffea arabica / C. canephora*	SNP	0.30	Deep species-level divergence
6	*Theobroma cacao*	SSR	0.35	High structure among Amazonian clusters
7	*Anacardium occidentale*	RAPD	0.20	Moderate divergence; potential for heterosis

26. Integration of Morphological and Molecular Data

Combining phenotypic and genotypic information enhances precision:

- » **Morpho-molecular correlations** identify QTL-linked descriptors (e.g., nut weight ↔ specific SSR allele in coconut).
- » **Multivariate data integration** through canonical correlation or Mantel tests reveals congruence between trait and marker-based distances.
- » **Molecular taxonomy** supports revision of species complexes (e.g., *Camellia* species boundaries).

Such integrative frameworks underpin **molecular-assisted germplasm management**.

Cytogenetics of Plantation Crops

28. Introduction to Cytogenetics in Plantation Crops

28.1 Concept and Scope

Cytogenetics is the branch of genetics that studies the structure, function, behavior, and evolution of chromosomes. In plantation crops, long-lived perennials with complex reproductive systems, cytogenetic knowledge is essential for understanding:

1. **Chromosome organization** and ploidy level.
2. **Mechanisms of inheritance** and gene expression.
3. **Hybrid fertility, sterility, and incompatibility phenomena.**

4. **Evolutionary relationships** among wild and cultivated taxa.

Because many plantation species are highly heterozygous, polyploid, or propagated vegetatively, classical Mendelian ratios rarely appear. Cytogenetic studies provide the foundation for successful hybridization, mutation breeding, and biotechnology interventions.

28.2 Historical Perspective

Early cytological investigations during the early 20th century focused on determining chromosome numbers:

» *Hevea brasiliensis* (2n = 36), *Cocos nucifera* (2n = 32), *Elaeis guineensis* (2n = 32), *Camellia sinensis* (2n = 30), *Coffea arabica* (2n = 44), *C. canephora* (2n = 22), *Theobroma cacao* (2n = 20), and *Anacardium occidentale* (2n = 42).*Advances in fluorescence staining, flow cytometry, and molecular cytogenetics (FISH, GISH) later clarified chromosomal rearrangements, genome size, and polyploidization events.

29. Chromosome Number and Karyotype Architecture

29.1 Chromosome Numbers of Major Plantation Crops

S. No.	Crop	Somatic Chromosome Number (2n)	Karyotype Features	Remarks
1	*Cocos nucifera L.*	32	8 pairs metacentric + 8 sub-metacentric	Stable diploid; no polyploidy recorded
2	*Elaeis guineensis Jacq.*	32	Mostly metacentric; large chromosomes	Diploid; minor size variation
3	*Areca catechu L.*	32	Medium-sized, symmetrical	High karyotype stability
4	*Hevea brasiliensis Muell. Arg.*	36	Metacentric–submetacentric	Diploid; some aneuploids
5	*Camellia sinensis (L.) O. Kuntze*	30	Metacentric–submetacentric	Diploid; triploid & tetraploid forms known
6	*Coffea arabica L.*	44	Allotetraploid (2n = 4x = 44)	Hybrid of *C. eugenioides* × *C. canephora*
7	*Coffea canephora Pierre ex Froehner*	22	Diploid	High heterozygosity
8	*Theobroma cacao L.*	20	Small, symmetric chromosomes	Stable diploid
9	*Anacardium occidentale L.*	42	Predominantly submetacentric	Diploid, relatively large genome

29.2 Karyotype Symmetry and Evolution

Karyotype studies reveal that most plantation crops possess **symmetrical karyotypes** (low asymmetry index), indicating chromosomal stability during evolution. Exceptions include *Camellia sinensis* and *Anacardium occidentale*, which show slight asymmetry, reflecting structural rearrangements contributing to diversification.

Karyotype symmetry correlates with evolutionary advancement: symmetrical karyotypes denote primitive, stable genomes, whereas asymmetry signals adaptive evolution via inversions or translocations.

30. Polyploidy and Genome Duplication

30.1 Significance of Polyploidy

Polyploidy, presence of more than two complete chromosome sets, is a central force in plant evolution, generating novel gene interactions and buffering deleterious mutations. In plantation crops, polyploidy contributes to:

1. Hybrid vigour and physiological robustness.
2. **Increased cell size**, enhancing leaf area or fruit volume.
3. **Enhanced stress tolerance and secondary-metabolite synthesis.**

30.2 Polyploidy in Coffea

» *Coffea arabica* is an **allotetraploid (2n = 4x = 44)** derived from natural hybridization between *C. canephora* (2n = 22) and *C. eugenioides* (2n = 22).
  - Combines high cup quality (from *C. eugenioides*) and disease resistance (from *C. canephora*).
  - Allopolyploid genome confers stable fertility but limits further cross-compatibility due to ploidy barriers.

30.3 Polyploidy in Camellia

Tea displays diploid (2n = 30), triploid (2n = 45), and tetraploid (2n = 60) forms. Triploids exhibit larger leaves, thicker cuticles, and higher caffeine content but partial sterility. Induced polyploidy through colchicine treatment has produced vigorous clones suitable for mechanized plucking.

30.4 Polyploidy in Hevea and Anacardium

Spontaneous or induced polyploids occur in *Hevea brasiliensis* and *Anacardium occidentale*. Tetraploid Hevea plants show increased latex-vessel number and thicker bark but reduced fertility, useful in clonal propagation. In cashew, colchicine-derived autotetraploids possess larger nuts and leaves with enhanced drought tolerance.

30.5 Absence of Polyploidy in Certain Crops

Cocos nucifera and *Elaeis guineensis* remain strictly diploid. Their reproductive biology and long life cycles restrict the survival of polyploid embryos. Instead, heterozygosity and hybridization, not polyploidy, maintain variability.

31. Chromosomal Aberrations and Structural Changes

31.1 Structural Aberrations

Rearrangements, translocations, inversions, duplications, and deletions, play adaptive roles:

» *Camellia sinensis* exhibits paracentric inversions linked with leaf-morphology variation.

» In *Hevea*, reciprocal translocations between homologous chromosomes occasionally produce partial sterility, detected through irregular bivalent formation during meiosis.

» *Coffea* hybrids reveal segmental exchanges stabilizing allopolyploid genomes.

31.2 Aneuploidy

Aneuploid forms (2n ± 1 or 2) occasionally arise in *Hevea* and *Camellia.* Though often sterile, they serve in gene-dosage studies. In *Hevea*, trisomics manifest altered leaf venation and latex-yield variability.

31.3 Chromosome Breakage and Mutagenesis

Ionizing radiation or chemical mutagens induce chromosomal breaks leading to translocations and deletions. Mutation breeding programs in *Coffea* and *Anacardium* exploit controlled breakage to generate novel variation.

32. Meiotic Behavior and Fertility Relationships

32.1 Normal Meiosis

Diploid plantation crops usually exhibit regular meiosis with 100 % bivalent formation and balanced segregation. For instance, *Cocos nucifera* shows 16 bivalents with rare abnormalities, explaining its high fertility and genetic stability.

32.2 Meiotic Irregularities in Hybrids

Hybridization between divergent species often produces meiotic irregularities:

» In *Hevea brasiliensis* × *H. benthamiana*, univalents and laggards reduce pollen viability (~35 %).

» *Coffea arabica* × *C. canephora* hybrids display quadrivalents and non-disjunction due to differing chromosome structures.

» In *Camellia*, inter-varietal crosses may yield triploids with sterile pollen.

32.3 Cytomixis and Chromatin Transfer

Cytomixis, inter-cellular chromatin migration during meiosis, has been reported in *Areca catechu* and *Camellia sinensis*. It may generate gametes with variable chromosome numbers, contributing to micro-variation and potential polyploid formation.

33. Molecular Cytogenetics and Genome Organization

33.1 Fluorescence *In Situ* Hybridization (FISH)

FISH mapping employs labeled DNA probes to visualize specific loci on chromosomes:

» rDNA (45S and 5S) sites in *Coffea arabica* clarify parental genome composition.

» Telomeric probes in *Hevea* reveal conserved terminal sequences without interstitial repeats.

» In tea, FISH identifies chromosomes carrying caffeine-synthase genes.

33.2 Genomic *In Situ* Hybridization (GISH)

GISH differentiates parental genomes in hybrids and polyploids:

» Applied to *Coffea arabica*, it confirmed genomic contributions from *C. eugenioides* and *C. canephora*.

» In *Hevea* interspecific hybrids, GISH distinguishes chromosomal introgression from wild species.

33.3 Flow Cytometry and Genome Size Estimation

Flow-cytometric analysis quantifies nuclear DNA content:

Crop	Approx. Genome Size (pg/2C)	Remarks
Cocos nucifera	5.7	Medium genome size
Elaeis guineensis	3.9	Compact diploid genome
Hevea brasiliensis	4.4	Moderate size; repetitive DNA ≈ 35 %
Camellia sinensis	6.2	Large genome with polyphenol-gene families
Coffea arabica	2.9	Small allotetraploid genome
Theobroma cacao	1.1	Compact diploid genome

Comparative genome sizes help calibrate sequencing projects and cytological mapping.

34. Cytoplasmic Inheritance and Organelle Genomes

34.1 Mitochondrial and Chloroplast DNA

Cytoplasmic genomes influence traits such as male sterility, photosynthetic efficiency, and disease tolerance. Maternally inherited chloroplast markers trace phylogeny:

» Coconut chloroplast haplotypes differentiate Pacific and Indo-Atlantic lineages.

» Coffee cpDNA identifies maternal parents in hybrids.

» In *Hevea*, mtDNA rearrangements associate with latex-flow physiology.

34.2 Cytoplasmic–Nuclear Interactions

Hybrid sterility sometimes arises from cytoplasmic–nuclear incompatibility. Crosses between distinct coffee species occasionally fail due to mismatched organelle–nucleus signaling, emphasizing the need for cytoplasmic characterization in breeding.

35. Applications of Cytogenetic Knowledge in Breeding

35.1 Hybrid Identification and Validation

Chromosome counts, meiotic pairing, and marker-aided cytology verify hybridity, ensuring purity of breeding populations (e.g., Dura × Pisifera oil-palm hybrids).

35.2 Induced Polyploidy and Mutation Breeding

Colchicine or oryzalin treatments generate polyploids with altered morphology. Cytological screening identifies stable chromosomal configurations suitable for commercialization.

35.3 Chromosome Engineering

Emerging techniques like **chromosome elimination, translocation breeding**, and **haploid induction** (through anther culture in tea) allow targeted manipulation of chromosome segments carrying valuable genes.

36. Cytogenetics of Specific Plantation Crops

36.1 Coconut (*Cocos nucifera L.*)

Stable diploid genome (2n = 32). Cytogenetic stability explains its long-term clonal consistency. Crosses between tall and dwarf forms exhibit normal meiosis, though slight pollen-fertility reduction occurs due to chromosome structural variation. Somaclonal variation from tissue culture rarely alters chromosome number, but sub-chromosomal changes affect oil content genes.

36.2 Oil Palm (*Elaeis guineensis Jacq.*)

Diploid (2n = 32) with uniform chromosomes. Cytological analysis reveals frequent heterochromatic regions associated with oil-biosynthetic gene clusters. Meiosis is regular in intraspecific crosses, but hybrids with *E. oleifera* (2n = 32) show delayed pairing and partial sterility, useful for introgressing disease resistance.

36.3 Rubber (*Hevea brasiliensis Muell. Arg.*)

Diploid (2n = 36). High chromosomal stability maintains clonal fidelity. However, interspecific hybrids often display multivalent formations, causing variable fertility. Latex yield differences correlate with chromosomal rearrangements affecting genes controlling laticifer development.

36.4 Tea (*Camellia sinensis (L.) O. Kuntze*)

Cytogenetically diverse: diploid, triploid, tetraploid forms. Triploids show 30 bivalents + 15 univalents in meiosis; seed set ≈ 20 %. Polyploidization increases leaf size and caffeine concentration. FISH mapping reveals caffeine-synthase genes on chromosomes 5 and 12. Triploids serve as rootstocks and breeding bridges.

36.5 Coffee (*Coffea arabica L., C. canephora Pierre ex Froehner*)

C. arabica – allotetraploid (2n = 44); exhibits regular bivalent pairing (22 II). *C. canephora* – diploid (2n = 22). Crosses between them yield triploids (2n = 33) with low fertility. GISH has confirmed chromosome homology between parents. Cytogenetic studies aid introgression of disease-resistance genes while minimizing sterility.

36. Cytogenetics of Specific Plantation Crops *(continued)*

36.6 Cocoa (*Theobroma cacao L.*)

Diploid (2n = 20) with very small chromosomes (≈1–2 μm), which long hampered classical karyotyping. Differential fluorochrome staining and FISH now reveal modest heterochromatin blocks distinguishing **Criollo** and **Forastero** types. Meiosis is generally regular (10 II), supporting high seed set in compatible matings. Somatic embryogenesis does not usually alter chromosome number, but prolonged callus phases may induce subtle segmental changes, warranting **flow cytometry checks** before field release. Inter-specific hybridization with *T. grandiflorum* (cupuaçu) requires embryo rescue; partial sterility reflects structural differences and post-zygotic barriers.

36.7 Cashew (*Anacardium occidentale L.*)

Diploid (2n = 42). Karyotype slightly asymmetric with several submetacentrics; satellites occur variably. Meiosis is regular in most varieties, but **hybrid sterility** can arise from asynchrony in floral phenology and sporophytic self-incompatibility (covered later). Colchicine-induced tetraploids show larger stomata and thicker leaves, with enhanced drought tolerance and nut size; **pollen fertility declines**, so such lines are used mainly as parents in **triploid breeding** or as vigor-inducing rootstocks.

37. Practical Cytogenetic Tools and Protocols

37.1 Chromosome Preparation and Staining

1. **Pretreatment:** ice-cold water or 0.002 M 8-hydroxyquinoline for 2–4 h to accumulate metaphases (root tips or young floral buds).
2. **Fixation:** Carnoy's solution (ethanol:acetic acid, 3:1) at 4 °C.
3. **Enzymatic maceration:** pectinase/cellulase mix for tough tissues (e.g., *Hevea* bark meristems).
4. **Staining:** aceto-orcein, aceto-carmine, Feulgen; DAPI or propidium iodide for fluorescence.
5. **Imaging:** epifluorescence microscopy with image analysis for arm-ratio and asymmetry indices.

Tip for tiny-chromosome species (cocoa): use **air-dry squash**, high-magnification objectives, and **FISH** to anchor rDNA or telomeric repeats.

37.2 Flow Cytometry for Ploidy/Genome Size

Nuclei isolated from young leaves are stained (propidium iodide) alongside an internal standard (e.g.,

Pisum sativum). DNA content histograms quantify **ploidy shifts** (triploids in tea, tetraploids in coffee callus) and detect **mixoploidy** in regenerated plants. Regular screening is advised for tissue-culture pipelines in coconut, tea, and cocoa.

37.3 FISH and GISH Workflows

- **FISH:** map 45S/5S rDNA, telomeres, and trait-linked BACs.
- **GISH:** differentiate parental genomes in *Coffea* allotetraploids or *Hevea* interspecific hybrids, confirming **introgression tracts**.
- **Applications:** verify wide crosses; detect cryptic translocations; guide backcrossing by tracking alien segments.

37.4 Meiotic Index and Fertility Metrics

- **Meiotic index (MI):** % normal tetrads; values >90% indicate cytological stability.
- **Pollen viability:** acetocarmine or FDA staining; **in vitro germination** on sucrose–boric acid media adds functional validation.
- **Use in breeding:** screen parents and F1s (e.g., *Hevea*, cashew) to anticipate seed set and field fertility.

38. Cytogenetic Constraints in Hybridization and Their Management

38.1 Homology and Pairing Control

Low homology leads to univalents, laggards, and bridges. **Strategies:**

- Choose **phylogenetically closer parents** (SSR/SNP distance + GISH pre-screen).
- Use **bridge crosses** (e.g., *Coffea eugenioides* bridging *arabica* × *canephora*).
- Employ **chromosome doubling** (colchicine/orzyalin) to restore pairing in wide hybrids.

38.2 Endosperm Imbalance and Post-zygotic Barriers

Distant crosses (e.g., *Theobroma cacao* × *T. grandiflorum*) fail due to endosperm imbalance. Rescue with:

- **Early embryo culture** (8–14 days post-pollination).
- **Nurse endosperm** or osmoticum-rich media.
- **Backcrossing + marker tracking** to retain target segments while restoring fertility.

38.3 Cytoplasmic–Nuclear Incompatibilities

Mitochondrial/chloroplast–nuclear mismatches can reduce vigor or cause male sterility (observed sporadically in *Coffea* inter-specifics). Maintain **cytoplasm-diagnostic markers** (cpSSR) to select optimal maternal parents and avoid maladapted cytoplasms.

39. Doubled Haploids and Androgenesis Prospects

Although tree crops are recalcitrant, **anther/ microspore culture** has shown success in **tea** (androgenic callus → green plants at low frequency). Doubled haploids (DH) would greatly accelerate **fixation** of recessives and **QTL mapping**, but optimization is crop-specific:

» Pretreatments: cold shock (4 °C), mannitol, heat shock (32–35 °C).

» Induction media: NAA + kinetin; activated charcoal for phenolic control (tea, cocoa).

» Chromosome doubling: antimitotics applied during callus proliferation.

40. Somaclonal Variation: Cytogenetic Basis and Use

Tissue culture can produce **chromosomal rearrangements** and **aneuploidy**, generating phenotypic novelties:

» **Coconut:** rare segmental variants affecting oil traits, screen via SSR + flow cytometry before release.

» **Hevea:** somaclones with altered laticifer density; validate with **meiotic index** and **latex protein profiles**.

» **Breeder guidance:** treat somaclonal lines as **pre-breeding material**; stabilize through clonal bulking after multi-site validation.

41. Integrating Cytogenetics with Genomics

41.1 Cytogenetic Anchoring of Genomes

High-quality genomes (oil palm, cacao, coffee) benefit from **chromosome-scale scaffolding** aligned with FISH markers (rDNA, telomeres) and **Hi-C** contact maps. Karyotype-informed assembly helps:

» Resolve **inversions/translocations** affecting recombination.

» Define **recombination deserts** (pericentromeres) vs **hotspots**, guiding marker placement and **MAS/GWAS**.

41.2 QTL–Cytology Convergence

» In *Hevea*, QTLs for latex yield coincide with regions showing **heterochromatin polymorphism**; breeders can target **high-pairing** haplotypes to enhance meiotic recombination around yield QTLs.

» In tea, caffeine-synthase loci mapped by FISH align with SNP-GWAS peaks for alkaloid content, useful for marker-assisted pyramiding without adverse cytological effects (e.g., translocation-induced sterility).

42. Breeding Playbooks Informed by Cytogenetics

42.1 Parent Choice and Crossing Blocks

» **Check ploidy and MI** of candidate parents; avoid low-MI clones as pollen donors.

» For **wide crosses** (cocoa, coffee, Hevea), perform **GISH pre-screens** to estimate homology; prefer combinations with >80% inferred homology.

» In coconut, maintain **Tall × Dwarf** crossing blocks with staggered phenology to maximize synchronous meiosis and viable pollen output.

42.2 Managing Hybrid Sterility

» Use **chromosome doubling** to restore bivalent pairing in inter-specific F1s (coffee).

» **Backcross** to the recurrent parent while tracking alien introgressions with SSR/SNPs; cull individuals showing **meiotic irregularities** (>10% univalents/bridges).

42.3 Clonal Deployment and Diversity

» Deploy **polyclonal mixtures** (Hevea, tea) to buffer against cytogenetic vulnerabilities; avoid single-clone landscapes that accentuate random aneuploid sectors.

» **Rotation of pollen parents** preserves recombination diversity across planting seasons.

43. Common Cytogenetic Pitfalls and Troubleshooting

S. No.	Problem	Likely Cytogenetic Cause	Actionable Fix
1	Poor seed set in wide crosses	Univalents/laggards; endosperm imbalance	Chromosome doubling; early embryo rescue
2	Off-type somaclones	Aneuploidy or translocations	Flow cytometry + SSR screening; shorten culture cycles
3	Low pollen viability	Meiotic irregularity; environmental stress during meiosis	Shift flowering window; select high-MI clones
4	Sterile triploids (tea)	Unbalanced pairing	Use as **male-sterile female** for controlled pollination; backcross with diploids
5	Unexpected male sterility in interspecifics	Cytoplasmic–nuclear incompatibility	Change maternal parent; cpDNA typing

44. Summary Table – Cytogenetic Features of Major Plantation Crops

S. No.	Crop	2n	Ploidy/ Cytotype	Notable Cytogenetic Traits	Breeding Implications
1	*Cocos nucifera*	32	Diploid	Very stable meiosis; rare aberrations	Reliable seed fertility; prioritize phenology for Tall × Dwarf
2	*Elaeis guineensis*	32	Diploid	Uniform karyotype; heterochromatin blocks	Hybrid purity checks; potential GISH in *E. oleifera* introgression
3	*Areca catechu*	32	Diploid	Occasional cytomixis	Screen pollen viability; maintain high-MI parents

4	*Hevea brasiliensis*	36	Diploid; aneuploids seen	Translocations in wide hybrids	GISH-guided introgression; monitor MI in F1s
5	*Camellia sinensis*	30	2x, 3x, 4x	Triploids with partial sterility; FISH-mapped caffeine genes	Use triploids for vigor/ mechanization; MAS for quality QTLs
6	*Coffea arabica*	44	Allotetraploid	Regular 22 II pairing; cytoplasm effects	Bridge crosses; cytoplasm typing
7	*C. canephora*	22	Diploid	High heterozygosity	Pollen donors to restore vigor and resistance
8	*Theobroma cacao*	20	Diploid	Tiny chromosomes; FISH resolves types	Embryo rescue for interspecifics; verify ploidy in somatic embryos
9	*Anacardium occidentale*	42	Diploid; induced 4x	Slight asymmetry; colchicine 4x lines	Use 4x as parents/ rootstocks; screen fertility

Variations and Natural Selection in Plantation Crops

46. Concept of Variation

46.1 Definition and Scope

Variation refers to the differences in morphological, physiological, biochemical, and genetic traits among individuals of the same species or between related species. In plantation crops, variation is expressed at multiple levels, from differences in leaf shape, flower structure, fruit size, and biochemical composition to molecular polymorphisms in DNA sequences. Because these crops are long-lived perennials, their cumulative exposure to environmental heterogeneity, mutations, and recombination across generations makes variation an indispensable asset for adaptation and breeding.

46.2 Types of Variation

1. **Somatic Variation** – Occurs in vegetative tissues; perpetuated through clonal propagation (e.g., bud sports in *Camellia sinensis*).
2. **Germinal Variation** – Arises in reproductive cells; heritable and the chief raw material for evolution.
3. **Continuous Variation** – Quantitative traits (yield, oil %, caffeine content) controlled by many genes.
4. **Discontinuous Variation** – Qualitative traits (flower colour, leaf pubescence) governed by major genes.
5. **Induced Variation** – Generated via mutagenesis, polyploidy, or tissue-culture stress.

Understanding the origin and maintenance of these variations enables scientists to manage genetic resources strategically.

47. Sources of Variation in Plantation Crops

47.1 Mutation

Mutations create new alleles through base substitutions, insertions, deletions, or chromosomal rearrangements. Examples:

- » Somatic mutations in *Hevea brasiliensis* have yielded clones such as RRIM 600 and PB 235 with superior latex flow.
- » Spontaneous dwarfism in *Cocos nucifera* generated the Chowghat Dwarf group.
- » Chemical mutagenesis in *Coffea arabica* produced rust-resistant Mokka-type mutants.

Mutation rates are low ($\approx 10^{-5} - 10^{-6}$ per locus per generation), yet their cumulative impact over centuries has enriched diversity.

47.2 Recombination

During meiosis, crossing-over between homologous chromosomes reshuffles alleles, producing new genotypic combinations. In highly heterozygous, cross-pollinated crops such as *Elaeis guineensis* and *Hevea*, recombination is a dominant evolutionary force. Controlled pollination in oil palm deliberately exploits recombination between *Dura* and *Pisifera* lines to create the heterotic *Tenera* hybrid.

47.3 Polyploidy and Chromosomal Variation

Polyploidy and aneuploidy alter gene dosage, enzyme activity, and phenotypic expression. Triploid tea shows higher biomass; autotetraploid cashew produces larger nuts. Such cytogenetic changes introduce novel variation while maintaining vigor through gene redundancy.

47.4 Epigenetic Variation

Epigenetic modifications, DNA methylation, histone acetylation, and small RNA activity, regulate gene expression without altering sequence. Environmental cues like drought or shading can induce heritable methylation patterns affecting caffeine metabolism in tea or oil biosynthesis in palm, representing a **reversible adaptive variation** crucial for rapid climate responses.

47.5 Somaclonal Variation

Plant tissue culture can trigger genetic and epigenetic instability. Somaclones of *Hevea* and *Cocos* often differ in growth rate or oil content. Though unpredictable, these variants serve as pre-breeding materials when screened cytologically and molecularly for stability.

48. Natural Selection as a Creative Force

48.1 Definition

Natural selection is the differential survival and reproduction of individuals due to heritable differences in fitness. It acts upon existing variation, increasing the frequency of advantageous alleles and reducing deleterious ones.

48.2 Forms of Natural Selection

1. **Stabilizing Selection** – Favours intermediate phenotypes; e.g., moderate nut size in coconut ensures both buoyancy and copra yield.
2. **Directional Selection** – Favours one extreme; e.g., drought-tolerant genotypes of cashew under arid stress.
3. **Disruptive Selection** – Favours both extremes, promoting polymorphism; e.g., tall vs. dwarf coconuts maintained under differing rainfall regimes.
4. **Balancing Selection** – Maintains multiple alleles; common in heterozygote-advantage systems like disease resistance in coffee.

48.3 Selective Agents in Plantation Systems

» **Abiotic:** temperature, rainfall, soil salinity, altitude.

» **Biotic:** pathogens (*Phytophthora palmivora* in cocoa, *Hemileia vastatrix* in coffee), pollinators, herbivores.

» **Anthropogenic:** management practices (fertilization, shade, pruning), chemical stress, and continuous clonal propagation.

49. Interaction Between Variation and Selection

Variation supplies raw diversity; selection filters it. In long-generation perennials, even slight selective advantages compound over decades.

Example: In a coconut population exposed to coastal salinity, seedlings with superior ion-exclusion genes survive and reproduce, gradually shifting allele frequencies toward salt tolerance, a micro-evolutionary adaptation observable within plantation timescales.

50. Case Studies of Variation and Selection

50.1 Coconut (*Cocos nucifera L.*)

» **Natural variation:** tall vs. dwarf ecotypes differ in flowering age, stature, and nut morphology.

» **Selection:** farmers historically retained palms with high copra yield; environmental selection favoured salt-tolerant rootstocks.

» **Outcome:** formation of regional cultivars such as West Coast Tall, Andaman Ordinary, and Chowghat Dwarfs representing ecological clines.

50.2 Oil Palm (*Elaeis guineensis Jacq.*)

» Natural recombination and farmer selection produced Dura, Pisifera, and Tenera fruit forms.

» Industrial breeding applied **recurrent selection**, increasing oil yield from 3 t ha^{-1} (1960s) to >6 t ha^{-1} today.

» Selection pressure under intensive management continues to favour high photosynthetic efficiency and low unsaturated-fat fractions.

50.3 Rubber (*Hevea brasiliensis Muell. Arg.*)

In its Amazonian home, natural selection shaped clones resistant to *Microcyclus ulei* (South American Leaf Blight). In Asia, breeding populations experienced **artificial directional selection** for latex yield, resulting in convergent phenotypes but narrower genetic bases , a cautionary example of **selection-induced erosion**.

50.4 Tea (*Camellia sinensis (L.) O. Kuntze*)

Continuous plucking constitutes a chronic selection regime favouring plants with rapid shoot regeneration and pest tolerance. Over centuries, these pressures differentiated **China**, **Assam**, and **Cambod** types, each optimized to distinct micro-climates.

50.5 Coffee (*Coffea arabica L.* and *C. canephora Pierre ex Froehner*)

Wild *C. arabica* evolved under shaded montane forests , traits like self-compatibility and limited branching enhance survival in low light. Human selection for cup quality maintained these characters even when cultivation shifted to open sun systems. Conversely, *C. canephora* remained an out-crosser adapted to humid lowlands with open canopies.

50.6 Cocoa (*Theobroma cacao L.*)

Pathogen pressure from *Phytophthora spp.* and *Moniliophthora roreri* has driven co-evolutionary selection for polygenic resistance. Modern "Trinitario" hybrids embody accumulated adaptive variation from both Criollo (flavour) and Forastero (vigor) ancestries.

50.7 Cashew (*Anacardium occidentale L.*)

Selection under drought and nutrient-poor soils yielded genotypes with high osmotic adjustment, waxy leaf cuticles, and efficient stomatal regulation. Farmers' preference for large nuts reinforced these drought-adapted phenotypes, illustrating **joint natural × artificial selection**.

51. Maintenance of Variation

51.1 Heterozygosity and Cross-Pollination

Most plantation crops are highly heterozygous due to cross-pollination, which maintains numerous alleles at low frequencies. In *Hevea* and *Cocos*, heterozygosity underpins yield stability and broad adaptability.

51.2 Gene Flow

Migration of pollen or seed across plantations maintains allelic continuity, preventing fixation and ensuring long-term evolutionary potential. However, excessive gene flow may homogenize populations, reducing local adaptation; thus managed **isolation blocks** are used in breeding gardens.

51.3 Mutation–Selection Balance

Low-frequency deleterious mutations persist because selection cannot purge them faster than they appear, a key source of latent variation that breeders can recombine into beneficial backgrounds.

51.4 Balancing and Frequency-Dependent Selection

In coffee, pathogen races vary; rare resistance alleles confer advantage until pathogens adapt, maintaining polymorphism, an ecological arms race ensuring enduring variation.

52. Environmental and Epigenetic Influences on Variation

52.1 Phenotypic Plasticity

Perennial crops display high **plasticity**, ability of a single genotype to produce different phenotypes under varying environments. For instance, the same coconut cultivar differs in leaf size, internode length, and nut yield when grown in coastal versus inland sites. Such plasticity mitigates environmental stress and enhances survival.

52.2 Epigenetic Memory and Transgenerational Effects

Stress-induced epigenetic marks (methylation) can persist through vegetative propagation. *Camellia sinensis* plants exposed to drought exhibit up-regulated methylation of aquaporin genes; the effect is retained in cuttings, constituting **epigenetic inheritance**, a rapid, non-genetic adaptation mechanism.

53. Artificial Selection and Domestication Revisited

Human choice interacts with natural selection to mould genetic architecture.

S. No.	Selective Force	Typical Target Trait	Example Crops	Outcome
1	Farmer mass selection	Fruit/nut size, yield	*Cocos*, *Anacardium*	Larger reproductive organs
2	Institutional breeding	Oil %, disease resistance	*Elaeis*, *Hevea*	Enhanced productivity
3	Clonal selection	Latex flow, flavour	*Hevea*, *Camellia*	Uniform elite clones
4	Molecular/ marker-assisted	Quality genes, stress alleles	*Coffea*, *Cocoa*	Targeted trait introgression

Artificial selection typically acts **directionally**, while natural selection maintains **stabilizing diversity**, their balance determines evolutionary resilience.

54. Quantitative Assessment of Variation and Selection

54.1 Statistical Indices

1. **Coefficient of Variation (CV %)** – quantifies phenotypic dispersion; >20 % indicates exploitable variability.
2. **Heritability (h^2)** – ratio of genetic to total variance; high h^2 traits respond swiftly to selection (e.g., oil content in palm).
3. **Genetic Advance (GA)** – expected improvement per selection cycle.

4. **Selection Differential (S)** – mean of selected individuals minus population mean.

Combined, these metrics guide breeders in predicting response to selection ($R = h^2 \times S$).

54.2 Molecular Indicators of Selection

Population-genetic parameters (π, Tajima's D, FST outlier tests) identify genomic regions under selection. In *Hevea*, QTLs for latex-flow enzymes (β-1,3-glucanase) show selective sweeps; in coffee, caffeine-biosynthesis genes display balancing selection reflecting consumer-driven preferences.

55. Adaptive Radiation and Speciation Trends

Over evolutionary timescales, accumulated variation and localized selection yield **adaptive radiation**, diversification into ecological niches:

» The genus *Camellia* expanded into >100 species across Asia, each adapted to distinct altitudes and light regimes.

» Wild *Coffea* species diversified across Africa following Pleistocene climatic oscillations.

» Hevea* species radiated along Amazonian river systems under hydrological selection gradients.

Such radiation underlies the modern gene pools used in breeding.

56. Conservation of Variability Under Selective Breeding

56.1 Risk of Genetic Erosion

Intensive clonal propagation (rubber, tea) and high-input monocultures compress genetic diversity, increasing vulnerability to epidemics , as illustrated by the catastrophic *Microcyclus ulei* outbreaks in uniform rubber stands.

56.2 Strategies to Retain Variation

1. **Recurrent Selection and Polyclonal Gardens** – periodic mixing of diverse clones.
2. **Participatory Breeding** – engaging farmers maintaining heterogeneous landraces.
3. **Dynamic Conservation** – rotational use of breeding populations allowing natural recombination.
4. **Molecular Monitoring** – annual SSR/SNP profiling to track allelic richness.

57. Table – Representative Forms of Variation and Selective Responses in Plantation Crops

S. No.	Crop	Principal Variation Type	Selective Force	Adaptive Outcome
1	*Cocos nucifera*	Tall vs. dwarf morphotypes	Salinity + farmer selection	Salt tolerance + precocity
2	*Elaeis guineensis*	Fruit shell thickness mutation	Human directional	Tenera hybrid yield gain
3	*Hevea brasiliensis*	Somatic mutation in latex cells	Artificial	High-latex clones

4	*Camellia sinensis*	Polyploid variation	Human + environmental	Triploid vigour
5	*Coffea arabica*	Allopolyploid origin	Natural hybridization	Quality + resistance
6	*Theobroma cacao*	Pathogen-driven allelic diversity	Co-evolution	Disease resistance
7	*Anacardium occidentale*	Morphological & biochemical	Drought + farmer choice	Drought hardiness + large nut

Variations and Natural Selection in Plantation Crops

46. Concept of Variation

46.1 Definition and Scope

Variation refers to the differences in morphological, physiological, biochemical, and genetic traits among individuals of the same species or between related species. In plantation crops, variation is expressed at multiple levels, from differences in leaf shape, flower structure, fruit size, and biochemical composition to molecular polymorphisms in DNA sequences. Because these crops are long-lived perennials, their cumulative exposure to environmental heterogeneity, mutations, and recombination across generations makes variation an indispensable asset for adaptation and breeding.

46.2 Types of Variation

1. **Somatic Variation** – Occurs in vegetative tissues; perpetuated through clonal propagation (e.g., bud sports in *Camellia sinensis*).
2. **Germinal Variation** – Arises in reproductive cells; heritable and the chief raw material for evolution.
3. **Continuous Variation** – Quantitative traits (yield, oil %, caffeine content) controlled by many genes.
4. **Discontinuous Variation** – Qualitative traits (flower colour, leaf pubescence) governed by major genes.
5. **Induced Variation** – Generated via mutagenesis, polyploidy, or tissue-culture stress.

47. Sources of Variation in Plantation Crops

47.1 Mutation

Mutations create new alleles through base substitutions, insertions, deletions, or chromosomal rearrangements. Examples:

- Somatic mutations in *Hevea brasiliensis* have yielded clones such as RRIM 600 and PB 235 with superior latex flow.
- Spontaneous dwarfism in *Cocos nucifera* generated the Chowghat Dwarf group.
- Chemical mutagenesis in *Coffea arabica* produced rust-resistant Mokka-type mutants.

Mutation rates are low ($\approx 10^{-5} - 10^{-6}$ per locus per generation), yet their cumulative impact over centuries has enriched diversity.

47.2 Recombination

During meiosis, crossing-over between homologous chromosomes reshuffles alleles, producing new genotypic combinations. In highly heterozygous, cross-pollinated crops such as *Elaeis guineensis* and *Hevea*, recombination is a dominant evolutionary force. Controlled pollination in oil palm deliberately exploits recombination between *Dura* and *Pisifera* lines to create the heterotic *Tenera* hybrid.

47.3 Polyploidy and Chromosomal Variation

Polyploidy and aneuploidy alter gene dosage, enzyme activity, and phenotypic expression. Triploid tea shows higher biomass; autotetraploid cashew produces larger nuts. Such cytogenetic changes introduce novel variation while maintaining vigor through gene redundancy.

47.4 Epigenetic Variation

Epigenetic modifications, DNA methylation, histone acetylation, and small RNA activity regulate gene expression without altering sequence. Environmental cues like drought or shading can induce heritable methylation patterns affecting caffeine metabolism in tea or oil biosynthesis in palm, representing a **reversible adaptive variation** crucial for rapid climate responses.

47.5 Somaclonal Variation

Plant tissue culture can trigger genetic and epigenetic instability. Somaclones of *Hevea* and *Cocos* often differ in growth rate or oil content. Though unpredictable, these variants serve as pre-breeding materials when screened cytologically and molecularly for stability.

48. Natural Selection as a Creative Force

48.1 Definition

Natural selection is the differential survival and reproduction of individuals due to heritable differences in fitness. It acts upon existing variation, increasing the frequency of advantageous alleles and reducing deleterious ones.

48.2 Forms of Natural Selection

1. **Stabilizing Selection** – Favours intermediate phenotypes; e.g., moderate nut size in coconut ensures both buoyancy and copra yield.
2. **Directional Selection** – Favours one extreme; e.g., drought-tolerant genotypes of cashew under arid stress.
3. **Disruptive Selection** – Favours both extremes, promoting polymorphism; e.g., tall vs. dwarf coconuts maintained under differing rainfall regimes.
4. **Balancing Selection** – Maintains multiple alleles; common in heterozygote-advantage systems like disease resistance in coffee.

48.3 Selective Agents in Plantation Systems

- **Abiotic:** temperature, rainfall, soil salinity, altitude.
- **Biotic:** pathogens (*Phytophthora palmivora* in cocoa, *Hemileia vastatrix* in coffee), pollinators, herbivores.
- **Anthropogenic:** management practices (fertilization, shade, pruning), chemical stress, and continuous clonal propagation.

49. Interaction Between Variation and Selection

Variation supplies raw diversity; selection filters it. In long-generation perennials, even slight selective advantages compound over decades.

Example: In a coconut population exposed to coastal salinity, seedlings with superior ion-exclusion genes survive and reproduce, gradually shifting allele frequencies toward salt tolerance , a micro-evolutionary adaptation observable within plantation timescales.

50. Case Studies of Variation and Selection

50.1 Coconut (*Cocos nucifera L.*)

- **Natural variation:** tall vs. dwarf ecotypes differ in flowering age, stature, and nut morphology.
- **Selection:** farmers historically retained palms with high copra yield; environmental selection favoured salt-tolerant rootstocks.
- **Outcome:** formation of regional cultivars such as West Coast Tall, Andaman Ordinary, and Chowghat Dwarfs representing ecological clines.

50.2 Oil Palm (*Elaeis guineensis Jacq.*)

- Natural recombination and farmer selection produced Dura, Pisifera, and Tenera fruit forms.
- Industrial breeding applied **recurrent selection**, increasing oil yield from 3 t ha^{-1} (1960s) to >6 t ha^{-1} today.
- Selection pressure under intensive management continues to favour high photosynthetic efficiency and low unsaturated-fat fractions.

50.3 Rubber (*Hevea brasiliensis Muell. Arg.*)

In its Amazonian home, natural selection shaped clones resistant to *Microcyclus ulei* (South American Leaf Blight). In Asia, breeding populations experienced **artificial directional selection** for latex yield, resulting in convergent phenotypes but narrower genetic bases , a cautionary example of **selection-induced erosion**.

50.4 Tea (*Camellia sinensis (L.) O. Kuntze*)

Continuous plucking constitutes a chronic selection regime favouring plants with rapid shoot regeneration and pest tolerance. Over centuries, these pressures differentiated **China**, **Assam**, and **Cambod** types, each optimized to distinct micro-climates.

50.5 Coffee (*Coffea arabica L.* and *C. canephora Pierre ex Froehner*)

Wild *C. arabica* evolved under shaded montane forests , traits like self-compatibility and limited branching enhance survival in low light. Human selection for cup quality maintained these characters even when cultivation shifted to open sun systems. Conversely, *C. canephora* remained an out-crosser adapted to humid lowlands with open canopies.

50.6 Cocoa (*Theobroma cacao L.*)

Pathogen pressure from *Phytophthora spp.* and *Moniliophthora roreri* has driven co-evolutionary selection for polygenic resistance. Modern "Trinitario" hybrids embody accumulated adaptive variation from both Criollo (flavour) and Forastero (vigor) ancestries.

50.7 Cashew (*Anacardium occidentale L.*)

Selection under drought and nutrient-poor soils yielded genotypes with high osmotic adjustment, waxy leaf cuticles, and efficient stomatal regulation. Farmers' preference for large nuts reinforced these drought-adapted phenotypes, illustrating **joint natural × artificial selection.**

51. Maintenance of Variation

51.1 Heterozygosity and Cross-Pollination

Most plantation crops are highly heterozygous due to cross-pollination, which maintains numerous alleles at low frequencies. In *Hevea* and *Cocos*, heterozygosity underpins yield stability and broad adaptability.

51.2 Gene Flow

Migration of pollen or seed across plantations maintains allelic continuity, preventing fixation and ensuring long-term evolutionary potential. However, excessive gene flow may homogenize populations, reducing local adaptation; thus, managed **isolation blocks** are used in breeding gardens.

51.3 Mutation–Selection Balance

Low-frequency deleterious mutations persist because selection cannot purge them faster than they appear, a key source of latent variation that breeders can recombine into beneficial backgrounds.

51.4 Balancing and Frequency-Dependent Selection

In coffee, pathogen races vary; rare resistance alleles confer advantage until pathogens adapt, maintaining polymorphism, an ecological arms race ensuring enduring variation.

52. Environmental and Epigenetic Influences on Variation

52.1 Phenotypic Plasticity

Perennial crops display high **plasticity**, the ability of a single genotype to produce different phenotypes under varying environments. For instance, the same coconut cultivar differs in leaf size, internode length, and nut yield when grown in coastal versus inland sites. Such plasticity mitigates environmental stress and enhances survival.

52.2 Epigenetic Memory and Transgenerational Effects

Stress-induced epigenetic marks (methylation) can persist through vegetative propagation. *Camellia sinensis* plants exposed to drought exhibit up-regulated methylation of aquaporin genes; the effect is retained in cuttings, constituting **epigenetic inheritance**, a rapid, non-genetic adaptation mechanism.

53. Artificial Selection and Domestication Revisited

Human choice interacts with natural selection to mould genetic architecture.

S. No.	Selective Force	Typical Target Trait	Example Crops	Outcome
1	Farmer mass selection	Fruit/nut size, yield	*Cocos*, *Anacardium*	Larger reproductive organs
2	Institutional breeding	Oil %, disease resistance	*Elaeis*, *Hevea*	Enhanced productivity
3	Clonal selection	Latex flow, flavour	*Hevea*, *Camellia*	Uniform elite clones
4	Molecular/ marker-assisted	Quality genes, stress alleles	*Coffea*, *Cocoa*	Targeted trait introgression

Artificial selection typically acts **directionally**, while natural selection maintains **stabilizing diversity**, their balance determines evolutionary resilience.

54. Quantitative Assessment of Variation and Selection

54.1 Statistical Indices

1. **Coefficient of Variation (CV %)** – quantifies phenotypic dispersion; >20 % indicates exploitable variability.
2. **Heritability (h^2)** – ratio of genetic to total variance; high h^2 traits respond swiftly to selection (e.g., oil content in palm).
3. **Genetic Advance (GA)** – expected improvement per selection cycle.
4. **Selection Differential (S)** – mean of selected individuals minus population mean.

Combined, these metrics guide breeders in predicting response to selection ($R = h^2 \times S$).

54.2 Molecular Indicators of Selection

Population-genetic parameters (π, Tajima's D, FST outlier tests) identify genomic regions under selection. In *Hevea*, QTLs for latex-flow enzymes (β-1,3-glucanase) show selective sweeps; in coffee, caffeine-biosynthesis genes display balancing selection reflecting consumer-driven preferences.

55. Adaptive Radiation and Speciation Trends

Over evolutionary timescales, accumulated variation and localized selection yield **adaptive radiation**, diversification into ecological niches:

- » The genus *Camellia* expanded into >100 species across Asia, each adapted to distinct altitudes and light regimes.
- » Wild *Coffea* species diversified across Africa following Pleistocene climatic oscillations.
- » Hevea* species radiated along Amazonian river systems under hydrological selection gradients.

Such radiation underlies the modern gene pools used in breeding.

56. Conservation of Variability Under Selective Breeding

56.1 Risk of Genetic Erosion

Intensive clonal propagation (rubber, tea) and high-input monocultures compress genetic diversity, increasing vulnerability to epidemics , as illustrated by the catastrophic *Microcyclus ulei* outbreaks in uniform rubber stands.

56.2 Strategies to Retain Variation

1. **Recurrent Selection and Polyclonal Gardens** – periodic mixing of diverse clones.
2. **Participatory Breeding** – engaging farmers maintaining heterogeneous landraces.
3. **Dynamic Conservation** – rotational use of breeding populations allowing natural recombination.
4. **Molecular Monitoring** – annual SSR/SNP profiling to track allelic richness.

57. Table Representative Forms of Variation and Selective Responses in Plantation Crops

S. No.	Crop	Principal Variation Type	Selective Force	Adaptive Outcome
1	*Cocos nucifera*	Tall vs. dwarf morphotypes	Salinity + farmer selection	Salt tolerance + precocity
2	*Elaeis guineensis*	Fruit shell thickness mutation	Human directional	Tenera hybrid yield gain
3	*Hevea brasiliensis*	Somatic mutation in latex cells	Artificial	High-latex clones
4	*Camellia sinensis*	Polyploid variation	Human + environmental	Triploid vigour
5	*Coffea arabica*	Allopolyploid origin	Natural hybridization	Quality + resistance
6	*Theobroma cacao*	Pathogen-driven allelic diversity	Co-evolution	Disease resistance
7	*Anacardium occidentale*	Morphological & biochemical	Drought + farmer choice	Drought hardiness + large nut

57. Controlled Pollination Techniques in Breeding *(continued)*

1. **Use of Male-Sterile or Self-Sterile Lines** – Naturally occurring or induced male sterility (e.g.,

triploid tea, interspecific coffee lines with sporophytic SI) enables efficient **controlled cross-pollination** without emasculation.

2. **Supplementary Pollination** – In oil palm and cocoa, **assisted pollen dusting** or **hand pollination** boosts fruit set where pollinators are limiting.
3. **Synchronization of Flowering** – Adjusting irrigation, pruning, or shade to align anthesis of parents (critical in *Cocos nucifera* Tall × Dwarf and in *Hevea*).
4. **Pollen Mixes and Diallel Designs** – In cross-pollinated crops (tea, Hevea, oil palm), **polycross** or **diallel** schemes increase genetic recombination; pollen identity can be verified by **marker paternity tests**.
5. **Hybrid Verification** – SSR/SNP fingerprints and morphological markers (shell thickness in oil palm; nut morphology in coconut) confirm hybridity before costly field evaluation.

58. Selfing, Outcrossing, and Mixed-Mating Systems

58.1 Selfing (Autogamy)

Advantages: genetic fixation, uniformity; Disadvantages: inbreeding depression in many perennials.

- **Prominent example:** *Coffea arabica* (largely autogamous, but outcrossing up to ~10–15% under pollinator abundance).
- **Breeding use:** Creation of **inbred lines** for mapping populations or to dissect recessive traits (rarely advanced to cultivars due to perennial load).

58.2 Outcrossing (Allogamy)

Advantages: heterozygosity, vigor, adaptability; Disadvantages: segregation load and heterogeneity.

- **Examples:** *Hevea brasiliensis*, *Elaeis guineensis*, *Camellia sinensis*, *C. canephora*, *Anacardium occidentale*, *Theobroma cacao*.
- **Breeding use:** Recurrent selection, half-sib and full-sib families, **clonal selection** of elite heterozygotes.

58.3 Mixed-Mating Systems

Some species show both selfing and outcrossing depending on ecology and variety (e.g., **tall vs. dwarf coconuts; arabica coffee** in bee-rich environments). Mixed systems are useful for **population improvement** while preserving local adaptation.

59. Specialized Pollination Syndromes

59.1 Weevil-Mediated Pollination (Cantharophily) in Oil Palm

- **Attraction cues:** phenolic volatiles; **thermogenesis** marginally elevates spadix temperature to enhance scent spread.
- **Practice tip:** Maintain **male inflorescence refugia** during breeding to sustain *Elaeidobius* populations.

59.2 Midge-Mediated Pollination in Cocoa

- Tiny **Forcipomyia** midges require **humid leaf-litter microhabitats**; shade trees and mulching dramatically increase fruit set.
- **Hand pollination protocol:** collect fresh anthers, dust onto receptive stigmas (mid-morning), tag and record; fruit set >80% under skilled hands.

59.3 Bee-Centric Systems in Cashew, Tea, and Robusta Coffee

- **Cashew:** long anthesis window benefits from **Apis** diversity; avoid pesticide sprays during bloom.
- **Tea:** bees enhance cross-set in SI clones used for seed gardens.
- **Robusta:** bee presence can **double** fruit set compared to pollinator-poor stands.

60. Pollen–Pistil Interactions and Pre-Zygotic Barriers

60.1 Stigma Receptivity

- Assessed by **hydrogen peroxide bubbling**, peroxidase staining, or β-galactosidase assays.
- Peak receptivity varies: **coconut (1–2 days), oil palm (24 h), cocoa (same day), tea (36–48 h)**.

60.2 Pollen Adhesion, Hydration, and Germination

- **Exine sculpturing** and **stigma exudates** mediate adhesion (sticky in cocoa, cashew).
- Controlled **humidity (60–80%)** is crucial in artificial pollinations (coffee/cocoa).

60.3 Guidance and Compatibility Signals

- **Calcium gradients, arabinogalactan proteins (AGPs),** and **lipid transfer proteins** steer tubes.
- **Self-recognition systems** (S-alleles in sporophytic/gametophytic SI) arrest incompatible tubes at stigma or style (details in Part 8).

61. Post-Pollination Events and Seed Development

1. **Double fertilization** timing (typical ranges): oil palm 12–24 h; coconut 24–36 h; cocoa 12–24 h; tea/rubber 36–72 h.
2. **Selective ovule abortion**: cocoa/cashew prune supernumerary ovules; usually one seed matures.
3. **Endosperm type**: nuclear → cellular (tea, coffee), scanty endosperm in **cashew; copious liquid endosperm** (coconut water) → **solid endosperm** (copra).
4. **Seed maturation**: extended in perennials, **arecanut/coconut** 10–12 months; **cocoa** 5–6 months; **coffee** 8–9 months.

62. Environmental Modulators of Pollination Success

- **Temperature:** High (>34 °C) shortens stigma receptivity (cashew) and dries exudates; cool/wet reduces bee activity (coffee/tea).

- **Humidity:** Essential for pollen hydration; extremes (<40% or >90%) reduce germination.
- **Wind/Rain:** Wind aids coconut; heavy rain impedes cocoa midges and bee foraging.
- **Agrochemicals:** Time sprays **post-anthesis** or at dusk; adopt pollinator-safe IPM.

63. Troubleshooting Low Fruit Set: Diagnostics Guide

S. No.	Symptom	Likely Cause	Rapid Remedy
1	Many flowers, few fruits (cocoa)	Pollinator scarcity	Hand pollination; conserve leaf-litter moisture
2	Pollen fails to germinate (tea)	Incompatible mating; pollen aged	Fresh pollen; choose compatible clone; adjust timing
3	Aborted nuts (cashew)	Heat stress during anthesis	Micro-sprinklers; shade nets; shift crossing window
4	Poor set (oil palm)	Low *Elaeidobius* density	Introduce weevils; maintain male inflorescences
5	Set but empty nuts (coconut)	Post-zygotic failure	Select high-MI parents; avoid nutrient/water stress

64. Quantifying Mating Systems and Gene Flow

- **Outcrossing rate (t_m):** multilocus estimates from codominant markers (SSR/SNP); critical for tea, Hevea, and oil palm breeding nurseries.
- **Paternity analysis:** assigns male parents in open-pollinated progenies (useful in **seed gardens** of tea/cashew).
- **Effective pollen clouds:** characterize pollinator movement; informs **isolation distance** and orchard design.

65. Table – Snapshot of Pollination & Fertilization Features in Major Plantation Crops

S. No.	Crop	Sexual System	Pollination Mode	Key Pollinators	Fertilization Window	Breeding Notes
1	*Cocos nucifera*	Monoecious; protandry	Wind + insects	Bees, flies, beetles	24–36 h	Tall × Dwarf hybrids; bagging feasible
2	*Elaeis guineensis*	Monoecious (male/female cycles)	Entomophily	*Elaeidobius* weevils	12–24 h	Controlled Dura × Pisifera; pollen storage long
3	*Areca catechu*	Monoecious; protandry	Entomophily	Bees, flies	~24 h	Manual pollination for hybrids
4	*Hevea brasiliensis*	Monoecious; central female	Entomophily	Thrips, small flies	36–48 h	Hand pollination standard; low natural set
5	*Camellia sinensis*	Hermaphrodite; SI	Entomophily	Bees, flies	36–48 h	Clonal crop; seeds for breeding only

6	*Coffea arabica*	Hermaphrodite; SC	Autogamy (insects aid)	Bees help	~48 h	Prevent selfing in hybrids
7	*C. canephora*	Hermaphrodite; SI	Entomophily	Bees (*Apis*, *Melipona*)	~48 h	Maintain pollinator habitat
8	*Theobroma cacao*	Hermaphrodite; self-compatible but poorly self-set	Entomophily	*Forcipomyia* midges	12–24 h	Hand pollination boosts set
9	*Anacardium occidentale*	Polygamous (H + M)	Entomophily	Bees, flies	≤24 h	Bagging/emasculation; short pollen life

Sterility and Incompatibility Systems in Plantation Crops

66. Introduction

66.1 Reproductive Barriers and Their Importance

Sterility and incompatibility mechanisms represent natural barriers to self-fertilization and hybrid formation. In plantation crops, long-lived perennials propagated through seed or vegetative means, these systems control the **extent of cross-pollination, genetic diversity, hybrid vigor, and seed set**. Understanding their biological basis helps breeders plan **controlled hybridization**, manage **clonal fertility**, and maintain **stable yields**.

66.2 Terminology

» **Sterility** – failure to produce or release viable gametes, pollen, or ovules.

» **Incompatibility** – failure of fertilization despite viable gametes, owing to biochemical rejection between pollen and pistil. Both act as **reproductive isolating mechanisms**, influencing genetic structure and breeding strategy.

67. Types of Sterility

67.1 Structural (Morphological) Sterility

Caused by malformation of floral organs, stigma exsertion, fused anthers, undeveloped ovary, or malformed pollen sacs.

Example: Certain *Hevea* clones show malformed pistils leading to non-functional stigmas.

67.2 Functional Sterility

Organs appear normal but are physiologically incapable of fertilization, often due to temperature or hormonal imbalance. *Example:* Excess heat (> 36 °C) during cashew anthesis in coastal India causes pollen desiccation and ovule abortion.

67.3 Cytoplasmic Sterility

Arises from mitochondrial or plastid mutations interfering with pollen development; often maternally inherited. While less documented in plantation crops, cytoplasmic sterility has been observed in *Camellia sinensis* hybrids and experimental *Coffea* lines.

67.4 Genetic (Nuclear) Sterility

Controlled by nuclear genes affecting gametogenesis, often single recessive mutations. In *Hevea*, segregants exhibiting complete pollen abortion have been traced to recessive nuclear sterility alleles.

67.5 Induced Sterility

Created by mutagens, irradiation, or polyploidization. For example, colchicine-induced triploid tea and tetraploid cashew display high pollen sterility; these lines are used as **female parents** in breeding.

68. Male Sterility in Plantation Crops

68.1 Definition and Relevance

Male sterility refers to failure of pollen formation or release, while female fertility remains intact. It is vital for hybrid-seed production because it obviates emasculation. In tree crops, natural male sterility is rare but invaluable when available.

68.2 Cytoplasmic Male Sterility (CMS)

Mitochondrial gene rearrangements cause abortion of the anther tapetum. Though well-established in cereals, CMS analogues occur in perennials too:

- In *Camellia sinensis*, a few accessions exhibit cytoplasmic sterility linked with aberrant mitochondrial open reading frames.
- In *Coffea*, experimental CMS lines are being explored for mechanized hybrid seed production.

68.3 Genetic Male Sterility (GMS)

Controlled by nuclear loci; used in *Hevea* and *Anacardium* breeding. GMS lines of *Hevea* display empty pollen sacs, useful for reciprocal crossing with fertile clones to produce hybrid seeds.

68.4 Environmental Male Sterility (EGMS)

Temperature- or photoperiod-sensitive sterility, documented experimentally in tea and cashew:

- Tea clones exposed to < 12 °C during meiosis show pollen abortion.
- Cashew inflorescences formed under > 38 °C have low viable pollen, reverting to fertility under cooler regimes.

68.5 Practical Exploitation

Breeders use male-sterile or triploid individuals as **seed parents**, ensuring cross-fertilization without emasculation (triploid tea × diploid pollen donors; sterile *Hevea* × fertile pollinator clones).

69. Female Sterility

Female sterility stems from ovule abortion, stigma non-receptivity, or embryo-sac degeneration.

Examples:

- *Hevea brasiliensis*: a high proportion of non-functional ovules in certain clones causes low seed set.

- *Theobroma cacao*: a fraction of flowers abort due to failure of pollen-tube penetration.
- *Coffea arabica*: occasionally shows pre-mature ovule degeneration under heat stress.

70. Incompatibility Systems

71.1 Concept

Self-incompatibility (SI) is a genetically controlled mechanism preventing self-fertilization in otherwise hermaphrodite species. It maintains heterozygosity and promotes outcrossing.

72.2 Types

1. **Gametophytic SI (GSI)** – pollen behaviour determined by its own haploid genotype. Common in *Camellia sinensis* and *Coffea canephora.*
2. **Sporophytic SI (SSI)** – pollen behaviour governed by diploid genotype of parent anther tissue; found in *Anacardium occidentale.*

73.3 Morphological Manifestations

- Failure of pollen germination on the stigma.
- Arrested tube growth within the style.
- Degeneration of pollen tubes before reaching ovule. Microscopic staining with aniline blue reveals fluorescent callose plugs marking rejection sites.

74. Crop-Specific Incompatibility and Sterility Patterns

74.1 *Camellia sinensis (L.) O. Kuntze*

- **System:** Gametophytic self-incompatibility.
- **Mechanism:** *S*-allele interaction between pollen and pistil; incompatible tubes callose-plugged in upper style.
- **Outcome:** Cross-pollination obligatory; self-pollinated flowers rarely form seeds.
- **Breeding practice:** Multiple compatible clones inter-planted in seed gardens; 20–30 m isolation ensures foreign pollen.
- **Biochemical markers:** High peroxidase and RNase activity during pollen rejection; pistil glycoproteins (S-RNases) implicated.

75.2 *Coffea arabica* L. and *C. canephora Pierre ex Froehner*

Species	Fertility System	Consequences for Breeding
C. arabica	Self-compatible, sometimes inbreeding depression minimal	True breeding possible; facilitates pure-line selection
C. canephora	Sporophytic SI, multiple S-alleles	Requires compatible pollinizers (two or more clones per block)

Molecular analyses show polymorphism at the **S-locus** (receptor kinase genes). Pollination between clones sharing identical S-alleles leads to **pollen-tube inhibition** at the stigma surface.

75.3 *Hevea brasiliensis Muell. Arg.*

» **Reproductive system:** Predominantly out-crossing.

» **Self-incompatibility:** Partial, post-zygotic type, selfed embryos abort early.

» **Other barriers:** High abscission of unpollinated and selfed flowers; low natural seed set (~ 1–2 %).

» **Implication:** Artificial hand-pollination mandatory for breeding.

75.4 *Cocos nucifera L.*

» **Fertility type:** Cross-pollinated tall forms, partial selfing in dwarfs.

» **Incompatibility:** No strict genetic SI, but mechanical and temporal (protandry) barriers restrict selfing.

» **Hybridization use:** Controlled bagging enables targeted Tall × Dwarf hybrids (heterosis > 25 %).

75.5 *Elaeis guineensis Jacq.*

» **System:** Functional self-incompatibility due to sequential male–female flowering (dioecious phases).

» **Male and female inflorescences** alternate; pollen from other plants is needed.

» **Weevil pollination** reinforces cross-fertilization. Controlled pollination with stored pollen yields nearly 100 % hybrid seed.

75.6 *Theobroma cacao L.*

» **Self-incompatibility:** Late-acting GSI in most Forastero types; compatible in some Criollo.

» **Mechanism:** Fertilization may occur, but the zygote fails post-zygotically.

» **Breeding implication:** Manual cross-pollination is obligatory for reliable pod set; controlled combinations avoid SI partners.

75.7 *Anacardium occidentale L.* (Cashew)

» **System:** Sporophytic SI.

» **Evidence:** Selfed pollen fails to germinate or tube terminates near the stigma.

» **Ratio of flowers:** High male: hermaphrodite (~ 20:1) amplifies the need for cross-pollination.

» **Breeding practice:** Compatible clones interplanted in orchards; bees ensure pollen transfer.

76. Genetic Control of Incompatibility

76.1 S-Locus Polymorphism

- Controlled by multi-allelic **S-gene complex** coding for pollen and pistil proteins.
- GSI – pollen S-determinant (S-RNase interaction); SSI – receptor kinase (SRK) and cysteine-rich ligand (SCR).
- **Camellia, Coffea,** and **Cashew** possess rich S-allele diversity (10–30 alleles per population).

76.2 Molecular Detection

- PCR primers targeting conserved *S-RNase* or *SRK* sequences diagnose compatibility before crossing.
- Marker-assisted selection enables pre-screening of compatible parental sets in tea and coffee breeding programs.

77. Physiological and Biochemical Basis

1. **Enzymatic Activity:** Rise in peroxidase, polyphenol oxidase, and RNase during incompatible interactions.
2. **Calcium Ion Flux:** Rapid Ca^{2+} accumulation in incompatible styles blocks tube elongation.
3. **Reactive Oxygen Species (ROS):** Burst in stigma tissue triggers pollen-tube apoptosis.
4. **Hormonal Imbalance:** Ethylene surge correlates with stigma senescence in selfed flowers.
5. **Programmed Cell Death (PCD):** Microscopic evidence of pollen-tube collapse due to PCD triggered by S-locus signalling.

78. Hybrid Sterility and Post-Zygotic Barriers

Even when fertilization occurs, hybrids between divergent species may be **sterile** because of:

1. **Chromosomal imbalance** (unequal pairing, as in *Coffea arabica* × *C. canephora*).
2. **Endosperm imbalance** (3:2 maternal: paternal genome ratio disturbed).
3. **Genic disharmony,** nuclear–cytoplasmic conflict causing embryo abortion (seen in *Hevea* × *H. benthamiana*).

Rescue measures: Embryo culture, backcrossing, polyploidization to restore pairing.

79. Environmental and Management Factors Affecting Fertility

Factor	Effect on Fertility	Mitigation Strategy
Temperature extremes	Pollen desiccation or ovule abortion	Adjust planting altitude; shading
Humidity < 40 %	Pollen dehydration	Timed pollination during early morning

Nutrient deficiency (B, Ca)	Pollen-tube growth inhibition	Foliar micronutrient sprays
Pollinator scarcity	Reduced cross-pollination	Pollinator gardens; IPM
Agrochemicals	Pollen kill	Avoid spraying during bloom

80. Exploiting Sterility and Incompatibility in Breeding

81.1 Hybrid-Seed Production

- **Triploid tea** (3 × = 45) × diploid fertile pollen donors → vigorous sterile progeny used for vegetative multiplication.
- **Male-sterile Hevea clones** × fertile pollen clones → hybrid seeds without emasculation.
- **Controlled compatibility mapping** in coffee and cashew optimizes crossing blocks.

82.2 Maintenance of Genetic Diversity

Self-incompatibility ensures cross-fertilization, preventing inbreeding depression and sustaining adaptive variation in natural populations of *Camellia* and *Coffea*.

83.3 Overcoming Incompatibility

1. **Bud Pollination:** Pollination at the immature stigma stage before SI expression.
2. **Mentor Pollen:** Mixing compatible pollen with self-pollen to stimulate tube growth.
3. **Stigma Washing / Heat Treatment:** Removes inhibitory proteins.
4. **In Vitro Fertilization:** Pollen germinated on artificial media; tubes guided into ovules under a microscope (experimental in *Camellia*).
5. **Mutation or Gene Editing:** CRISPR/Cas knock-out of *S-RNase* or *SRK* genes to break SI (future research frontier).

84. Table – Summary of Sterility and Incompatibility Patterns in Major Plantation Crops

S. No.	Crop	Sterility Type	Incompatibility System	Expression Level	Breeding Implication
1	*Cocos nucifera*	Functional (protandry)	None genetic	Temporal barrier	Partial selfing in dwarfs; cross in talls
2	*Elaeis guineensis*	Functional (separate inflorescences)	None genetic	Sequential flowering	Mandatory cross-pollination
3	*Areca catechu*	Functional	None genetic	Protandry	Manual pollination for hybrids
4	*Hevea brasiliensis*	Genetic/ Functional	Partial post-zygotic	Embryo abortion	Low seed set; hand pollination

5	*Camellia sinensis*	Cytoplasmic/ Functional	Gametophytic SI	Strong	Cross required; clonal propagation
6	*Coffea arabica*	,	Self-compatible	Weak	Selfing possible
7	*C. canephora*	,	Sporophytic SI	Strong	Two or more clones per block
8	*Theobroma cacao*	Functional + Genetic	Late-acting GSI	Variable	Hand pollination necessary
9	*Anacardium occidentale*	Functional	Sporophytic SI	Moderate	Cross-compatible orchard design

85. Molecular and Genomic Insights

- Transcriptomics in *Camellia* reveals up-regulation of S-RNase, ubiquitin ligases, and defense-related genes during incompatible pollinations.
- Coffee genome sequencing has identified S-locus receptor kinases analogous to Brassica SRK, confirming evolutionary conservation.
- In cashew, differential expression of CaSRK-like and CaSCR-like transcripts corresponds with compatible versus incompatible reactions.
- These discoveries enable molecular tagging of compatibility for predictive crossing.

86. Overcoming Sterility for Hybridization

86.1 Hormonal Manipulation

Gibberellin (20–50 ppm) or NAA sprays enhance anther dehiscence and stigma receptivity in *Hevea* and *Camellia*, partially restoring fertility.

87.2 Polyploidization

Doubling chromosome number restores pairing in sterile interspecific hybrids, e.g., synthetic tetraploid *Coffea* hybrids fertile compared with their diploid × tetraploid parents.

88.3 Embryo Rescue and Tissue Culture

Excising immature embryos from incompatible crosses (cocoa, Hevea) before abortion allows survival in vitro, bypassing post-zygotic barriers.

Chapter - 6

Breeding Methods

1. Introduction and Principles of Plant Breeding in Plantation Crops

1.1 Definition and Scope

Plant breeding is the purposeful manipulation of genetic variability to develop cultivars possessing improved yield, quality, resistance, and adaptability. In **plantation crops**, long-lived perennials such as *Cocos nucifera* (coconut), *Elaeis guineensis* (oil palm), *Hevea brasiliensis* (rubber), *Camellia sinensis* (tea), *Coffea arabica* (Arabica coffee), *Theobroma cacao* (cocoa), and *Anacardium occidentale* (cashew), the breeder's objective is not merely yield maximization but also **genetic resilience and economic sustainability** over decades of cultivation. Because these crops have extended juvenile phases, large plant size, and complex reproductive biology, their breeding strategies differ fundamentally from those of annuals. Progress, therefore, depends on integrating **classical selection, hybridization, clonal evaluation, mutation, and molecular technologies** within long-term programs.

1.2 Historical Development

Phase	Approx. Period	Major Features	Representative Achievements
I – Pre-scientific	before 1900	Farmer and planter selection of elite trees and palms	Traditional *Cocos* landraces (West Coast Tall, Tiptur Tall)
II – Exploratory	1900–1950	Introductions, acclimatization, simple selection	Early *Hevea* clones RRIM 501, PB 86; seedling tea gardens
III – Systematic	1950–1990	Controlled hybridization, clone testing, seed gardens	Tall × Dwarf coconut hybrids, Arabica × Robusta coffee
IV – Molecular Era	1990 → present	Marker-assisted and genomic breeding	SSR-fingerprinting in tea, SNP maps in oil palm

DOI: 10.1201/9781042010332-7

This gradual evolution shows the shift from phenotype-based selection to genomics-enabled design breeding.

1.3 Objectives of Breeding in Plantation Crops

1. **Yield improvement** – higher copra, latex, oil, caffeine, or bean output.
2. **Quality enhancement** – desirable biochemical composition (fatty-acid profile, aroma, polyphenol content).
3. **Resistance breeding** – to major pests and pathogens such as *Hemileia vastatrix* (coffee rust) and *Phytophthora palmivora* (cocoa pod rot).
4. **Abiotic-stress tolerance** – salinity, drought, and temperature extremes.
5. **Precocity and uniformity** – reduced juvenile phase and synchronized flowering.
6. **Architectural modification** – compact canopies, reduced height, easy harvest.
7. **Clonal stability** – high rooting percentage and fidelity in vegetative propagation.
8. **Resource-use efficiency and sustainability** – superior photosynthetic and nutrient-use efficiency under climate variability.

1.4 Genetic Base and Breeding Constraints

Many plantation crops were domesticated from limited introductions, producing **narrow genetic bases** and heightened vulnerability to diseases.

S. No.	Crop	Primary Centre of Origin	Genetic Status in Cultivation
1	*Cocos nucifera*	Indo-Pacific region	Moderate diversity; tall × dwarf hybrids broaden base
2	*Elaeis guineensis*	West Africa	Narrow; few Dura–Pisifera founders
3	*Hevea brasiliensis*	Amazon basin	Very narrow (< 20 founder trees in Asia)
4	*Camellia sinensis*	China & Indo-Burma	Wide; China and Assam types inter-crossed
5	*Coffea arabica*	Ethiopia	Narrow; tetraploid origin from two species
6	*Theobroma cacao*	Upper Amazon	Moderate; Criollo, Forastero, Trinitario groups
7	*Anacardium occidentale*	Brazil	Narrow; few introductions to India and Africa

The challenge, therefore, is to **expand variability** through germplasm exploration, wild-species introgression, and biotechnological intervention.

1.5 Fundamental Principles of Breeding

1. **Variation** – source of heritable differences arising from mutation, recombination, and polyploidy.
2. **Selection** – preferential propagation of superior genotypes.

3. **Hybridization** – creation of novel gene combinations via controlled crossing.
4. **Evaluation and Testing** – multi-location and multi-season trials for stability.
5. **Propagation and Maintenance** – vegetative multiplication of elite genotypes.
6. **Genetic Purity and Documentation** – use of DNA fingerprints, barcoding, and clone registers.

1.6 Breeding Systems and Reproductive Behaviour

Reproductive System	Typical Crops	Breeding Implication
Cross-pollinated (allogamous)	*Hevea*, *Cocos tall*, *Elaeis*, *Camellia*, *C. canephora*	Maintain heterozygosity; exploit heterosis; clonal propagation
Self-pollinated (autogamous)	*Coffea arabica*, dwarf coconut	Allows fixation; pure-line selection possible
Asexual (clonal)	Tea, rubber, cashew clones	Elite heterozygotes perpetuated unchanged
Partial apomixis	Cashew and related species	Seed formation without fertilisation is possible

Recognizing these systems dictates the choice of selection, crossing, and propagation strategies.

1.7 Breeding Cycle and Time Management

Because one breeding cycle may require **5–10 years**, time-saving innovations are vital:

» **Early-selection indices** (enzyme markers for latex flow, leaf biochemistry for oil content).

» **Vegetative propagation of hybrids** to test clones without waiting for seed set.

» **Off-season flower induction** via nutrient or hormonal treatments.

» **Overlapping generations**, simultaneous evaluation and crossing of young and mature plants.

1.8 Infrastructure for Breeding Programmes

1. **Isolated breeding gardens** for controlled pollination.
2. **Clonal repositories and gene banks** under strict quarantine.
3. **Seed gardens**, biclonal or polyclonal, to supply hybrid seed (tea, coffee).
4. **Molecular labs** for DNA fingerprinting and marker analysis.
5. **Data management systems** linking field phenotypes with genomic information.

Emerging **digital phenotyping** (drones, AI-based image analytics) now complements manual scoring.

1.9 Integration of Conventional and Molecular Approaches

Conventional Practice	Molecular Complement
Phenotypic selection	Marker-assisted selection (MAS)
Pedigree breeding	QTL mapping and genomic selection
Clonal evaluation	SSR/SNP identity verification

Mutation breeding	TILLING and whole-genome resequencing
Quality assessment	Metabolomics and transcriptomics

This convergence shortens breeding cycles and increases precision while retaining empirical field wisdom.

2 – Selection Methods in Plantation Crops

2. Introduction

2.1 Concept of Selection

Selection is the foundation of all plant-breeding programmes. It is the process of identifying and preserving individuals possessing superior genetic combinations for further propagation. In plantation crops, mostly perennial, heterozygous, and cross-pollinated, selection assumes greater complexity and long-term importance than in annuals, because a single elite tree or palm can influence commercial plantations for decades through clonal multiplication.

Selection exploits **natural variation** generated by mutation, recombination, and environmental interaction. Over time, the continuous selection and propagation of elite phenotypes lead to **genetic improvement of populations** without necessarily changing their chromosomal constitution.

2.2 Historical Perspective

Early planters practiced empirical selection long before the advent of genetics. By visually identifying trees with higher yields or better quality, they established landraces such as *Cocos nucifera* 'West Coast Tall' and *Camellia sinensis* 'Assam Jat'. With the rediscovery of Mendel's laws, breeders introduced systematic procedures, mass selection, progeny testing, and recurrent selection to accelerate improvement.

3. Types of Selection

3.1 Natural Selection

In wild or semi-cultivated populations, environmental pressures such as drought, salinity, or pathogen attack favour the survival of specific genotypes. For example:

» Coastal coconut populations naturally selected for salt tolerance.

» *Hevea brasiliensis* populations in humid zones evolved resistance to *Microcyclus ulei*.

Breeders exploit such naturally adapted populations as **primary germplasm pools**.

3.2 Artificial Selection

Deliberate human intervention distinguishes artificial from natural selection. It involves identifying superior genotypes based on measurable traits. Major forms include:

1. Mass Selection
2. Pure-Line Selection
3. Progeny Selection

4. Clonal Selection
5. Recurrent Selection
6. Marker-Assisted and Genomic Selection

Each method suits particular reproductive systems and breeding objectives.

4. Mass Selection

4.1 Definition and Principle

Mass selection involves choosing a number of superior individuals from a heterogeneous population, bulking their propagules, and growing them collectively to form an improved population. The assumption is that selected phenotypes possess desirable genotypes with additive effects.

4.2 Procedure

1. Survey natural or cultivated stands.
2. Record phenotypic superiority (yield, quality, vigour, disease resistance).
3. Collect nuts, seeds, or cuttings from the best 1–5 % trees.
4. Raise progeny collectively.
5. Evaluate the composite population under uniform management.

4.3 Applications in Plantation Crops

Crop	Selection Criteria	Achievements
Cocos nucifera	Copra yield, nut size, disease tolerance	West Coast Tall, East Coast Tall, Laccadive Tall
Hevea brasiliensis	Latex yield, girth increment	RRIM 500 series clones
Camellia sinensis	Shoot density, leaf size	Assam Jat selections
Anacardium occidentale	Nut weight, apple color	Vengurla 1–5 selections

4.4 Advantages

» Simple, inexpensive, field-based.

» Maintains population variability.

» Effective when heritability is high.

4.5 Limitations

» Ineffective for low-heritability traits.

» Phenotypic superiority may reflect environment, not genotype.

» Heterozygosity remains high, limiting uniformity.

Mass selection thus serves as a **first step** in population improvement and germplasm characterization.

5. Pure-Line Selection

5.1 Definition

Pure-line selection isolates a homozygous line from a self-pollinated population by selfing and selection of a single superior plant whose progeny are genetically uniform.

5.2 Relevance to Plantation Crops

Although most plantation crops are cross-pollinated, pure-line selection is applicable to **autogamous** or **self-compatible species** such as *Coffea arabica* and some dwarf coconuts. In these, homozygosity enables the stable inheritance of desired traits.

5.3 Procedure

1. Identify promising individual plants in a population.
2. Harvest and sow seeds from each separately.
3. Evaluate progeny for uniformity and performance.
4. Select the best pure line for release.

5.4 Example

» In *Coffea arabica*, pure-line selection produced cultivars like 'Kent', 'S.795', and 'Selection 9'.

» In dwarf coconut, uniform self-pollination led to development of Chowghat Dwarf Green and Orange.

5.5 Advantages and Limitations

Advantages	Limitations
Fixes desirable genes permanently	Reduces genetic variability
Produces uniform cultivars	Inbreeding depression in perennials
Suitable for self-pollinated crops	Long generation time delays progress

6. Progeny Selection

6.1 Concept

Progeny selection evaluates offspring derived from known parents to estimate breeding value. It separates genetic from environmental effects and is especially useful in cross-pollinated species.

6.2 Types

1. **Half-sib selection** – progenies share one parent (usually the female).
2. **Full-sib selection** – progenies share both parents from controlled crosses.

6.3 Implementation

1. Collect controlled pollination seed or open-pollinated progeny.
2. Grow and evaluate each family separately.
3. Estimate variance among families.
4. Select superior families and individuals within them.
5. Recombine selected individuals to form a new cycle (recurrent selection).

6.4 Applications

» In *Hevea brasiliensis*, progeny testing distinguishes high-latex clones early in breeding cycles.

» In *Elaeis guineensis*, progeny selection identified superior *Dura* × *Pisifera* combinations for oil yield.

» In *Camellia sinensis*, seedling progeny trials identify elite mother bushes for clonal release.

7. Clonal Selection

7.1 Principle

Clonal selection is the most dominant method for perennial plantation crops. It relies on the **vegetative propagation of superior heterozygous individuals** so that their genetic identity remains unchanged.

7.2 Procedure

1. Survey existing plantations for outstanding trees based on yield, quality, or disease tolerance.
2. Collect vegetative material (budwood, cuttings, grafts).
3. Establish **clonal evaluation trials** in replicated designs.
4. Compare performance over multiple years.
5. Release superior clones for commercial multiplication.

7.3 Major Achievements

Crop	Examples of Clones	Key Traits
Hevea brasiliensis	RRIM 600, PB 235, GT 1	High latex yield, disease tolerance
Camellia sinensis	TV 1–29, UPASI 3, UPASI 9	Quality, yield, drought tolerance
Coffea canephora	S.274, S.274 B, CxR composites	Vigour, rust resistance
Anacardium occidentale	Vengurla 4, NRCC Selection 2	Nut yield, apple quality

7.4 Advantages

» Captures heterosis in fixed form.

» Rapid multiplication of elite genotype.

- Ensures genetic uniformity of plantation.
- Allows evaluation under identical management.

7.5 Limitations

- Vulnerability to epidemics due to uniformity.
- Accumulation of somatic mutations with age.
- Low multiplication rate in woody perennials.
- High initial cost for nursery establishment.

Clonal selection remains the **cornerstone of breeding strategy** in crops like tea, rubber, and cashew.

8. Recurrent Selection

8.1 Concept

Recurrent selection cyclically improves a population by selecting superior individuals each generation and inter-mating them to recombine favorable alleles. It is particularly effective in **cross-pollinated species** where additive genetic variance is high.

8.2 Cycle Steps

1. Selection of superior individuals.
2. Controlled or open inter-mating among selections.
3. Evaluation of progeny in the next generation.
4. Re-selection and recombination.
5. Repetition over cycles.

8.3 Forms

- **Half-sib recurrent selection** – uses open-pollinated seed of selected mothers.
- **Full-sib recurrent selection** – based on controlled crosses between selected parents.
- **Reciprocal recurrent selection** – simultaneous improvement of two heterotic populations.

8.4 Applications

- In *Hevea*, recurrent selection among high-yielding clones expands breeding base and captures additive gains.
- In *Cocos nucifera*, inter-se crossing of elite Tall and Dwarf types creates superior hybrid progenies with enhanced combining ability.
- In *Camellia sinensis*, recurrent selection among regional seedling populations improves yield and cup quality.

8.5 Advantages and Limitations

Advantages	Limitations
Accumulates favourable genes over cycles	Long generation interval
Maintains variability	Requires large populations
Enhances adaptability	Controlled pollination laborious

9. Marker-Assisted and Genomic Selection

9.1 Marker-Assisted Selection (MAS)

MAS uses DNA markers closely linked to desirable genes or QTLs to select genotypes at the seedling stage.

» In *Hevea*, SSR markers associated with latex biosynthetic enzymes aid early prediction of yield.

» In *Camellia*, QTLs for catechin content enable quality-oriented selection.

» In *Coffea*, rust-resistance genes (*SH1–SH9*) are tracked with specific markers.

9.2 Genomic Selection (GS)

GS estimates the genomic breeding value (GEBV) of individuals using genome-wide SNP data, allowing prediction without phenotyping every generation. This approach, though recent, can drastically reduce generation time in long-cycle crops when combined with tissue culture and early flowering induction.

9.3 Advantages

» Selection independent of environment.

» Accelerated breeding cycles.

» Simultaneous improvement of multiple traits.

» Useful for complex polygenic characters like yield and drought tolerance.

10. Comparative Summary of Selection Methods

S. No.	Method	Reproductive System Best Suited	Examples of Crops	Key Outcome
1	Mass selection	Cross-pollinated, high heritability	*Cocos, Cashew*	Population improvement
2	Pure-line selection	Self-pollinated	*Coffea arabica*	Uniform cultivars
3	Progeny selection	Cross-pollinated	*Hevea, Oil palm*	Accurate breeding-value estimate
4	Clonal selection	Vegetatively propagated	*Tea, Rubber, Cashew*	Fixation of heterosis

5	Recurrent selection	Cross-pollinated	*Hevea, Coconut*	Cumulative additive gain
6	Marker/Genomic selection	All systems	All major plantation crops	Precision and cycle reduction

11. Integration of Selection Methods

Modern breeding rarely relies on a single method. Instead, breeders adopt **integrated selection frameworks**, combining field evaluation with molecular and statistical tools:

» **Phenotypic + Molecular Selection:** field performance validated by DNA markers.

» **Clonal + Progeny Testing:** clonal trials followed by sexual recombination.

» **Participatory Selection:** involving growers for trait prioritization.

» **Digital Selection Tools:** AI-based image analysis for canopy vigour, fruit load, or disease scoring.

3 – Chimeras and Clonal Selection in Plantation Crops

1. Introduction

1.1 Vegetative Propagation and Genetic Fixation

Most plantation crops are perennial and highly heterozygous. They are therefore **propagated vegetatively**, through cuttings, grafts, or tissue culture, so that elite genotypes are reproduced **identically**. This practice, while preserving desired combinations, also exposes the breeder to the phenomena of **chimerism** and **somaclonal variation**, both of which can generate useful novelty or unwanted instability. The **cellular organization** of clonal tissues and the mechanisms that create **genetic mosaics** is essential for rational clonal improvement and for maintaining genetic fidelity in large-scale propagation.

2. Chimeras – Concept and Classification

2.1 Definition

A **chimera** is an individual plant composed of two or more genetically distinct cell layers or tissues originating from a single zygote or from spontaneous mutation within meristematic layers. Each sector retains its own genotype but remains physically united with others. Chimeras often arise **spontaneously** in perennial trees and can be detected through changes in morphology or physiology such as leaf colour, thorn number, or yield attributes.

2.2 Origin of Chimeras

1. **Somatic Mutation:** Most common origin , a mutation in one cell layer of the apical meristem gives rise to a genetically distinct lineage that co-exists with normal tissue.

2. **Graft Union:** Exchange of cells at graft junctions can produce **graft chimeras**, particularly in clonal seed orchards of *Coffea* or *Hevea*.
3. **Polyploidization Events:** Endopolyploidy or chromosome doubling in limited tissues leads to **cytochimeras** (mixoploids).
4. **Adventitious Shoot Formation:** Regeneration from multiple callus sectors in tissue culture can yield **sectorial chimeras** with mosaic genotypes.

2.3 Meristematic Structure and Histogenic Layers

The shoot apical meristem typically has three histogenic layers:

Layer	Position / Function	Cell Fate
L-I	Outer epidermal layer	Epidermis, stomata
L-II	Sub-epidermal	Gametes, leaf mesophyll
L-III	Inner corpus	Vascular tissues, cortex, pith

Mutations restricted to one layer yield different forms of chimeras depending on how mutated cells are distributed.

3. Types of Chimeras

3.1 Periclinal Chimera

» Mutant layer completely surrounds normal tissue in one histogenic layer (e.g., entire L-I).

» **Most stable** type; maintained through vegetative propagation.

» Example: Variegated leaves in *Camellia sinensis* where chlorophyll mutation in L-I forms white-green patterns.

» Used to develop ornamental tea accessions valued for pigmentation studies.

3.2 Mericlinal Chimera

» Mutated cells occupy only part of a single histogenic layer.

» **Unstable**, often reverts or becomes periclinal over time.

» Observed occasionally in *Hevea brasiliensis* leaves showing patchy anthocyanin expression.

3.3 Sectorial Chimera

» Mutation spans several layers in a sector of the meristem.

» Common in woody perennials after somatic mutation.

» **Example:** Branches of *Coffea arabica* differing in fruit colour (red vs. yellow) on the same tree.

3.4 Cytochimera (Mixoploid)

» Different ploidy levels coexist in tissues (diploid + tetraploid).

- Produced by chromosome doubling in limited zones or in vitro culture.
- *Camellia* and *Anacardium* have shown mixoploids with enlarged stomata and thicker leaves.

3.5 Graft Chimera

- Results from cellular fusion between stock and scion at graft junction.
- Rare but recorded in *Hevea* and *Coffea* experiments.
- May combine disease resistance of stock with quality traits of scion, though stability is uncertain.

4. Detection and Characterization of Chimeras

4.1 Morphological Indicators

- Sectorial differences in colour, leaf thickness, or trichome density.
- Variable fruit shape or seed number on different branches.

4.2 Cytological Analysis

- Determination of ploidy by counting chromosomes in root-tip or anther cells.
- Flow-cytometry detects mixoploidy rapidly.

4.3 Molecular Markers

- SSR, AFLP, or SNP markers reveal intra-plant polymorphism.
- Example: Somatic mutation in *Hevea brasiliensis* clone RRIM 600 detected by RAPD band variation between sectors.

4.4 Physiological and Biochemical Tests

- Isozyme or metabolite profiling distinguishes mutant tissues (e.g., altered caffeine biosynthesis in *Coffea* chimeras).
- Chlorophyll and phenol analyses reveal partial loss-of-function mutants in tea leaves.

5. Utilization of Chimeras in Breeding

5.1 As Sources of Novel Variation

Chimeras often carry **unique mutations** for colour, disease tolerance, or growth habit. By isolating and stabilizing such tissues, breeders can derive **bud sports** that serve as new clones or parents.

- ***Camellia sinensis:*** Bud sports yielding "purple-leaf" or "anthocyanin-rich" varieties used for functional-tea production.
- ***Hevea brasiliensis:*** Bark chimeras with improved latex-vessel distribution selected as superior clones.
- ***Coffea arabica:*** Yellow-berry mutants derived from red-berry trees via periclinal mutation; used in breeding for cup-quality diversity.

5.2 Stabilization Through Clonal Propagation

Once identified, chimeras must be propagated through methods that **preserve their histogenic structure**:

Propagation Method	Suitability for Chimera Maintenance
Budding / Grafting	Maintains periclinal arrangement
Stem Cuttings	Usually preserves L-I + L-II but may lose L-III
Tissue Culture (organogenesis)	Often breaks chimera structure
Somatic Embryogenesis	Produces uniform, non-chimeric regenerants

Therefore, conventional vegetative propagation is preferred when the chimera phenotype is desired.

5.3 Breeding Value of Chimeras

While most chimeras are valuable mainly for clonal release, some have been successfully converted into **sexually stable mutants** through:

1. **Gametogenic Inclusion:** If the mutation resides in L-II (gamete-forming layer), the variant gene may be transmitted to progeny.
2. **Recurrent Selection from Sport-Derived Seedlings:** Used in tea and cashew to capture favourable somatic variants.

6. Clonal Selection in Plantation Crops

6.1 Definition and Rationale

Clonal selection refers to the identification, evaluation, and multiplication of superior individual genotypes (clones) from genetically variable populations, ensuring faithful replication through vegetative propagation. It remains the **most successful breeding method** for plantation crops, enabling rapid dissemination of outstanding genotypes without waiting for sexual recombination.

6.2 Steps in Clonal Selection

1. **Exploratory Survey:** Locate high-yielding or disease-resistant trees in existing plantations.
2. **Preliminary Selection:** Evaluate yield records, morphological descriptors, and farmer feedback.
3. **Collection of Propagules:** Budwood, cuttings, or grafts taken from elite mother plants.
4. **Clonal Evaluation Trials:** Replicated field testing under uniform conditions for 4–6 years.
5. **Multi-Location Testing:** Assessment of genotype × environment interaction.
6. **Release and Registration:** Naming, documentation, and large-scale multiplication.

6.3 Evaluation Criteria

Criterion	Parameters Measured	Illustrative Crop
Productivity	Latex yield (g/tapping), nut number, bean weight	*Hevea*, *Cocos*, *Coffea*
Quality	Oil composition, polyphenol content, flavour	*Elaeis*, *Camellia*, *Coffea*
Stress Tolerance	Drought, salinity, temperature extremes	*Cashew*, *Coconut*
Disease Resistance	Incidence % of leaf/pod pathogens	*Hevea*, *Cocoa*
Morphology	Tree architecture, rooting ability	All perennials

6.4 Achievements of Clonal Selection

Crop	Selected Clones / Varieties	Key Features
Hevea brasiliensis	RRIM 600, PB 235, RRII 105	High latex yield, disease tolerance
Camellia sinensis	TV 1–29, UPASI 3, UPASI 9	Quality, drought tolerance
Coffea canephora	S.274, CxR composites	Vigour, rust resistance
Cocos nucifera	Kera Sagara, Kera Chandra	High copra recovery
Anacardium occidentale	VRI-3, Vengurla 4, BPP-8	High nut yield, bold kernels

Each released clone represents years of rigorous field testing and contributes directly to productivity gains.

6.5 Advantages of Clonal Selection

1. **Fixation of Heterosis:** Superior heterozygous combinations preserved indefinitely.
2. **Uniform Plantations:** Facilitates mechanized harvesting and quality consistency.
3. **Shortened Breeding Cycle:** No need for sexual generation turnover.
4. **Early Testing:** Young clonal trials predict mature performance accurately.
5. **Compatibility with Modern Propagation Systems:** Cuttings, grafting, micropropagation.

6.6 Limitations and Risks

1. **Genetic Uniformity → Vulnerability:** Monoclonal plantations risk epidemic losses (e.g., *Hevea* and *Microcyclus ulei*).
2. **Somatic Mutation Accumulation:** Aging clones may exhibit yield decline or off-types.
3. **Propagation Bottlenecks:** Limited rooting/grafting success restricts dissemination.
4. **Environmental Interaction:** Clones may perform differently under varying microclimates.
5. **Loss of Genetic Base:** Over-reliance on a few clones narrows diversity.

6.7 Maintenance of Clonal Fidelity

To ensure genetic integrity, the following are implemented:

- **Source Verification:** Certified mother plants with traceable lineage.
- **Periodic Molecular Fingerprinting:** SSR/SNP assays to detect off-types.
- **Quarantine and Indexing:** Virus- and pathogen-free propagation stock.
- **Controlled Nursery Environment:** Standardized hormone and substrate regimes.
- **Replacement of Aging Mother Plants:** After 10–12 years to prevent somatic drift.

7. Integration of Chimera and Clonal Studies with Modern Biotechnology

7.1 Somaclonal Variation

Tissue-culture-derived clones occasionally exhibit heritable variation ("somaclonal variation"). While undesirable for mass multiplication, it can yield novel traits such as dwarfism or disease tolerance.

- *Hevea brasiliensis*: Somaclonal lines with altered latex flow identified via RAPD markers.
- *Camellia sinensis*: Micropropagated plantlets showing enhanced catechin content stabilized through successive pruning.

Breeders must distinguish between **useful somaclonals** (for selection) and **ephemeral epigenetic changes** (to be eliminated).

7.2 Molecular Monitoring of Clones

High-throughput genotyping (SSR, SNP, GBS) enables:

1. **Identity verification** among large clone collections.
2. **Detection of mutation hotspots** responsible for somaclonal change.
3. **Epigenetic profiling** (methylation patterns) correlating with yield or stress response

4 – Inter-generic, Inter-specific and Inter-varietal Hybridization

1. Introduction

1.1 Definition and Purpose

Hybridization is the deliberate or natural crossing of genetically dissimilar individuals to combine desirable traits in one genotype. When parents belong to different **varieties of the same species**, the cross is **inter-varietal**; when they belong to **different species**, it is **inter-specific**; and when they originate from **different genera**, it is **inter-generic**.

In **plantation crops**, hybridization has been the most powerful instrument for:

1. Combining complementary characters (yield × quality × resistance).
2. Exploiting heterosis for productivity and vigour.
3. Transferring resistance genes from wild or related species.

4. Generating novel biochemical or architectural types suitable for mechanized or climate-resilient production.

1.2 Historical Perspective

Hybridization in perennial crops began early in the 20th century when natural hybrids were observed in *Coffea* and *Camellia*. Controlled crossing later gave rise to hybrid varieties that revolutionized productivity.

Period	Milestone in Hybridization of Plantation Crops	Key Achievements
1900 – 1930	Recognition of natural hybrids in *Camellia sinensis* and *Coffea*	Foundation of breeding programmes in Asia and Africa
1930 – 1960	First controlled inter-varietal crosses in tea, coconut and coffee	Early hybrids such as 'Kent × S.288' in *Arabica*
1960 – 1990	Expansion of inter-specific and inter-generic research	*Tenera* oil-palm hybrids; *Arabusta* coffee; Hevea crosses
1990 → present	Molecular assisted hybridization and wild introgressions	QTL-assisted crossing; DNA marker confirmation

2. Inter-varietal Hybridization

2.1 Concept and Objectives

Inter-varietal hybridization involves crossing different varieties or ecotypes within a species. Its goal is to combine high yield and quality with disease resistance or adaptation from diverse local gene pools. Because most plantation species exhibit wide intra-specific variability, this approach underlies modern cultivar development.

2.2 Examples and Achievements

Crop	Cross Combination	Objective and Result
Cocos nucifera	Tall × Dwarf (e.g., West Coast Tall × Chowghat Dwarf)	Early bearing, high copra yield; released as 'Kera Chandra' and 'Kera Sagara'
Hevea brasiliensis	RRIM 501 × PB 86	Higher latex yield, vigour, wind-resistance
Camellia sinensis	Assam × China types	Hybrid tea clones (TV series) combining yield + quality
Coffea arabica	Kent × S.288	Rust resistance with cup quality
Anacardium occidentale	Vengurla 4 × NRCC Sel-2	Larger nut, compact canopy

2.3 Procedure

1. **Parent selection** based on complementary traits.

2. **Controlled pollination** under bags or isolation cages.
3. **Hybrid seed collection and raising** in nurseries.
4. **Field evaluation** for yield, quality, and resistance.
5. **Clonal propagation** of elite hybrids.

2.4 Significance

» Simple and economical.

» Explores intra-specific variation before resorting to wide crosses.

» Source of heterosis and recombinants adapted to regional conditions.

3. Inter-specific Hybridization

3.1 Definition and Rationale

Inter-specific hybridization crosses individuals of different species within the same genus. It widens the genetic base and introduces traits absent in the cultivated species, particularly **disease resistance**, **abiotic-stress tolerance**, or **quality modifiers**.

3.2 Reproductive Barriers

1. **Pre-fertilization barriers:** pollen-stigma incompatibility, pollen tube arrest, different chromosome numbers.
2. **Post-fertilization barriers:** embryo abortion, endosperm failure, hybrid sterility.
3. **Overcoming barriers:** embryo rescue, bridge crosses, chromosome doubling, mentor pollination, and somatic hybridization.

3.3 Notable Examples

3.3.1 *Coffee (Coffea arabica* × *C. canephora)*, "Arabusta"

» **Objective:** Combine cup quality of *arabica* with vigour and disease resistance of *robusta.*

» **Technique:** Hand pollination, embryo rescue to prevent abortion.

» **Outcome:** Triploid hybrids ($2n = 3x = 33$) showing intermediate fertility and superior growth under low elevations.

» **Use:** as bridging parents in tetraploid breeding or for direct planting in Africa.

3.3.2 *Hevea brasiliensis* × *H. benthamiana* / *H. nitida*

» **Goal:** introgression of resistance to South-American Leaf Blight.

» **Barrier:** chromosomal imbalance ($2n = 36$ vs 34); solved by colchicine-induced tetraploidy and backcrossing.

» **Result**: hybrids with partial fertility and enhanced disease tolerance; foundation for future resistant lines.

3.3.3 *Theobroma cacao* × *T. grandiflorum*

- **Purpose:** Introduce flavour and disease resistance.
- **Method:** Controlled pollination and embryo culture.
- **Hybrid:** Sterile F_1 but useful for mapping and gene transfer.

3.3.4 *Camellia sinensis* × *C. assamica spp. lasiocalyx*

- **Objective:** Cold hardiness and flavour balance.
- Outcome: intermediate leaf size and improved cup quality; basis of China × Assam hybrids (TV series).

3.3.5 *Cocos nucifera* × *C. zeylanica*

- **Objective:** Improved drought tolerance and short stature.
- **Status:** Experimental; embryo culture required to rescue abortive hybrids.

3.4 Techniques to Overcome Hybridization Barriers

Technique	Purpose	Application in Crops
Embryo rescue	Prevent hybrid embryo death	*Hevea*, *Coffea*, *Cocoa*
Chromosome doubling	Restore fertility in sterile hybrids	*Hevea*, *Coffee*
Mentor pollination	Stimulate pollen-tube growth with irradiated compatible pollen	*Camellia*, *Coffea*
Bridge crossing	Use intermediate species for gene transfer	*Hevea* complex
Somatic hybridization	Protoplast fusion across species / genera	Experimental in *Coffea*, *Citrus*, *Camellia*

4. Inter-generic Hybridization

4.1 Definition and Scope

Inter-generic hybridization crosses individuals belonging to different genera within a family. It is extremely difficult due to severe genomic incompatibility but offers potential to transfer **novel resistance or physiological traits** unavailable within the genus.

4.2 Notable Attempts

Cross Combination	Objective	Outcome / Remarks
Hevea brasiliensis × *Manihot esculenta*	Transfer of resistance to leaf blight and root rot	Somatic hybrids produced via protoplast fusion; partial chlorophyll retention
Camellia sinensis × *C. oleifera*	High seed-oil content and cold tolerance	Sexual hybrids rare; embryo culture improves success

Coffea × *Psilanthus*	Introgression of drought tolerance	Confirmed via DNA markers; fertility low
Cocos nucifera × *Syagrus romanzoffiana*	Dwarfness and ornamental traits	Experimental in Brazil; no fertile offspring
Theobroma cacao × *Herrania purpurea*	Flavour and disease resistance	Embryo rescue successful; limited backcross fertility

4.3 Somatic Hybridization (Inter-generic via Protoplast Fusion)

Somatic hybridization circumvents sexual barriers by fusing protoplasts from two species or genera.

- **Procedure:** Enzymatic isolation of protoplasts → PEG or electrofusion → hybrid callus formation → regeneration.
- **Advantages:** Combines nuclear and cytoplasmic traits; overcomes incompatibility.
- **Applications:** Experimental *Coffea arabica* + *Psilanthus molecularis* hybrids showing enhanced water-use efficiency; *Camellia sinensis* + *C. oleifera* fusion produced oil-rich somatic hybrids.

Such hybrids often require chromosome doubling or embryo rescue to attain fertility.

5. Hybrid Evaluation and Selection

5.1 Early Screening

Hybrid seedlings are screened for:

- Morphological markers (leaf size, stem colour).
- Biochemical profiles (caffeine, catechins, oil composition).
- Molecular confirmation using SSR/SNP markers to verify true hybridity.

5.2 Field Evaluation

Selected hybrids undergo multi-year testing for yield, quality, and adaptability. Crosses showing consistent superiority are propagated clonally to stabilize performance.

5.3 Hybrid Maintenance

Because F_1 hybrids are heterozygous and often sterile, breeders maintain them vegetatively or reconstruct crosses periodically in seed gardens (e.g., oil palm Dura × Pisifera).

6. Breeding Significance of Hybridization

Type	Primary Goal	Key Outcome in Plantation Crops
Inter-varietal	Combine yield, quality, resistance	Kera Chandra (Coconut), TV 1 (Tea)
Inter-specific	Introduce new genes across species	Arabusta (Coffee), Hevea × benthamiana
Inter-generic	Transfer distant traits (resistance, oil quality)	Coffea × Psilanthus somatic hybrids

7. Challenges and Limitations

1. **Long Juvenile Phase:** Evaluation of F_1 takes years (5–10 years in Hevea and Cocos).
2. **Incompatibility and Sterility:** Most wide crosses require embryo rescue or chromosome doubling.
3. **Low Hybrid Fertility:** Limits recombination and introgression.
4. **Environmental Influence:** Hybrid expression varies with climate and management.
5. **Limited Infrastructure:** Need for pollination control facilities and molecular labs.

5 – Heterosis Breeding in Plantation Crops

1. Introduction

1.1 Concept of Heterosis

Heterosis, or **hybrid vigour**, refers to the phenomenon wherein the F_1 hybrid between genetically divergent parents exhibits superior performance in growth, yield, quality, or stress tolerance compared with either parent. It represents the positive expression of dominance and over-dominance effects of favorable alleles that mask deleterious recessives in heterozygous combinations.

In plantation crops, long-lived, mostly cross-pollinated and heterozygous species, heterosis breeding is a logical extension of inter-varietal and inter-specific hybridization programmes. The goal is to exploit **hybrid vigour** for traits such as yield, precocity, vigour, disease resistance and adaptability, while fixing this vigour through clonal propagation.

1.2 Historical Perspective of Heterosis Exploitation

Although heterosis was scientifically described by Shull (1908) and East (1909) in maize, its principle was recognized earlier by planters of perennial crops who observed the superior performance of naturally hybrid seedlings.

Period	Milestone in Heterosis Breeding	Key Crop Examples
1900–1930	Observation of vigourous seedlings from mixed stands	*Cocos nucifera*, *Camellia sinensis*
1930–1960	Systematic Tall × Dwarf crosses and interspecific hybrids	*Coconut*, *Coffee*
1960–1990	Hybrid seed gardens and clonal evaluation	Oil palm (*Dura* × *Pisifera*), Rubber (*Hevea crosses*)
1990 →	Molecular characterization of heterotic groups	Tea, Cocoa, Cashew

2. Genetic Basis and Theories of Heterosis

2.1 Dominance Hypothesis

Superior hybrid performance is due to masking of deleterious recessive alleles by dominant ones. Example: In *Coffea arabica*, hybrid plants show higher yield and disease tolerance because dominant resistant alleles from *robusta* mask recessive susceptibilities.

2.2 Over-dominance Hypothesis

Heterozygotes at certain loci have superior performance to either homozygote (e.g., enzyme activity or stress response loci). Observed in *Hevea brasiliensis* where hybrids show enhanced latex flow.

2.3 Epistasis or Interaction Hypothesis

Cumulative interaction among non-allelic genes creates novel metabolic balances. Examples include hybrid tea clones with optimized polyphenol biosynthesis.

2.4 Cytoplasmic and Maternal Effects

In perennials propagated via seed, maternal influences (e.g., cytoplasmic organelles and endosperm ratios) can contribute to hybrid vigour, especially in *Cocos* and *Coffea* crosses.

3. Measurement of Heterosis

Type of Heterosis	Formula (%)	Interpretation
Mid-parent		Average gain over both parents
Better-parent (heterobeltiosis)		Gain over better parent
Standard heterosis		Gain over standard commercial variety

Quantitative traits like yield show high heterosis (> 50 %) in *Cocos* and *Elaeis* hybrids, while quality traits often show moderate heterosis.

4. Heterosis Breeding Approaches in Major Plantation Crops

4.1 Coconut (*Cocos nucifera*)

Biological Basis: Tall types are cross-pollinated and vigorous but late-bearing; dwarf types are self-pollinated, early and precocious but less productive.

Strategy: Crossing Tall × Dwarf (T × D) and D × T to combine tall's yield with dwarf's precocity.

Hybrid	Parentage	Key Features
Kera Chandra	West Coast Tall × Chowghat Dwarf Orange	High copra yield, early bearing (3 yrs)
Kera Sagara	Lakshadweep Ordinary × Chowghat Dwarf Green	High oil content, shorter stature
Kalpa Samrudhi	West Coast Tall × Chowghat Dwarf Green	Superior fruit set under stress

Outcome: T × D hybrids yield 20–30 % more copra per palm and start bearing 2–3 years earlier than talls. Clonal seed gardens maintain these combinations through controlled pollination.

4.2 Oil Palm (*Elaeis guineensis*)

Parent Types: Dura (shell-thick), Pisifera (shell-less), and Tenera (thin shell).

Crossing Scheme: Dura × Pisifera → Tenera hybrid (F_1).

Result: Tenera fruits possess a higher mesocarp-to-kernel ratio, yielding 30–40 % more oil. This is the global standard for commercial oil-palm production.

Trait	Dura	Pisifera	Tenera (F_1)
Shell thickness (mm)	2–8	0	0.5–1.0
Oil % in bunch	15–18	18–20	23–25
Mesocarp ratio %	50	60	75

Hybrid seed production requires isolation gardens and trained pollination teams. Global programmes in Malaysia and India have standardized Dura × Pisifera seed gardens to sustain hybrid vigour.

4.3 Hevea brasiliensis (Rubber Tree)

Objective: Enhance latex yield, wind resistance, and disease tolerance.

Approach: Inter-clonal crosses among elite clones such as PB 86, RRIM 501, GT 1 produce F_1 seedlings with superior girth increment and latex flow.

Hybrid Combination	Yield Gain over Parent (%)	Notable Trait
RRIM 501 × PB 86	+35	Latex productivity
PB 217 × RRIM 600	+25	Wind tolerance
GT 1 × PB 235	+30	Early bark renewal

Elite seedlings are cloned for commercial release, thus fixing heterosis permanently.

4.4 Tea (*Camellia sinensis*)

Background: Tea is cross-pollinated; seed populations show wide variation.

Strategy: Controlled crosses between China and Assam types ("Assam Jat × China Jat") exploit hybrid vigour for yield and quality.

Outcome: Clones such as TV 1, TV 9, UPASI 3 originated from seedling hybrids showing 30–40 % yield gain. Once selected, these are maintained as clones to preserve heterosis.

4.5 Coffee (*Coffea arabica* and *C. canephora*)

1. **Intra-specific Hybrids:** Arabica × Arabica crosses (e.g., Kent × S.288 → S.795) combine yield and rust resistance.
2. **Inter-specific Hybrids:** Arabica × Robusta ("Arabusta") for vigour and adaptability at low altitudes.
3. **Triploid Hybrids:** Show larger leaves and berries, useful for vegetative propagation.

Result: S.795 and Selection 9 exhibit 25–30 % higher yield and cup quality.

4.6 Cocoa (*Theobroma cacao*)

Hybridization between Criollo and Forastero types produces the **Trinitario** group , vigourous, disease-resistant, and high-flavour. Controlled hybrids such as ICG 156 and ICS 95 combine Forastero vigour with Criollo flavour.

4.7 Cashew (*Anacardium occidentale*)

Hybrid vigour is expressed in nut size and apple weight. Crosses among elite clones (Vengurla 4 × NRCC Sel-2) produce F_1 seedlings with 20–25 % higher yield. Promising hybrids are grafted for clonal release.

5. Exploitation and Fixation of Heterosis

5.1 Clonal Fixation

Because plantation crops are perennial and vegetatively propagated, heterosis can be fixed by cloning the F_1 hybrid. Examples include Tenera oil palm and hybrid tea clones. This is analogous to F_1 seed use in annuals but requires clonal multiplication systems.

5.2 Hybrid Seed Gardens

Where seed propagation is used (*Cocos*, *Elaeis*), heterosis is maintained by reconstructing specific parental crosses each generation under controlled pollination to produce uniform F_1 seed.

5.3 Recurrent Selection for Combining Ability

Long-term programmes in Hevea and Cocos evaluate parental combinations for general (GCA) and specific (SCA) combining ability. High-GCA parents form heterotic pools for sustained gain.

6. Molecular and Physiological Bases of Heterosis

6.1 Molecular Mechanisms

» **Increased gene expression breadth** in hybrids.

» **Epigenetic reprogramming** leading to enhanced metabolism.

» **Additive mitochondrial and chloroplast function** improving photosynthesis.

» **Allelic complementation** of stress-response genes.

Genomic studies in *Coffea* and *Camellia* show heterotic F_1s have higher expression of genes related to carbon fixation, sugar transport, and antioxidant systems.

6.2 Physiological Manifestations

» Greater leaf area and chlorophyll density.

» Enhanced root volume for nutrient uptake.

» Higher photosynthetic rate and transpiration efficiency.

» Improved enzyme activities (latex biosynthesis, oil accumulation).

7. Advantages and Limitations of Heterosis Breeding

Advantages	Limitations / Challenges
Combines complementary traits rapidly	Requires controlled pollination and isolation
Fixes heterosis via cloning	High cost and labour for hand pollination
Produces uniform, high-yielding clones	Limited availability of divergent parents
Improves stress tolerance and adaptability	Long juvenile period delays evaluation
Increases economic return per unit area	Sterility in wide hybrids reduces breeding scope

8. Examples of Commercial Heterotic Hybrids in Plantation Crops

S. No.	Crop	Hybrid Name / Designation	Parentage	Gain over Parent (%)	Propagation Mode
1	Coconut	Kera Chandra	West Coast Tall × Chowghat Dwarf Orange	+30	Seed
2	Oil Palm	Tenera	Dura × Pisifera	+40	Seed
3	Rubber	RRIM 501 × PB 86	Inter-clonal cross	+35	Clone
4	Tea	TV 1 / TV 9	Assam × China	+30	Clone
5	Coffee	S.795	Kent × S.288	+25	Seed
6	Cocoa	Trinitario	Criollo × Forastero	+20	Seed / Clone
7	Cashew	Vengurla 4 × NRCC Sel-2	Intra-varietal	+25	Graft

6 Mutation and Polyploidy Breeding in Plantation Crops

1. Introduction

1.1 Rationale

Genetic improvement in plantation crops is constrained by their long juvenile phase, high heterozygosity, and limited sexual recombination. To overcome these limitations, **mutation** and **polyploidy breeding** offer unique, non-conventional routes for inducing heritable variation and generating new genotypes **without hybridization.**

Both methods are particularly valuable where:

- Crossing barriers restricts gene flow.
- Clonal uniformity has narrowed variability.
- Novel ideotypes (dwarf, early, disease-tolerant) are required rapidly.

2. Mutation Breeding

2.1 Definition

A **mutation** is a sudden heritable change in the genetic material. When such alterations are artificially induced through physical or chemical mutagens to create useful variability, the process is known as **mutation breeding**.

2.2 Types of Mutations

Category	Nature of Change	Examples / Remarks
Gene / Point mutation	Change in one or few nucleotides	Alters single enzyme or pigment gene (e.g., leaf colour mutants in *Camellia*)
Chromosomal mutation	Deletion, duplication, inversion, translocation	Structural variants affecting fertility and growth
Genome mutation	Gain or loss of entire sets of chromosomes	Basis for polyploidy (treated separately)
Somatic mutation (bud sport)	Mutation in vegetative cells; produces chimeras	Common in *Coffea*, *Camellia*, *Hevea*

2.3 Mutagenic Agents

2.3.1 Physical Mutagens

» **Gamma rays** (^{60}Co source): most widely used for perennial crops.

» **X-rays**: Effective for seeds and buds.

» **Fast neutrons**: Produce large deletions.

» **UV radiation**: Causes pyrimidine dimers in tissue culture.

2.3.2 Chemical Mutagens

» **Ethyl methane sulphonate (EMS)** – alkylates DNA bases; highly efficient.

» **Sodium azide** – induces point mutations at low concentrations.

» **Diethyl sulphonate (DES), Nitroso guanidine (NTG)** – potent but require strict safety controls.

2.4 Methodology of Mutation Breeding

1. **Selection of Material** – seeds, pollen, buds, or tissue-culture calli.
2. **Optimization of Dose** – determining LD_{50} (dose causing 50 % mortality).
3. **Treatment and Recovery** – exposure to mutagen; recovery of viable propagules.
4. **Raising M_1 Generation** – identification of putative mutants; often chimeric.
5. **Stabilization (M_2–M_4)** – fixation of mutations by clonal propagation or selfing.
6. **Evaluation and Release** – field testing for stability and performance.

2.5 Examples of Mutation Breeding in Plantation Crops

S. No.	Crop	Mutagen Used	Mutant Developed / Trait
1	*Camellia sinensis*	Gamma rays (20 Gy)	'TRI 2043' – early sprouting and drought tolerance
2	*Hevea brasiliensis*	Gamma rays on budwood	Clone RRIM 703 – improved latex yield
3	*Coffea arabica*	EMS treatment of seeds	Rust-resistant mutant lines
4	*Theobroma cacao*	UV irradiation	Pod-shape and anthocyanin mutants
5	*Anacardium occidentale*	Gamma rays (10–15 Gy)	Compact canopy, dwarf habit
6	*Cocos nucifera*	Chemical mutagenesis in embryos	Experimental dwarfs with early bearing

2.6 Applications and Utility

1. **Induction of Novel Morphotypes:** dwarf, compact, early-maturing types suited for high-density planting.
2. **Quality Improvement:** altered caffeine, theobromine, or polyphenol content in beverage crops.
3. **Stress Tolerance:** gamma-induced drought-resistant clones in *Camellia.*
4. **Disease Resistance:** mutants resistant to leaf spot and rust in *Coffea* and *Hevea.*
5. **Functional Foods:** pigment-enriched mutants (anthocyanin-rich tea leaves).

2.7 Mutation Breeding via Tissue Culture (In Vitro Mutagenesis)

Combining mutation with micropropagation accelerates the generation of stable variants.

Procedure

» Treat callus or somatic embryos with EMS or gamma rays.

» Regenerate plantlets via organogenesis or somatic embryogenesis.

» Screen at the cellular level using molecular or biochemical assays.

Advantages

» Produces large populations in limited space.

» Reduces chimerism; mutants arise from single cells.

» Enables selection for biochemical traits (secondary-metabolite content, enzyme profiles).

Example: In *Camellia sinensis*, gamma-irradiated callus lines yielded clones with increased catechin levels and drought tolerance.

2.8 Advantages and Limitations of Mutation Breeding

Advantages	Limitations / Cautions
Creates novel alleles absent in germplasm	Many mutations deleterious or sterile
Rapid and cost-effective	Requires intensive screening
Applicable to clones where crossing is difficult	May cause chimeras in treated buds
Preserves parental genotype except at mutated locus	Limited predictability of outcome

3. Polyploidy Breeding

3.1 Definition and Concept

Polyploidy is the condition of possessing more than two complete sets of chromosomes. Polyploids may arise spontaneously or be induced artificially through agents such as **colchicine** or **oryzalin.**

Polyploidy modifies morphology, physiology, and biochemistry by increasing cell size, altering gene dosage, and enhancing tolerance to stress. Many plantation crops naturally include polyploids (*Coffea arabica* – $2n = 4x = 44$; *Camellia sinensis* – $2n = 30, 45, 60$), making polyploidy breeding a practical route for improvement.

3.2 Types of Polyploidy

Type	Origin	Example / Significance
Autopolyploid	Duplication of same genome within species	Induced in *Camellia* and *Cashew* for larger leaves and fruits
Allopolyploid	Combination of genomes from different species	*Coffea arabica* (from *C. canephora* × *C. eugenioides*)
Segmental Allopolyploid	Partial pairing among homologues	*Hevea* species hybrids
Aneuploid / Mixoploid	Irregular chromosome numbers	Often unstable; useful for cytogenetic studies

3.3 Induction of Polyploidy

3.3.1 Chemical Methods

» **Colchicine:** inhibits spindle formation, doubling chromosomes in meristem cells.

» **Oryzalin / Trifluralin:** microtubule inhibitors effective at lower, safer concentrations.

Procedure:

1. Treat apical buds, seeds, or callus with colchicine (0.05–0.2 %).
2. Rinse, recover, and grow under controlled conditions.
3. Confirm doubling via chromosome counts or flow cytometry.

3.3.2 In Vitro Induction

Tissue culture systems permit precise treatment of explants; regenerated polyploids are screened for morphology and DNA content.

3.4 Cytological Confirmation

Indicator	Method	Expected Change in Polyploid
Chromosome count	Microscopic analysis of root tips	Doubling (e.g., 2n → 4n)
Stomatal guard-cell size	Micrometry	Larger (1.3–1.6×) than diploid
DNA content	Flow cytometry	Twofold increase
Leaf thickness and chlorophyll	Morphometry	Increased values

3.5 Polyploid Breeding Achievements

Crop	Type of Polyploid	Method / Trait	Result / Significance
Coffea arabica	Natural allotetraploid (2n = 4x = 44)	Combines genomes of *C. canephora* & *C. eugenioides*	Superior flavour, self-fertility
Camellia sinensis	Induced tetraploids (4x = 60)	Colchicine on shoot tips	Larger leaves, higher biomass
Hevea brasiliensis	Synthetic tetraploids (4x = 72)	Colchicine treatment of buds	Used as male parent to produce triploids
Cocos nucifera	Triploids (3x = 48)	Dwarf × tetraploid cross	Compact palms with high copra yield
Anacardium occidentale	Autotetraploids	Oryzalin on seedlings	Bold nut, thick apple
Theobroma cacao	Triploids	2x × 4x crossing	Seedless pods for confectionery use

3.6 Triploid Breeding in Rubber and Coconut

In Hevea, tetraploid clones (e.g., RRIM 703) crossed with diploids produce **triploid progeny** (3x = 54) showing thick bark, higher latex flow, and wind resistance. In *Cocos nucifera*, controlled 4x × 2x crosses yield **triploid hybrids** with intermediate height and precocity, with potential for high-density plantations.

3.7 Physiological and Biochemical Consequences of Polyploidy

1. **Increased Cell Size:** larger leaves, fruits, and stomata.
2. **Reduced Growth Rate:** often slower but more robust.
3. **Enhanced Drought Tolerance:** due to reduced transpiration.
4. **Modified Secondary Metabolism:** higher caffeine or polyphenols.

5. **Altered Fertility:** sterility in odd-ploid (triploid) hybrids useful for seedless fruits.

3.8 Applications of Polyploidy Breeding

1. **Improvement of Yield Components:** polyploids with larger organ size and thicker leaves.
2. **Creation of Seedless Varieties:** triploid *Cacao* with parthenocarpy potential.
3. **Breeding Bridge for Interspecific Crosses:** tetraploids facilitate pairing and gene flow between species (e.g., *Hevea* and *H. benthamiana*).
4. **Development of Ornamental Types:** polyploid *Camellia* with novel flower forms and colours.
5. **Enhancement of Secondary Metabolites:** tetraploid tea lines richer in catechins and theaflavins.

3.9 Integration with Modern Tools

Flow Cytometry: rapid screening of thousands of regenerants for ploidy.

Molecular Markers: confirm genetic stability of induced polyploids.

Transcriptomics: reveals differential gene expression controlling cell size and metabolite pathways.

CRISPR-Cas9 Editing: now explored to induce targeted chromosomal duplications for polyploid creation without chemical mutagens.

3.10 Advantages and Limitations

Advantages	Limitations / Challenges
Creates instant variability in clones	High frequency of sterility and abnormalities
Enhances organ size and product yield	Slow growth and delayed flowering
Improves stress tolerance	Complex cytogenetics complicates breeding
Facilitates distant hybridization	Requires cytological expertise
May produce seedless or parthenocarpic forms	Difficult to stabilize odd-ploid forms

4. Comparative Overview: Mutation vs Polyploidy Breeding

Aspect	Mutation Breeding	Polyploidy Breeding
Genetic change	Gene or chromosome alteration	Whole-genome duplication
Predictability	Random	Semi-directed
Generation time	One or two cycles	Multiple cycles for stabilization
Applicability	Both annuals and perennials	Mainly perennials
Common purpose	Induce new alleles	Enlarge organs / bridge species
Example	Gamma-induced rust-resistant coffee	Triploid coconut

5. Future Prospects

1. **High-throughput Mutation Libraries:** catalogued mutant collections in *Camellia* and *Coffea* for functional-genomics studies.
2. **Targeted Mutagenesis:** CRISPR-based knock-outs replacing random chemical mutagenesis.
3. **Synthetic Polyploid Platforms:** controlled chromosome doubling for wide hybrid stabilization.
4. **Integration with Omics:** metabolomics + transcriptomics to identify desirable gene-dosage effects.
5. **Genome Editing-Induced Polyploidy:** precision duplication of specific chromosomes for trait enhancement.

7 – Resistance Breeding to Biotic and Abiotic Stresses in Plantation Crops

1. Introduction

1.1 Importance of Resistance Breeding

Plantation crops, such as *Cocos nucifera* (coconut), *Elaeis guineensis* (oil palm), *Hevea brasiliensis* (rubber), *Camellia sinensis* (tea), *Coffea arabica* (coffee), *Theobroma cacao* (cocoa), and *Anacardium occidentale* (cashew), are long-lived perennials exposed to numerous **biotic and abiotic stresses** over their productive lifespan. Diseases, pests, drought, temperature extremes, and salinity can cumulatively cause yield reductions of 30–70%.

Resistance breeding aims to develop cultivars capable of **withstanding or escaping these stresses** through genetic improvement rather than external chemical or management inputs, thereby ensuring **sustainable production, reduced pesticide dependence, and environmental safety.**

2. Concept and Principles of Resistance Breeding

2.1 Definitions

- **Resistance:** The ability of a plant to restrict the establishment, development, or damage caused by a pathogen or pest.
- **Tolerance:** Ability to endure infestation or stress while maintaining acceptable yield.
- **Immunity:** Complete absence of infection (rare in perennials).
- **Escape:** Avoidance of disease due to morphological or phenological traits.

2.2 Genetic Basis

Resistance may be controlled by:

1. **Major (R) genes** – provide vertical or specific resistance (e.g., SH genes in coffee rust).
2. **Minor (polygenic) genes** – provide horizontal or non-specific resistance.
3. **Gene-for-gene interactions** – pathogen avirulence gene matches plant resistance gene.

2.3 Breeding Strategies

Approach	Principle	Suitable Crops
Conventional selection	Phenotypic screening for resistance	*Hevea, Cashew*
Hybridization and introgression	Transfer resistance genes from wild relatives	*Coffee, Cocoa*
Mutation breeding	Induction of resistant mutants	*Tea, Cashew*
Marker-assisted selection (MAS)	DNA marker-linked R-gene tracking	*Coffee, Cocoa, Hevea*
Transgenic / genome editing	Gene insertion or modification	Experimental in *Hevea, Camellia*

3. Breeding for Resistance to Biotic Stresses

3.1 Major Diseases of Plantation Crops

S. No.	Crop	Major Pathogen / Pest	Type of Damage
1	*Coffea arabica*	*Hemileia vastatrix* (coffee leaf rust)	Defoliation, yield loss
2	*Hevea brasiliensis*	*Microcyclus ulei* (South American Leaf Blight)	Epidemic leaf fall
3	*Camellia sinensis*	*Exobasidium vexans* (blister blight)	Leaf lesions, reduced photosynthesis
4	*Theobroma cacao*	*Phytophthora palmivora* (black pod), *Moniliophthora perniciosa* (witch's broom)	Pod rot, twig deformation
5	*Cocos nucifera*	*Phytophthora katsurae* (bud rot)	Crown death
6	*Anacardium occidentale*	Tea mosquito bug (*Helopeltis antonii*)	Flower and nut loss
7	*Elaeis guineensis*	*Ganoderma boninense* (basal stem rot)	Vascular decay, death

3.2 Sources of Resistance

1. **Wild Relatives:** *Coffea canephora* – donor of rust resistance genes (*SH1–SH9*).*Hevea benthamiana* – donor of leaf-blight tolerance.*Theobroma grandiflorum* – donor of disease resistance in cocoa.*Camellia irrawadiensis* – donor of blister-blight resistance.
2. **Landraces:** Traditional *Cocos* and *Cashew* landraces adapted to local environments often display field tolerance to multiple pathogens.
3. **Induced Mutants:** Gamma-irradiated *Camellia* and *Coffea* mutants with enhanced tolerance to leaf and root pathogens.

3.3 Mechanisms of Disease Resistance

Type	Description	Example
Structural	Cuticular thickness, lignified tissues, trichomes	Cashew nuts with hard shells resist *Helopeltis*
Biochemical	Phenolics, tannins, alkaloids inhibit pathogen growth	High catechin in tea resists *Exobasidium*
Physiological	Rapid hypersensitive response, leaf abscission	*Coffea arabica* resists rust by HR reaction
Molecular	Activation of pathogenesis-related (PR) proteins	PR-2 and PR-5 genes in *Camellia* under infection

3.4 Case Studies of Biotic-Resistance Breeding

3.4.1 Coffee Leaf Rust – *Hemileia vastatrix*

» The most devastating disease of *Coffea arabica.*

» Resistance derived from *C. canephora* and *C. liberica.*

» Genes SH1–SH9 govern specific resistance; pyramiding using MAS has created multiline varieties.

» Example: 'Selection 9' and 'S.795' – high yield + rust resistance.

3.4.2 Blister Blight in Tea – *Exobasidium vexans*

» Severe defoliating disease in humid zones.

» Resistant donors: *Camellia irrawadiensis, C. assamica subsp. lasiocalyx.*

» Screening through detached-leaf assay; phenolic content correlates with tolerance.

» Resistant clones: TV 26, UPASI 9, TRI 2043 (mutant-derived).

3.4.3 South American Leaf Blight – *Microcyclus ulei* in Hevea*

» Endemic in South America; potential threat elsewhere.

» Resistance from *H. benthamiana* introgressed into *H. brasiliensis* via inter-specific hybridization and polyploid bridge.

» Molecular markers for R genes now used to monitor crosses.

3.4.4 Cocoa Pod Rots – *Phytophthora palmivora* and *M. perniciosa*

» Resistance found in wild *Theobroma grandiflorum* and *T. subincanum.*

» Polygenic inheritance; screening under artificial inoculation.

» Resistant hybrids: ICS 95, Scavina 6, TSH 1188.

» Genomic tools identify QTLs on linkage groups 1, 3, and 9 associated with resistance.

3.4.5 Cashew Tea Mosquito Bug (*Helopeltis antonii*)

- Major pest in India; causes 30–40 % yield loss.
- Field screening identified tolerant genotypes: VRI-3, NRCC Sel-2.
- Tolerance correlated with higher phenol and tannin content.

3.4.6 Basal Stem Rot in Oil Palm – *Ganoderma boninense*

- Fungal disease causing root decay and mortality.
- Tolerant *Elaeis oleifera* used in interspecific crosses with *E. guineensis*.
- QTL mapping identified markers linked to tolerance genes.

3.5 Screening and Evaluation Methods

Technique	Purpose	Crop Examples
Artificial inoculation	Controlled disease pressure for selection	*Cocoa, Tea*
Detached-leaf or leaf-disc assay	Rapid evaluation of resistance	*Hevea, Camellia*
Field screening	Multi-season confirmation	*Cocos, Cashew*
Biochemical assays	Quantify phenolics, PR proteins	*Coffee, Tea*
Molecular markers	Confirm resistant alleles	*Coffee* (SH genes), *Cocoa* (QTLs)

4. Breeding for Resistance to Abiotic Stresses

4.1 Nature and Impact

Abiotic stresses such as **drought, salinity, temperature extremes, and flooding** severely limit productivity of perennials that remain exposed for decades. With climate change, breeding for **multi-stress resilience** has become a top priority.

4.2 Genetic and Physiological Basis

Stress Type	Plant Response / Mechanism	Key Genetic Control
Drought	Osmotic adjustment, deep roots, waxy leaves	Polygenic; QTLs for water-use efficiency
Heat	Heat-shock proteins, canopy temperature regulation	Inducible HSP genes
Salinity	Ion exclusion, Na^+/K^+ homeostasis	Transporter genes (NHX1, HKT1)
Flooding	Aerenchyma formation, ethylene signaling	Adaptive regulatory genes
Cold	Antifreeze proteins, membrane stabilization	CBF transcription factors

4.3 Crop-Specific Efforts

4.3.1 Drought Tolerance in Coconut

- *West Coast Tall* and *Laccadive Tall* tolerant due to deep root system and osmotic adjustment.
- Breeding goal: develop hybrids (e.g., WCT × Chowghat Dwarf) with drought tolerance + early bearing.
- Physiological markers: leaf water potential, proline accumulation, stomatal conductance.

4.3.2 Drought and Heat Tolerance in Tea

- Clones UPASI 3 and TRI 2023 maintain high leaf water potential.
- Screening under field drought conditions for survival and recovery rate.
- QTLs for drought tolerance mapped on linkage groups 4 and 9 using SSR markers.

4.3.3 Salinity Tolerance in Cashew and Coconut

- Coastal populations of *Cocos nucifera* naturally adapted to saline soils.
- Salt-tolerant selections (Benaulim Tall, Lakshadweep Ordinary) used as parents.
- Cashew clones Vengurla 6 and NRCC Sel-2 tolerate EC up to 8 dS/m due to ion exclusion.

4.3.4 Cold Tolerance in Tea and Coffee

- High-altitude *Camellia sinensis var. sinensis* and *Coffea arabica var. typica* exhibit cold hardiness.
- Controlled hybridization with hardy types used to expand cultivation limits.
- Cold tolerance is associated with the accumulation of soluble sugars and unsaturated lipids.

4.3.5 Flooding Tolerance in Hevea and Oil Palm

- *Hevea spruceana* and *H. nitida* exhibit tolerance to waterlogging.
- Used as donors in inter-specific breeding for low-lying regions.
- Oil palm hybrids with *E. oleifera* show improved root aeration and survival.

4.4 Physiological and Biochemical Markers for Abiotic-Stress Tolerance

Marker	Mechanistic Relevance	Applicable Crops
Leaf relative water content (RWC)	Indicates dehydration avoidance	*Tea, Cashew*
Proline accumulation	Osmotic adjustment	*Hevea, Coconut*
Antioxidant enzymes (SOD, CAT, POD)	ROS scavenging	*Coffee, Cocoa*
Chlorophyll stability index	Heat tolerance indicator	*Camellia, Cashew*
Electrolyte leakage	Membrane injury	All perennials

4.5 Breeding and Screening Techniques

1. **Controlled Environment Screening** – growth chambers simulate drought or heat stress for early selection.
2. **Field Phenotyping Networks** – multi-locational trials in contrasting stress zones.
3. **Marker-Assisted Selection (MAS)** – identification of QTLs associated with stress tolerance.
4. **Physiological Selection Indices** – integration of physiological and morphological traits into composite scores.

5. Integration of Biotic and Abiotic Resistance Breeding

5.1 Need for Combined Resistance

Because perennials face multiple stresses simultaneously, breeding for **combined resistance** (biotic + abiotic) provides durable resilience.

Example: *Coffea arabica* 'Selection 9' combines rust resistance (SH genes) with drought tolerance (deep root system).

5.2 Breeding Framework

Stage	Activity	Tools Used
Germplasm screening	Identify tolerant sources	Pathogen assays, physiological tests
Genetic analysis	Determine inheritance	Segregating populations
Hybridization	Combine tolerance genes	Interspecific / intervarietal crosses
Molecular validation	Confirm introgression	SSR, SNP, QTL mapping
Field testing	Evaluate durability	Multi-location trials

6. Role of Biotechnology in Resistance Breeding

6.1 Marker-Assisted Breeding

- SH gene markers in *Coffea* enable pyramiding of rust resistance.
- SSR-linked markers for Phytophthora tolerance in *Cocoa*.
- SNP-based selection for drought-tolerance QTLs in *Tea*.

6.2 Transgenic Approaches

- **Antifungal genes** (chitinase, glucanase) introduced into *Hevea* to combat leaf diseases.
- **Osmoprotectant genes** (e.g., *P5CS*, *DREB*) transferred into *Camellia* to improve drought tolerance (experimental).
- **RNA interference (RNAi)** strategies for insect resistance in *Cashew*.

6.3 Genomic and Omics Integration

1. **Transcriptomics:** identifies stress-responsive gene networks (e.g., PR-proteins, HSPs).
2. **Proteomics:** characterizes defense enzymes and antioxidants.
3. **Metabolomics:** quantifies phenolics, flavonoids, and alkaloids contributing to resistance.
4. **Genome Editing:** CRISPR/Cas used for targeted insertion or activation of resistance genes in *Hevea* and *Camellia* (pilot studies).

7. Challenges and Constraints

1. **Long Breeding Cycles:** Evaluation for disease/stress resistance may require 10–15 years.
2. **Complex Polygenic Inheritance:** Difficult to fix resistance genes in heterozygous clones.
3. **Pathogen Variability:** Emergence of new virulent strains may overcome resistance.
4. **Trade-offs:** Resistance is sometimes associated with yield or quality penalties.
5. **Resource Limitation:** High cost of screening and molecular validation in woody perennials.

8. Breeding for Improving Quality in Plantation Crops

1. Introduction

1.1 Importance of Quality Improvement

In plantation crops, *quality* determines not only consumer preference and market price but also the **industrial utility and export competitiveness** of the produce. While yield drives economic output, **quality defines sustainability and brand value**.

Each plantation crop has a distinctive biochemical composition, flavour, aroma, fatty-acid profile, caffeine, polyphenols, latex composition, or kernel properties, governed by its **genetic makeup** and **environmental interactions**.

Breeding for quality therefore, involves improving the **chemical, physical, and sensory attributes** of economic parts (beans, nuts, leaves, latex, or fruits) without compromising yield or resistance.

1.2 Determinants of Quality

Crop	Primary Quality Parameters	Determinants
Camellia sinensis (tea)	Catechins, theaflavins, aroma	Polyphenol biosynthesis, leaf age
Coffea arabica / C. canephora (coffee)	Caffeine, chlorogenic acids, sugars	Species, altitude, roasting potential
Theobroma cacao (cocoa)	Fat %, theobromine, flavour precursors	Genetic type, fermentation quality
Cocos nucifera (coconut)	Oil %, medium-chain fatty acids	Genotype, maturity

Elaeis guineensis (oil palm)	Palmitic–oleic–linoleic ratio	Fruit form, mesocarp oil genes
Hevea brasiliensis (rubber)	Hydrocarbon chain quality (cis-1,4-polyisoprene)	Enzyme activity, clonal genotype
Anacardium occidentale (cashew)	Kernel size, shelling %, tannin content	Genetics, orchard environment

2. Genetic Basis of Quality Traits

2.1 Nature of Inheritance

Quality traits in plantation crops are often **quantitative,** governed by multiple genes influenced by the environment. Some, however, show **simple Mendelian control**.

Type	Example	Inheritance
Monogenic	Caffeine biosynthesis gene (*CaXMT1*) in coffee	Simple dominant/recessive
Oligogenic	Fatty-acid composition in coconut	2–3 loci interaction
Polygenic	Polyphenol and flavour traits in tea, cocoa	Additive and epistatic effects

2.2 Genetic–Biochemical Pathways

1. **Phenylpropanoid Pathway** – governs catechins, lignins, flavonoids in tea.
2. **Methylxanthine Pathway** – synthesizes caffeine, theobromine, theophylline in coffee and cocoa.
3. **Lipid Biosynthesis Pathway** – controls oil composition in coconut, palm, and cashew.
4. **Isoprenoid Pathway** – produces latex polymers in *Hevea.*

3. Breeding Objectives for Quality Improvement

1. **Enhance desirable chemical constituents** (flavour, aroma, oil composition, latex quality).
2. **Reduce undesirable components** (bitterness, high caffeine, high saturated fat).
3. **Stabilize quality across environments** (high heritability, low genotype × environment interaction).
4. **Develop specialty products** (low-caffeine coffee, high-oleic palm oil, antioxidant-rich tea).
5. **Combine quality with yield and resistance** for overall genetic gain.

4. Breeding Approaches for Quality Improvement

4.1 Conventional Selection and Hybridization

Phenotypic selection remains the first step when heritability is high. In perennial crops, **clonal selection** is most effective for fixing quality traits.

Crop	Method	Example / Outcome
Camellia sinensis	Clonal selection	TV 1, TV 9 – high polyphenols and flavour
Coffea arabica	Inter-varietal hybridization	S.795 – superior cup quality and aroma
Theobroma cacao	Recurrent selection	ICS 95, TSH 1188 – balanced flavour and yield
Cocos nucifera	Hybridization (T × D)	'Kera Chandra' – high oil %, early maturity
Elaeis guineensis	Selection within *Tenera*	High-oleic, low-palmitic genotypes

4.2 Mutation Breeding for Quality

» Gamma-ray-induced mutants in tea (TRI 2043) with higher catechin levels.

» EMS-induced low-caffeine mutants in *Coffea arabica.*

» UV-induced flavour variants in cocoa. Mutation modifies specific metabolic genes without altering gross morphology.

4.3 Hybridization and Heterosis for Quality

Objective: Combine biochemical superiority of one parent with yield or adaptability of another.

» *C. arabica* × *C. canephora* (Arabusta) hybrids show improved body and flavour.

» *Camellia sinensis* × *C. assamica* hybrids yield higher theaflavin ratios.

» *Cocos Tall* × *Dwarf* hybrids maintain higher lauric acid content.

Heterosis for quality is moderate but cumulative selection stabilizes it over cycles.

4.4 Marker-Assisted Selection (MAS)

DNA markers linked to biochemical traits enable **early, non-destructive selection.**

Trait / Gene	Marker Type	Crop / Application
Catechin biosynthesis (LAR, ANR genes)	SSR, SNP	*Camellia sinensis*
Caffeine metabolism (N-methyltransferases)	CAPS	*Coffea arabica*
Fatty-acid desaturase genes	SNP	*Cocos nucifera, Elaeis guineensis*
Latex biosynthesis (Hev b genes)	SSR	*Hevea brasiliensis*
Kernel flavour and sugars	AFLP	*Anacardium occidentale*

4.5 Biotechnological Approaches

4.5.1 Tissue Culture and Somaclonal Variation

Somaclonal variants may exhibit altered secondary metabolism. **Example:** micropropagated *Camellia* clones with enhanced theaflavin:catechin ratio.

4.5.2 Genetic Transformation

- Antisense suppression of caffeine synthase genes yields **low-caffeine coffee**.
- Over-expression of phenylalanine ammonia-lyase (PAL) enhances polyphenols in tea.
- Expression of *Hev b 1* and *Hev b 7* genes regulates latex biosynthesis in *Hevea*.

4.5.3 Metabolic Engineering

Redirects carbon flux towards desired metabolites using pathway gene manipulation.

4.5.4 Genome Editing (CRISPR/Cas)

Targeted knock-out of undesirable genes (e.g., caffeine synthase) or activation of quality-enhancing alleles (e.g., lipid desaturase) is now under experimental validation.

5. Crop-Specific Quality Breeding

5.1 Tea (*Camellia sinensis*)

Key Traits

- Polyphenol content (25–30 % of dry leaf).
- Ratio of catechins: theaflavins: thearubigins.
- Caffeine (~2–4 %).
- Aroma volatiles (linalool, geraniol).

Breeding Strategies

- Selection of clones with high catechin content (TV 22, UPASI 9).
- Hybridization of Assam and China types for flavour balance.
- Mutation breeding for early flush and reduced bitterness.
- MAS targeting *ANR*, *LAR*, and *DFR* genes.

Outcome: Development of functional teas rich in antioxidants, anthocyanin teas, and green-tea cultivars with low caffeine.

5.2 Coffee (*Coffea arabica* and *C. canephora*)

Quality Components

- Caffeine (0.8–2.4 %).
- Chlorogenic acid (CGA), trigonelline, and sucrose.
- Cup quality (flavour, aroma, acidity, body).

Breeding Approaches

- **Interspecific hybridization:** Arabica × Robusta to blend flavour and robustness (e.g., Arabusta).

- **Low-caffeine mutants:** EMS-induced lines with suppressed *CaDXMT* gene.
- **MAS:** Identification of alleles controlling caffeine synthase and sucrose accumulation.
- **Clonal selection:** Elite Arabica lines (Selection 9, S.795) with balanced quality profiles.

Achievements: Stable low-caffeine, high-flavour hybrids for premium coffee segments.

5.3 Cocoa (*Theobroma cacao*)

Key Traits

- Fat content (55–60 %).
- Theobromine: caffeine ratio.
- Flavour precursors (amino acids, sugars).

Strategies

- Inter-varietal hybridization between Criollo (flavour) and Forastero (vigour).
- QTL mapping for flavour compounds and fat content (LG 1, 3, 5).
- Selection for fermentation quality and reduced bitterness.

5.4 Coconut (*Cocos nucifera*)

Quality Traits

- Oil content (65–72 % copra).
- Lauric acid proportion (45–55 %).
- Kernel thickness, sweetness of tender nut water.

Breeding Methods

- Hybridization (Tall × Dwarf) for early, high-oil palms.
- Clonal selection for high lauric-acid types (Kera Sagara).
- Molecular markers linked to *FAD2* genes for oil unsaturation.

Goal: Produce genotypes suitable for **virgin coconut oil (VCO)** and **functional beverage industries**.

5.5 Oil Palm (*Elaeis guineensis*)

Quality Traits

- Fatty-acid composition: 40 % palmitic, 45 % oleic, 10 % linoleic acids.
- Iodine value (unsaturation index).
- Vitamin E (tocotrienols), carotenoids.

Breeding Progress

- Intra-specific selection for **high-oleic (low-saturated)** lines.

- *E. guineensis* × *E. oleifera* hybrids – reduced palmitic acid and higher unsaturation.
- Marker discovery for *FAD2*, *DGAT*, *KASII* genes governing lipid biosynthesis.

Outcome: Development of **heart-healthy palm oil** with improved nutritional profile.

5.6 Rubber (*Hevea brasiliensis*)

Quality Determinants

- Hydrocarbon chain length and purity of cis-1,4-polyisoprene.
- Rubber-to-resin ratio and ash content.

Breeding Approach

- Clonal selection for latex viscosity and elasticity (RRIM 600, PB 235).
- Molecular tagging of *Hev b 1* and *Hev b 3* genes related to biosynthesis.
- Mutation and tissue-culture variants with improved polymer properties.

Goal: Improve latex quality for specialized industrial and medical applications.

5.7 Cashew (*Anacardium occidentale*)

Quality Components

- Kernel weight and size (W210–W240 grades).
- Oil content (45–50 %).
- Apple juice sugar: acid ratio.
- Cashew-nut shell liquid (CNSL) phenolic composition.

Breeding Efforts

- Clonal selection for large kernel and high shelling % (Vengurla 4, BPP-8).
- Hybridization for high TSS apples.
- Mutation breeding for low tannin content.
- Molecular screening for fatty-acid genes to improve oil quality.

6. Quality Evaluation and Phenotyping

Parameter	Analytical Method	Purpose / Crop
Polyphenols	Folin–Ciocalteu assay	Tea, coffee, cocoa
Caffeine	HPLC, spectrophotometry	Coffee, tea
Fatty acids	Gas chromatography	Coconut, palm, cashew
Theobromine	TLC / HPLC	Cocoa
Latex viscosity	Rheometer / Mooney viscometer	Rubber
Flavour volatiles	GC–MS	Tea, coffee
Kernel grading	BIS/ISO standards	Cashew

Phenotyping laboratories increasingly use **non-destructive NIR spectroscopy** for rapid screening of chemical traits.

7. Integration of Quality Breeding with Genomic and Metabolomic Tools

7.1 Quantitative Trait Loci (QTL) Mapping

- QTLs for catechin and caffeine in *Camellia sinensis* on LG 2, 4, 9.
- Fat composition QTLs in *Elaeis* and *Cocos* mapped via SNP arrays.
- Theobromine:caffeine ratio QTLs in *Theobroma cacao*.

7.2 Genomic Selection

Predictive models using genome-wide SNPs estimate breeding values for quality traits, especially useful in perennials where phenotyping is costly and slow.

7.3 Metabolomics and Systems Biology

Metabolomic profiling identifies biochemical signatures associated with superior sensory quality (e.g., linalool in tea, methylxanthines in coffee). Integration with transcriptomics helps elucidate **gene–metabolite networks** controlling flavour and aroma.

8. Challenges in Quality Breeding

1. **Complexity and Polygenic Control** – difficult to fix in segregating generations.
2. **Environment × Genotype Interaction** – climate and processing strongly influence quality.
3. **Trade-off with Yield and Resistance** – selection for one may reduce the other.
4. **Long Evaluation Time** – years required to assess sensory and biochemical traits.
5. **Standardization of Quality Testing** – need for international calibration protocols.

9. Genetics of Important Traits and Their Inheritance Patterns in Plantation Crops

1. Introduction

1.1 Significance of Genetic Understanding

An in-depth knowledge of the **genetics and inheritance patterns** of economically important traits is the backbone of effective breeding strategies. In perennial plantation crops, *Camellia sinensis*, *Coffea arabica*, *Cocos nucifera*, *Hevea brasiliensis*, *Elaeis guineensis*, *Theobroma cacao*, and *Anacardium occidentale*, traits such as yield, quality, disease resistance, and stress tolerance are often **quantitatively inherited**, with strong **genotype × environment interactions**. Because of their long generation cycles and vegetative propagation, the elucidation of gene action helps breeders design **efficient selection schemes** and predict **hybrid performance**.

2. Genetic Architecture of Plantation Crops

Parameter	Typical Nature in Plantation Crops	Implications for Breeding
Ploidy	Mostly diploid (2n = 2x = 36 – 54); *Coffea arabica* is tetraploid	Complicates segregation analysis
Mating system	Predominantly cross-pollinated or mixed	High heterozygosity; need for clonal fixation
Generation time	Long (4–10 years)	Requires early-selection markers
Propagation	Vegetative (budding, grafting, cloning)	Enables fixation of heterozygotes
Genetic variation	Broad due to out-crossing and hybridization	Exploitable for recurrent selection

3. Categories of Economically Important Traits

1. **Yield and Yield Components** – number of nuts, berries, pods, or latex volume.
2. **Quality Traits** – biochemical constituents (oil %, caffeine, catechins, fatty acids).
3. **Morphological Traits** – plant height, canopy shape, leaf size, bark thickness.
4. **Resistance Traits** – to diseases, pests, and environmental stresses.
5. **Physiological Traits** – photosynthetic efficiency, drought tolerance, water-use efficiency.

4. Inheritance of Yield and Yield Components

4.1 Polygenic Nature

Yield in all plantation crops is **quantitative**, governed by **polygenes with additive + non-additive effects**. Heritability is generally **moderate to high** under clonal evaluation.

Crop	Main Yield Determinants	Genetic Control / Heritability
Camellia sinensis	Number of pluckable shoots, shoot density	Additive + dominance ($h^2 \approx 0.60$)
Coffea arabica	Fruit set %, berry weight, bean yield	Additive (0.55 – 0.70)
Cocos nucifera	Number of nuts, copra yield	Polygenic; maternal influence (0.40 – 0.65)
Hevea brasiliensis	Girth increment, latex flow	Over-dominance; high h^2 (0.75)
Elaeis guineensis	Bunch weight, oil content	Additive + dominance; heritable (0.60 – 0.80)
Anacardium occidentale	Nuts per panicle, nut size	Additive; moderate h^2 (0.50)

4.2 Gene Action

- **Additive gene action** predominates for stable, heritable traits (e.g., bean weight in coffee).
- **Dominance and epistasis** influence heterosis in *Cocos* and *Hevea.*
- **Maternal effects** pronounced in nut crops where endosperm genotype influences kernel size.

5. Inheritance of Quality Traits

5.1 Tea (*Camellia sinensis*)

Trait	Genetic Control	Notes
Catechin content	Polygenic; additive + epistatic	High heritability (0.70 – 0.80)
Theaflavin ratio	Dominant epistasis	Determines liquor colour
Caffeine content	Controlled by single major gene (*TCS1*)	Recessive mutants = low caffeine
Aroma volatiles	Quantitative inheritance	Moderate heritability (0.45 – 0.60)

5.2 Coffee (*Coffea arabica / C. canephora*)

Trait	Gene Action	Remarks
Caffeine	One or two major genes (*CaXMT1, DXMT*)	Additive + dominance
Cup quality	Polygenic	Moderate h^2 (0.50)
Chlorogenic acid	Additive	Linked with disease resistance
Bean shape and size	Quantitative	Heritable (0.55 – 0.65)

5.3 Cocoa (*Theobroma cacao*)

- **Theobromine :caffeine ratio** – governed by two genes; additive effects.
- **Fat content** – polygenic, moderate heritability (0.60).
- **Flavour precursors** – controlled by multiple QTLs (LG 1, 3, 5).

5.4 Coconut and Oil Palm

Trait	Genetic Control	Implications
Oil % in copra	Polygenic	Selectable in breeding populations
Lauric acid %	Controlled by *FAD2* and *FAD3*	Marker assisted selection feasible
Fatty-acid unsaturation	Quantitative inheritance	Environmental influence high

5.5 Hevea brasiliensis

- **Latex yield** – over-dominant and polygenic.
- **Dry rubber content** – additive effects.

» **Viscosity and polymer length** – polygenic; high heritability.

5.6 Cashew (*Anacardium occidentale*)

» **Kernel weight and shelling %** – additive control.

» **Tannin content** – monogenic recessive.

» **Apple sweetness (TSS)** – polygenic; environment-sensitive.

6. Inheritance of Morphological Traits

Trait	Crop	Type of Inheritance	Notes
Plant height	*Cocos nucifera*	Polygenic + maternal effect	Dwarf vs tall types
Leaf size and shape	*Camellia sinensis*	Quantitative	Correlated with yield
Bark thickness	*Hevea brasiliensis*	Additive	Affects tapping efficiency
Canopy spread	*Cashew*	Polygenic	Key for high-density planting
Fruit length / breadth	*Coffea, Cocoa*	Quantitative	Stable heritability > 0.60

7. Inheritance of Disease and Pest Resistance

7.1 Major Gene Resistance

Crop	Pathogen / Pest	Genes Involved	Inheritance Pattern
Coffea arabica	*Hemileia vastatrix* (rust)	SH1 – SH9	Major dominant genes
Hevea brasiliensis	*Microcyclus ulei*	R genes polygenic	Partial dominance
Camellia sinensis	*Exobasidium vexans*	Polygenic control	Quantitative tolerance
Theobroma cacao	*Phytophthora palmivora*	QTLs on LG 1, 3 & 9	Additive + dominant
Cashew	*Helopeltis antonii*	Polygenic	Additive heritability 0.55

7.2 Gene-for-Gene Relationship

The classical **Flor's hypothesis** holds for *Coffea arabica–H. vastatrix* interaction: each host R-gene corresponds to a pathogen avirulence gene. Breakdown of resistance occurs when new pathogen races evolve.

8. Inheritance of Abiotic-Stress Tolerance

Stress Type	Crop	Mechanism / Gene Action
Drought	*Cocos, Camellia, Cashew*	Polygenic; additive heritability ≈ 0.40–0.60
Salinity	*Cocos, Cashew*	Quantitative; selection for ion exclusion
Heat	*Tea, Hevea*	Dominant gene effects; inducible HSPs

Cold	*Camellia, Coffea*	Polygenic tolerance; CBF transcription factors
Flooding	*Hevea, Elaeis*	Additive effects; ethylene-regulated genes

9. Quantitative Genetic Parameters

Parameter	**Meaning / Use**	**Typical Range in Plantation Crops**
Genotypic variance (σ^2g)	True genetic variation	20 – 50 % of phenotypic variance
Heritability (h^2)	Proportion of phenotypic variance due to genetics	0.45 – 0.85 (depending on trait)
Genetic advance (GA)	Expected progress under selection	10 – 40 % of mean
GCV / PCV ratio	Indicates environmental influence	0.6 – 0.9 (desirable > 0.7)
Correlation coefficients	Association between traits	Positive between yield and size traits

These parameters guide the intensity and timing of selection, especially when early screening is needed before maturity.

10. Molecular and Genomic Insights into Inheritance

10.1 QTL Mapping

» *Coffea arabica*: QTLs for rust resistance (SH genes), caffeine content, drought tolerance.

» *Camellia sinensis*: QTLs for catechin and flavour on LG 2 & 9.

» *Hevea brasiliensis*: QTLs for latex yield on linkage groups 5 & 7.

» *Elaeis guineensis*: Oil quality QTLs linked to *SAD* and *FAD2*.

» *Cocoa*: Disease-resistance QTLs on LG 1 & 9.

10.2 Genome-Wide Association Studies (GWAS)

Recent GWAS in *Coffea* and *Camellia* have identified **SNP markers** linked to polyphenol and flavour genes. In *Hevea*, association mapping detected candidate genes regulating **cis-1,4-polyisoprene** polymerization.

10.3 Transcriptomics and Epigenetics

» Gene expression profiling under stress reveals **differential activation** of HSPs and antioxidant enzymes.

» **DNA methylation patterns** influence clonal phenotypic stability in tea and rubber.

» Epigenetic inheritance contributes to **trans-generational tolerance** in drought-prone perennials.

11. Genotype × Environment Interaction (G × E)

Quality and yield traits show significant G × E effects in plantation crops due to diverse ecologies.

Crop	Trait	Variance Component of G × E (%)	Breeding Implication
Camellia	Polyphenols	25	Multi-location testing needed
Coffea	Cup quality	35	Stable clone selection
Hevea	Latex yield	20	Clone × environment interaction important
Cocos	Copra yield	28	Evaluate hybrids across zones

Breeders use **AMMI** and **GGE biplot analyses** to identify stable genotypes across environments.

12. Cytogenetics and Chromosomal Behaviour

» *Coffea arabica* (2n = 4x = 44) – allotetraploid with regular meiosis and self-compatibility.

» *Hevea brasiliensis* (2n = 36) – diploid; irregular pairing in interspecific hybrids.

» *Camellia sinensis* (2n = 30–60) – segmental allopolyploid; variable fertility.

» *Elaeis guineensis* (2n = 32) – normal meiosis; stable F_1 fertility.

» *Cocos nucifera* (2n = 32) – disomic inheritance; partial self-compatibility. Cytogenetic analysis aids in identifying **chromosomal rearrangements**, ensuring hybrid stability.

13. Statistical Tools in Genetic Analysis

1. **Diallel and Line × Tester Analysis** – partition GCA and SCA effects.
2. **North Carolina Design I/II/III** – estimate additive and dominance variances.
3. **Generation Mean Analysis** – identify gene actions (additive, dominant, epistatic).
4. **Path Coefficient Analysis** – disentangle direct and indirect effects on yield.
5. **Bayesian and mixed-model approaches** – handle unbalanced clonal trials.

14. Implications for Breeding Strategy

Inheritance Pattern	Recommended Breeding Approach	Example Crop / Trait
Additive	Clonal or recurrent selection	Yield in tea, coffee
Dominance / Over-dominance	Hybrid breeding / heterosis exploitation	*Cocos, Hevea*
Polygenic Quantitative	Recurrent selection with index selection	Quality traits in tea and cocoa
Major Gene Resistance	Backcross / gene pyramiding	SH genes in coffee
Epigenetic Inheritance	Clonal propagation to maintain state	Tea and rubber clones

15. Emerging Genetic Concepts

1. **Genomic Selection (GS):** Predicts breeding values using dense SNP markers; shortens selection cycle.
2. **Pangenomics:** Captures core and dispensable genes across species (e.g., *Coffea* complex).
3. **Epigenome Breeding:** Uses DNA-methylation markers for stable phenotypes.
4. **Systems Genetics:** Integrates transcriptomic, proteomic, and metabolomic networks for trait prediction.
5. **Speed-breeding in perennials:** In vitro flowering and early selection to accelerate

10. Molecular, Transgenic, and Other Biotechnological Tools in Plantation-Crop Improvement

1. Introduction

1.1 Why biotechnology for plantation crops?

Perennial plantation crops, *Camellia sinensis*, *Coffea arabica*, *Cocos nucifera*, *Elaeis guineensis*, *Hevea brasiliensis*, *Theobroma cacao*, *Anacardium occidentale*, pose **three classic bottlenecks**: long juvenility, high heterozygosity, and complex reproduction (SI, dichogamy). Modern biotechnology compresses time and increases precision via **markers, tissue culture, doubled ploidy, transgenics, and genome editing**, while multi-omics enables **mechanistic selection** for quality, resilience, and yield.

2. DNA markers and genotyping frameworks

2.1 Marker systems

S. No.	Marker class	Key features	Typical use in plantation crops
1	SSR/microsatellite	Co-dominant, multi-allelic	Clone ID (tea, rubber), hybridity tests (coconut, cocoa)
2	SNP arrays/GBS	High density, genome-wide	Genomic selection (oil palm), GWAS (coffee/tea)
3	CAPS/KASP	Simple, trait-linked assays	Gene tracking (caffeine synthase, SH rust genes)
4	AFLP/RAPD (legacy)	Dominant; low cost	Preliminary diversity, nursery screening

Best practice: Anchor breeding pipelines on SSR/SNPs for **identity, purity, parentage**, and **foreground selection**; adopt low-cost KASP panels for routine nursery triage.

3. Marker-assisted selection (MAS) and genomic selection (GS)

3.1 MAS playbook

1. Discover QTL/gene → 2) validate across backgrounds → 3) design assay → 4) integrate with stage-gates (seedling, pre-field, field).
2. **Use-cases:**
   » *Coffee:* SH1–SH9 rust-resistance gene pyramiding.
   » *Tea:* catechin-pathway genes (*ANR*, *LAR*, *DFR*) for quality.
   » *Cocoa:* Phytophthora-tolerance QTLs (LG1/3/9).
   » *Hevea:* latex-yield and tapping-panel traits.

3.2 GS for long-cycle perennials

Train a model (SNPs + historical phenotypes) → predict **GEBV** for new seedlings → advance only the top 10–20%. **Payoff:** 1–2 selection cycles saved per decade; ideal for **oil palm bunch traits**, tea yield/quality indices, and rubber latex flow.

4. Tissue culture, micropropagation, and somatic embryogenesis

4.1 Clonal multiplication at scale

Crop	Preferred route	Notes
Camellia	Nodal cuttings; microshoots	Maintain clonal fidelity; periodic SSR checks
Hevea	Budding (field); somatic embryogenesis (R&D)	SE helpful for elite, hard-to-root clones
Coffea	Somatic embryogenesis (leaf explants)	Industrial pipelines exist; acclimatization is critical
Cocos	Zygotic embryo culture; SE (pilot)	Enables safe movement of germplasm
Cocoa	Somatic embryos from floral tissues	Used for disease-free, uniform planting material
Cashew	Softwood grafting; in vitro shoots (developing)	Juvenility rejuvenation needed

Quality gates: genetic fidelity (SSR/SNP), **virus indexing**, ploidy (flow cytometry), and nursery hardening SOPs.

4.2 Doubled haploids (DH) and gametic culture

Not yet routine in most plantation crops; experimental anther culture in *Camellia* and *Coffea* shows potential to shorten trait fixation for mapping populations.

5. Mutation discovery and targeted mutagenesis

5.1 TILLING / EcoTILLING

Reverse-genetics scans for point variants within elite clones. Useful to find **loss-of-function** in anti-quality or susceptibility genes (e.g., **low-caffeine** alleles in coffee).

5.2 CRISPR/Cas mutagenesis (non-transgenic routes)

Transient RNP delivery (protoplasts/meristems) can generate **edited, DNA-free plants**, regulatory-friendly in several jurisdictions. Candidate edits:

» *Coffee*: **DXMT/XMT** knock-downs → low-caffeine lines.

» *Tea*: edit **CsANR/LAR** for tailored catechin spectra.

» *Oil palm*: fine-tune **FAD2/SAD** for high-oleic oil.

» *Hevea*: regulatory edits in **cis-1,4-polyisoprene** pathway genes to affect latex quality.

6. Transgenic approaches (GMO), status and design

Note: Deploy only where permitted and with full biosafety compliance.

6.1 Trait stacks

Trait	Gene concept	Target crop	Breeding value
Fungal resistance	**Chitinases, β-1,3-glucanases, *PR* proteins**	*Hevea*, *Cocoa*, *Tea*	Reduce chemical use, stabilize yield
Insect resistance	**Bt toxins/RNAi against** *Helopeltis*	*Cashew*	Protect flowering/fruit set
Abiotic stress	*DREB*, *HSP*, *P5CS*	*Camellia*, *Coffea*	Drought/heat resilience
Quality modulation	**Antisense** *caffeine synthase*	*Coffee*	Naturally, "decaf" beans

6.2 Transformation platforms

» *Coffea/Cacao*: **Agrobacterium** on somatic embryos.

» *Hevea*: particle bombardment; Agrobacterium on friable callus (challenging).

» *Camellia*: Agrobacterium on juvenile tissues.

» *Cashew*: needs improved regeneration for stable integration.

Validation: event copy number (qPCR), insertion site (TLA/NGS), expression (RT-qPCR), trait bioassay, and **multi-season field trials**.

7. Molecular diagnostics, purity, and IP management

1. **DNA barcoding/fingerprinting**, clone identity, DUS testing, and **seed-garden policing** (e.g., *Cocos* T×D hybridity checks).

2. **Paternity analysis,** confirm pollination in open orchards (tea/cashew seed gardens).
3. **Freedom-to-operate (FTO),** marker panels and edited alleles must be IP-cleared; use open assays where possible.

8. Multi-omics for trait dissection

8.1 Transcriptomics

Stress-responsive networks (HSPs, ROS enzymes) in tea/coffee; latex biosynthesis modules in rubber; **oil biosynthesis regulons** in palm/coconut.

8.2 Metabolomics

Cup quality (coffee) = CGA, sugars, volatiles; tea liquor = catechin→theaflavin conversion; cocoa flavour precursors; **link metabolite QTLs to sensory panels.**

8.3 Proteomics and phospho-proteomics

Pinpoint post-translational switches in tapping response (*Hevea*) and fermentation enzymes (cocoa).

8.4 Systems integration

eQTL + mQTL + pQTL → **causal candidates**; feed into GS to raise prediction accuracy (hybrid performance, quality indices, stress scores).

9. Bioinformatics & decision support

» **Breeding databases:** genotype–phenotype–environment warehouses; immutable pedigree chains.

» **Predictive crossing tools:** genomic relationship matrices, **MateSel-style** optimizers to balance gain vs. inbreeding.

» **Field phenomics:** drone RGB/multispectral for canopy vigour; mobile NIR for oil, polyphenols; on-tree yield sensors (oil palm bunch count via CV).

10. Containment, biosafety, and ethics

1. **Pathogen-free pipelines** (thermo-therapy + meristem culture) for clonal health.
2. **Biosafety trial design:** refuge planting, pollen barriers, compliance logs.
3. **Ethics & access:** Nagoya Protocol, traceability and **benefit-sharing** when using wild relatives (e.g., Amazonian *Theobroma* or Ethiopian *Coffea*).
4. **Data stewardship:** privacy of genomic data; open science where feasible.

11. End-to-end biotech breeding workflow (template)

1. **Define targets** (yield/quality/resilience indices).
2. **Mine diversity** (wild relatives, landraces; sequence core sets).

3. **Map/QTL/GS** (build marker panels; train GS).
4. **Create novelty** (crosses, mutagenesis, edits).
5. **Fast-track propagation** (SE/microcuttings; fidelity checks).
6. **Stage-gated trials** (nursery → multi-site METs; AMMI/GGE stability).
7. **Release & stewardship** (identity assays, seed-garden SOPs, IP/FTO).

12. Quick-reference tables

Table 1. Biotech toolkits by crop

S. No.	Crop	Routine tools	Advancing tools
1	*Camellia sinensis*	SSR/SNP MAS; microcuttings	CRISPR (quality/stress), GS
2	*Coffea arabica/ canephora*	SE pipelines; MAS for SH genes	Low-caffeine edits; GS cup models
3	*Theobroma cacao*	SE + Agrobacterium; Phytophthora QTL MAS	Multi-omics flavour GS
4	*Hevea brasiliensis*	SSR ID; SE (R&D)	Latex-pathway editing; proteomics
5	*Elaeis guineensis*	SNP arrays; GS for oil traits	HOSO (high-oleic) edits
6	*Cocos nucifera*	Embryo culture; hybridity PCR	SE scale-up; oil QTL MAS
7	*Anacardium occidentale*	Grafting; SSR ID	Regeneration breakthroughs; insect-resistance RNAi

Table 2. Example trait→marker/assay pairs

S. No.	Trait	Assay	Crop
1	Rust resistance (SH genes)	KASP/CAPS	Coffee
2	Catechin profile	SNPs in *ANR/LAR*	Tea
3	Phytophthora tolerance	SSR-QTL (LG1/3/9)	Cocoa
4	Oil unsaturation	SNPs in *FAD2/SAD*	Oil palm/coconut
5	Latex yield potential	SSRs near pathway candidates	Rubber

13. Practical guidance for deployment

» **Start small, scale fast:** pilot MAS on 2–3 high-value traits, then expand panels.

» **QC at every step:** molecular identity and ploidy checks prevent costly mix-ups.

» **Blend phenomics + genomics:** sensor-derived indices (NDVI, canopy temp) sharpen GS.

» **Diversify clonally:** release **multi-clonal varietal packages** to reduce epidemic risk.

» **Plan seed-garden genetics:** paternity tracking and spatial designs to maintain **hybridity purity**.

Chapter - 7

Ideotype Breeding, Breeding Problems and Achievements in Plantation Crops

Part 1 – Introduction and General Breeding Objectives

1. Introduction

1.1 Concept and Scope

Plantation crops form a distinct category of long-duration perennial species cultivated mainly for exportable commodities such as **beverages, oils, latex, and nuts.** They include *Camellia sinensis* (tea), *Coffea arabica* and *C. canephora* (coffee), *Theobroma cacao* (cocoa), *Cocos nucifera* (coconut), *Elaeis guineensis* (oil palm), *Hevea brasiliensis* (rubber), and *Anacardium occidentale* (cashew).

These crops are integral to the **economic, social, and environmental fabric** of tropical agriculture, providing livelihood to millions, foreign exchange to producing nations, and valuable raw materials to industries. However, they exhibit several breeding challenges, long juvenile phases, high heterozygosity, limited seed set, and complex reproductive biology that necessitate **specialized, long-term, and multi-disciplinary breeding strategies**. Breeding achievements in plantation crops, therefore, represent decades of systematic work in **selection, hybridization, clonal propagation, and biotechnological interventions,** aimed at increasing productivity, improving quality, and enhancing resilience to biotic and abiotic stresses.

1.2 Historical Perspective of Breeding in Plantation Crops

The genetic improvement of plantation crops can be traced to the early 20th century, when colonial research stations began **clonal selection** from naturally occurring superior trees. The evolution of breeding can broadly be divided into four phases:

Phase	Period	Approach	Examples / Achievements
Exploratory Phase	1900–1940	Collection, introduction, and selection of superior clones	Selection of Assam Jat and China Jat tea, Kent coffee, high-oil palms

DOI: 10.1201/9781042010332-8

Hybridization Phase	1940–1970	Intra- and inter-specific hybridization to combine desirable traits	Tall × Dwarf coconut hybrids, Arabica × Robusta "Arabusta"
Systematic Breeding Phase	1970–2000	Multi-locational trials, disease resistance breeding, molecular analysis	Clonal series in tea (TV, UPASI), RRIM and PB clones in rubber
Genomic and Biotech Phase	2000–present	Marker-assisted and genomic selection, molecular characterization, transgenic approaches	Rust-resistant coffee (SH genes), oil palm genomic selection, CRISPR studies in tea

1.3 Importance of Breeding Achievements

1. **Economic Significance:** Genetic gains in yield (20–50%) and quality have increased national revenues and export competitiveness.
2. **Sustainability:** Disease-resistant and stress-tolerant varieties reduce dependence on chemical inputs and ensure ecological stability.
3. **Adaptation to Climate Change:** Breeding for drought, heat, and salinity tolerance is crucial for maintaining productivity under variable climates.
4. **Quality and Consumer Demand:** High-value attributes like flavour, aroma, antioxidant content, oil composition, and kernel size ensure premium pricing and market expansion.

1.4 Need for Ideotype-Oriented Breeding

Traditional breeding focused primarily on yield and disease resistance. However, the modern concept of **ideotype breeding** emphasizes the development of an **optimal plant type** that integrates yield potential, physiological efficiency, stress resilience, and product quality.

For example:

» Compact, high-density, early-bearing coconut palms.

» Tea ideotypes with short internodes, high leaf-area index, and rapid shoot turnover.

» Rubber clones with thin bark, high latex flow, and wind resistance.

» Coffee plants with compact growth, high fruit density, and rust resistance.

2. General Breeding Objectives in Plantation Crops

The ultimate goal of plantation crop breeding is to develop **high-yielding, high-quality, stress-resilient genotypes** that remain productive and profitable for decades. These objectives are multifaceted, addressing **biological, industrial, and environmental dimensions**.

2.1 Increasing Yield and Productivity

Yield is a complex quantitative trait influenced by several physiological and morphological characters. Breeding aims to maximise yield through:

1. Higher photosynthetic efficiency and biomass accumulation.
2. **Improved reproductive efficiency** – higher fruit-set and nut-set rates.
3. **Reduction of unproductive growth** – compact plant types for high-density planting.
4. **Early bearing and sustained productivity** over the plantation lifespan.

Examples

» *Cocos nucifera* (Tall × Dwarf hybrids) yielding 25–30% more copra.

» *Hevea brasiliensis* clones (RRIM 600, PB 86) showing latex yield gains up to 35%.

» *Elaeis guineensis* Tenera hybrid producing 25% more oil than parent forms.

2.2 Enhancing Quality and Industrial Value

Quality improvement involves biochemical and sensory enhancement of the harvested product:

Crop	Key Quality Traits	Breeding Focus
Camellia sinensis	Catechin, theaflavin, aroma volatiles	Flavour and antioxidant richness
Coffea arabica	Caffeine, chlorogenic acids, cup flavour	Balance between quality and disease resistance
Theobroma cacao	Flavour precursors, fat %, theobromine ratio	Trinitario flavour types
Cocos nucifera / Elaeis guineensis	Oil %, lauric/oleic ratio	Health-oriented oil profile
Hevea brasiliensis	Latex viscosity, rubber hydrocarbon %	Industrial-grade polymer quality
Anacardium occidentale	Kernel weight, apple TSS	Premium export kernels

2.3 Improving Resistance to Biotic Stresses

Resistance to diseases and pests ensures sustainability and reduced input costs.

Key Goals

» Develop genotypes resistant/tolerant to major diseases and insect pests.

» Introduce durable resistance through gene pyramiding or polygenic inheritance.

Examples

» *Coffea arabica* resistant to leaf rust (*Hemileia vastatrix*) via SH genes.

» *Camellia sinensis* clones resistant to blister blight (*Exobasidium vexans*).

» *Theobroma cacao* hybrids are tolerant to black pod and witches' broom.

» *Hevea brasiliensis* hybrids showing field resistance to South American Leaf Blight.

2.4 Enhancing Tolerance to Abiotic Stresses

Plantation crops are often grown under marginal conditions, high temperatures, drought, salinity, or flood-prone soils. Breeding objectives include:

1. **Drought tolerance** – efficient stomatal regulation and root systems (*Cocos*, *Camellia*).
2. **Temperature tolerance** – heat- or cold-hardiness (*Coffea arabica*, *Camellia sinensis*).
3. **Salinity tolerance** – ion exclusion and osmotic adjustment (*Cocos nucifera*, *Cashew*).
4. **Flood tolerance** – root aeration and anaerobic metabolism (*Hevea*, *Oil palm*).

Molecular marker-assisted screening for stress-related QTLs (e.g., *HSP*, *DREB*, *NHX*) now accelerates selection for abiotic resilience.

2.5 Shortening the Juvenile Phase

One of the greatest constraints in perennial crop breeding is the **long juvenile period**, delaying evaluation and release of new varieties.

Breeding and biotechnological approaches aim to:

» Introduce early-flowering alleles.

» Use dwarfing genes and growth regulators.

» Apply molecular markers for early selection.

» Employ tissue culture and somatic embryogenesis for clonal propagation of early bearers.

For example, Tall × Dwarf coconut hybrids flower 2–3 years earlier than pure tall types.

2.6 Developing Ideal Plant Architecture

Optimized plant architecture ensures higher photosynthetic efficiency and suitability for high-density or mechanized management systems.

Desirable Architectural Traits

» Compact and uniform canopy (cashew, coffee).

» Short internodes and erect leaves (tea).

» Reduced height and strong trunk (rubber, coconut).

» Balanced root: shoot ratio for efficient nutrient uptake (oil palm).

2.7 Enhancing Nutritional and Functional Quality

Modern breeding integrates **nutritional and nutraceutical improvement**, enhancing antioxidant, polyphenol, and micronutrient levels to meet health-conscious markets.

» **Tea:** catechin-rich clones for antioxidant teas.

» **Coffee:** low-caffeine, high-CGA lines for functional beverages.

» **Coconut and oil palm:** high-oleic and tocotrienol-rich oils for cardiovascular health.

» **Cocoa:** flavonoid-rich varieties for functional chocolate production.

2 – Ideotype Breeding in Plantation Crops

1. Introduction to Ideotype Breeding

1.1 Concept of Ideotype Breeding

The concept of an **ideotype** (from the Greek "ideal type") was first articulated by Donald (1968), who defined it as *"a biological model which is expected to perform or behave in a predictable manner within a defined environment."*

Ideotype breeding in plantation crops aims to design a *biological prototype* possessing an integrated set of morphological, physiological, biochemical, and genetic traits that confer **high productivity, superior quality, resilience, and sustainability** over a long lifespan. Unlike conventional breeding, which relies on empirical selection of superior phenotypes, ideotype breeding uses a **blueprint-based approach**, specifying the desirable plant attributes before initiating the breeding program.

1.2 Importance of Plantation Crops

Ideotype breeding is particularly relevant to plantation crops because they:

1. Exhibit **long juvenile phases**, making early selection critical.
2. Have **heterogeneous populations** with high environmental variability.
3. Require **precise adaptation** to specific agro-climatic conditions.
4. Demand traits suited to **mechanization, high-density planting, and climate resilience**.

Hence, ideotype models act as *guiding frameworks* to channel breeding efforts toward a defined goal, such as:

- » Compact canopy and early maturity in coconut.
- » High photosynthetic efficiency and antioxidant-rich leaves in tea.
- » Wind-resistant and high-latex clones in rubber.
- » Flavour-rich and disease-tolerant hybrids in coffee and cocoa.

1.3 Components of an Ideotype

An ideal plant type integrates multiple components:

Category	Representative Traits	Functional Significance
Morphological	Plant stature, leaf size, canopy shape	Determines light interception, harvest ease
Physiological	Photosynthetic rate, WUE, nutrient use efficiency	Governs yield and resilience
Biochemical	Quality metabolites (polyphenols, caffeine, oil composition)	Enhances market value
Genetic	Stable gene combinations, resistance alleles	Ensures heritability and adaptability
Ecological	Root architecture, wind/heat tolerance	Improves sustainability

Socio-economic	Ease of harvest, labour requirement, product uniformity	Increases adoption and profitability

1.4 Steps in Ideotype Breeding

1. **Define the target environment and production system.** – Rainfed vs. irrigated, smallholder vs. estate cultivation.
2. **List desirable traits quantitatively** – e.g., leaf angle 45°, compact height <2.5 m, high WUE.
3. **Identify donor sources.** – Wild relatives, mutants, elite clones.
4. **Design crossing and selection strategy.** – Recurrent selection, hybridization, marker-assisted introgression.
5. **Validate ideotype performance.** – Multi-environment testing and physiological evaluation.
6. **Refine model continuously** as production ecology and market demands evolve.

2. Ideotype Models for Major Plantation Crops

2.1 Tea (*Camellia sinensis*) Ideotype

2.1.1 Desired Traits

Trait Category	Specific Features	Rationale
Morphology	Bushy, compact canopy; short internodes; erect leaves	Maximizes shoot density and harvest frequency
Physiology	High photosynthetic rate; rapid shoot renewal; efficient nutrient uptake	Sustains continuous plucking cycles
Biochemistry	High catechin and theaflavin ratio; low caffeine	Ensures flavour, antioxidant value
Stress Tolerance	Drought and blister blight resistance	Adaptation to climate variability
Root System	Deep, branched roots	Drought endurance and nutrient absorption

2.1.2 Ideotype Description

The **"Tea Ideotype"** envisages a medium-sized bush with a **flat plucking table**, strong lateral branching, high leaf-area index (LAI ≈ 6–7), and efficient photosynthetic conversion. It should combine the **flavour profile of *C. sinensis*** with the **vigour of *C. assamica*.**

2.1.3 Achievements

Clonal series like **TV 1, TV 9, UPASI 3, TRI 2025** embody these ideotype principles, high yield, balanced quality, and stress tolerance.

2.2 Coffee (*Coffea arabica, C. canephora*) Ideotype

2.2.1 Desirable Characteristics

Trait Type	Specific Features	Functional Benefit
Architecture	Compact canopy, short internodes, high fruit density	Suitable for mechanization and high density
Physiology	Early bearing, efficient photosynthesis, high WUE	Productivity under marginal zones
Biochemistry	Low caffeine, high sugars and flavour volatiles	Premium-quality beverage
Resistance	Tolerance to rust and drought	Durable performance
Root Traits	Deep root system	Drought resilience

2.2.2 Ideotype Concept

The **"Compact Arabica Ideotype"** aims for a self-pruning plant of 1.5–2.0 m height with 60–70 fruits per node, resistant to rust (*Hemileia vastatrix*) and nematodes, producing beans with balanced flavour.

2.2.3 Realized Ideotypes

Hybrids such as **Selection 9, S.795,** and **Catimor (C. arabica × C. canephora)** are practical examples combining quality and resilience.

2.3 Cocoa (*Theobroma cacao*) Ideotype

Trait Domain	Ideotype Feature	Rationale
Canopy	Narrow, upright branches	Higher planting density
Pod Traits	Thin pod wall, high bean count	Processing efficiency
Biochemical	Balanced theobromine:caffeine ratio, high fat %	Flavour and quality
Resistance	Black pod and witches'-broom tolerance	Stable yield
Root	Deep taproot	Drought tolerance

The **"Trinitario Ideotype"** integrates flavour from *Criollo* and vigour/disease tolerance from *Forastero*. Modern hybrids like **TSH 1188** and **ICS 95** approximate this design.

2.4 Coconut (*Cocos nucifera*) Ideotype

2.4.1 Desirable Traits

Trait Class	Characteristic	Utility
Architecture	Semi-tall (10–12 m), compact canopy	Wind resistance, ease of harvest
Reproduction	Early flowering (3–4 years)	Reduces non-bearing period

Yield Traits	>120 nuts/palm/year; high copra/oil	Economic sustainability
Stress Traits	Drought and salinity tolerance	Climate adaptation
Quality	High lauric acid and tender nut water sweetness	Dual-purpose use

2.4.2 Ideotype Vision

A semi-dwarf palm combining the **yield of tall types** and **precocity of dwarfs**, manageable, high-yielding, with resistance to root wilt and environmental stress.

Realized Models: *Kera Chandra*, *Kalpa Sagara*, *Kera Sree*, and *Kalpa Surya* embody these traits.

2.5 Oil Palm (*Elaeis guineensis*) Ideotype

Parameter	Ideal Feature	Goal
Height	Slow-growing trunk (<30 cm/year)	Reduces harvesting difficulty
Leaf	Long, erect leaves with high LAI	Efficient photosynthesis
Fruit	High mesocarp:kernel ratio	Oil yield and quality
Physiology	High bunch number, short internode	Increased yield
Composition	High oleic, low palmitic content	Healthy oil profile

Modern Ideotype: *Tenera* (Dura × Pisifera) type with high oil extraction (25–28%) and compact crown suited for high-density plantations.

Advanced breeding aims for **"climate-smart Tenera"** with slow height increment and high unsaturated oil.

2.6 Rubber (*Hevea brasiliensis*) Ideotype

2.6.1 Desired Traits

Trait Type	Ideotype Feature	Benefit
Morphology	Medium height, flexible trunk, strong wind resistance	Safety in storm-prone regions
Latex Physiology	High latex-flow rate, thick bark renewal	Increased productivity
Disease Resistance	Tolerance to SALB and Corynespora leaf fall	Yield stability
Root System	Deep, spreading roots	Drought and wind tolerance
Maturity	Early tappability (6–7 years)	Reduced immature period

2.6.2 Ideotype Form

The **"High-Yield, Wind-Resistant Rubber Clone"** is envisioned as a deep-rooted, medium-sized tree with strong girth (≥50 cm at 6 years), high latex flow (≥50 g/tree/tapping), and rapid bark regeneration.

Examples: *RRIM 600*, *PB 217*, *GT 1*, and *RRII 105* fulfil major ideotype attributes.

2.7 Cashew (*Anacardium occidentale*) Ideotype

Ideotype Component	Ideal Trait	Justification
Tree Form	Compact canopy (3–4 m), semi-spreading	Suited for high-density orchards
Flowering	High perfect-flower ratio, synchronized flowering	Enhanced nut set
Nut / Kernel	Bold nuts (≥7 g), high shelling %	Export value
Pest Resistance	Tolerance to *Helopeltis antonii*	Reduced pesticide dependence
Quality	Low tannin apple, high TSS	Beverage and processing potential

Realized Ideotypes: *Vengurla 4*, *NRCC Sel-2*, and *BPP-8* are compact, high-yielding, and pest-tolerant selections approximating the ideal model.

3. Comparative Ideotype Framework

Crop	Plant Architecture	Physiology	Quality Focus	Stress/Resistance Target
Tea	Compact bush	High photosynthesis	Catechin:theaflavin balance	Blister blight, drought
Coffee	Compact tree	Early bearing	Flavour, low caffeine	Rust, drought
Cocoa	Upright canopy	High fruit set	Flavour balance	Pod rot
Coconut	Semi-dwarf palm	Early flowering	Oil, water sweetness	Drought, wind
Oil palm	Compact crown	High bunch index	High oleic oil	Basal stem rot
Rubber	Medium height	High latex flow	Polymer quality	SALB, wind
Cashew	Short, dense canopy	High flowering ratio	Kernel weight	Helopeltis tolerance

4. Physiological and Biochemical Foundations of Ideotype Design

1. **Photosynthetic Efficiency:** Ideotypes emphasize high chlorophyll density and balanced leaf area. Rubber and oil palm clones with higher Rubisco activity show enhanced productivity.
2. **Source–Sink Balance:** Tea and coffee ideotypes focus on rapid leaf regeneration (source) and efficient reproductive sink allocation (buds, fruits).
3. **Water-Use Efficiency (WUE):** Drought-tolerant ideotypes exhibit lower stomatal conductance, deeper roots, and osmolyte accumulation (proline, sugars).
4. **Biochemical Pathways:** Tea and coffee ideotypes integrate desirable pathways, phenylpropanoid (flavonoids), methylxanthine (caffeine), for enhanced quality.

5. **Canopy Architecture and Light Utilization:** Narrow, upright leaves and balanced branching maximize photosynthetically active radiation (PAR) interception in cocoa and rubber ideotypes.

5. Modern Tools in Ideotype Breeding

Tool / Approach	Purpose	Example
Physiological modeling	Predicts yield response under varying environments	Tea ideotype modeling (LAI vs. yield)
Genomic prediction	Early ideotype selection based on genomic values	Coffee, oil palm
Marker-assisted ideotype selection	Tagging quality and resistance genes	Tea (catechin genes), Coffee (SH genes)
Phenomic platforms	Drone-based canopy imaging	Coconut and rubber field evaluation
CRISPR gene editing	Targeted modification of key traits	Caffeine and fatty-acid pathways

6. Advantages of Ideotype Breeding

1. Provides **clear direction** for long-term breeding programmes.
2. Enables **targeted improvement** of physiological efficiency.
3. Facilitates **integration of multi-disciplinary data** (physiology, genetics, environment).
4. Reduces the risk of **undesirable linkages** by designing trait combinations beforehand.
5. Suits **clonal crops**, where ideotypes can be fixed and multiplied directly.

7. Limitations and Challenges

1. Complexity of multi-trait integration and long evaluation period.
2. Difficulty in quantifying physiological parameters accurately.
3. Environmental variability altering ideotype performance.
4. Limited genomic information for some crops (e.g., cashew, coconut).
5. Institutional and infrastructural constraints for modeling-based breeding.

Despite these, ideotype breeding remains the most forward-looking approach for sustainable improvement of plantation crops.

3 – Breeding Problems and Constraints in Plantation Crops

1. Introduction

Plantation crops differ sharply from annuals in their breeding behavior: they are long-lived, highly heterozygous, and mostly perennial, with intricate reproductive systems. While these characteristics underpin their ecological success, they also impose **serious limitations on breeding progress**.

Understanding these constraints is essential for planning effective breeding pipelines and for interpreting the rate of genetic gain realized so far.

2. Biological and Genetic Constraints

2.1 Long Juvenile Phase

Most plantation crops require **4–10 years** to reach the first reproductive phase.

Crop	Juvenile Period (yrs)	Impact on Breeding
Hevea brasiliensis	6–7	Delays tapping trials
Coffea arabica	3–4	Late evaluation of yield and quality
Cocos nucifera	5–8	Slow generation turnover
Camellia sinensis	3–5	Delayed selection of pluckable shoots

Consequence: low selection intensity, delayed turnover, and high maintenance cost.

Remedies: Use of **early-flowering genes, growth-regulators**, and **molecular markers** for early selection.

2.2 Heterozygosity and Perennial Nature

Perennial crops preserve heterozygosity through outcrossing and vegetative propagation.

» Crossing two heterozygotes leads to segregation and unpredictable recombination.

» Fixation of traits through inbreeding causes **inbreeding depression** (loss of vigor and fertility). Thus, breeding progress depends on **clonal selection** or **recurrent selection** rather than simple pedigree lines.

2.3 Complex Pollination Mechanisms

Mechanism	Crops Affected	Problem
Dichogamy (temporal separation of sexes)	*Cashew, Coconut*	Selfing difficult → uncontrolled pollination
Self-incompatibility	*Camellia, Coffea canephora*	Hinders inbreeding and line development
Protandry / Proterogyny	*Hevea, Cocoa*	Limits controlled crossing efficiency
Entomophily	*Cashew, Coconut*	Pollination depends on insects and weather

Pollination control demands intensive manual bagging, tagging, and monitoring, often resulting in **low fruit-set (<20 %)** in hybridization experiments.

2.4 Polyploidy and Chromosomal Complexity

Several species exhibit variable ploidy levels (*Camellia sinensis* $2n = 30–60$, *Coffea arabica* $2n = 4x = 44$). Consequences include:

» Irregular meiosis and reduced fertility.

- » Segregation distortion and genetic imbalance.
- » Difficulty in defining inheritance ratios. Cytogenetic studies and flow-cytometry screening are therefore indispensable before hybridization.

2.5 Low Seed Set and Poor Viability

Low pollination success and embryo abortion reduce the number of viable hybrid seeds. For instance, in *Hevea*, cross-pollinated fruit set rarely exceeds 10 %, while *Cocos* crosses yield < 20 % hybrid nuts. Embryo-rescue and in vitro germination techniques are increasingly used to recover hybrid progenies.

2.6 Lack of Sexual Reproduction in Clonal Species

Crops such as *Hevea* and *Camellia* are maintained clonally for uniformity. However, **sexual sterility or incompatibility** in elite clones restricts recombination and development of new genetic combinations. Breeders rely on **induced mutations, somaclonal variation, or polyploidy** to introduce novelty.

3. Environmental and Agro-climatic Constraints

3.1 Climate Sensitivity

Being long-lived, plantation crops face cumulative exposure to **climate variability**, temperature extremes, drought, cyclone, and uneven rainfall.

- » **Tea** suffers from winter dormancy and heat stress.
- » **Coffee** experiences flower abortion under erratic rains.
- » **Coconut and cashew** suffer nut fall during dry spells.

Stable performance requires breeding for **climate-resilient ideotypes** with high water-use efficiency and stress-responsive genes.

3.2 Soil and Topographical Limitations

Shallow lateritic soils, low fertility, and poor drainage constrain growth and influence genotype × environment interactions. Because these crops occupy land for decades, soil exhaustion and acidity gradually reduce productivity, making **rootstock and nutrient-efficient genotypes** vital breeding targets.

3.3 Biotic Stress Complexity

Pathogen populations co-evolve with crops. Resistance breakdown is frequent due to:

1. **Monoclonal cultivation** (e.g., single clones in tea and rubber).
2. **Emergence of new pathogen races** (*Hemileia vastatrix* race XXIII in coffee).
3. **Limited resistance genes** in cultivated gene pools.

Continuous introgression from wild relatives and molecular gene pyramiding are needed to restore durability.

4. Technical and Institutional Constraints

4.1 Inadequate Genetic Resources and Characterization

Many gene banks exist but remain under-characterized at molecular and physiological levels.

Crop	Germplasm Collections (approx.)	Gaps
Camellia sinensis	> 3 000 accessions worldwide	Limited trait tagging, redundancy
Coffea arabica & *C. canephora*	> 10 000 accessions	Incomplete phenotypic data
Cocos nucifera	> 2 500	Poor pollen storage, viability
Hevea brasiliensis	> 1 000 clones	Low wild diversity
Cashew	~ 400	Narrow base, limited wild introgression

Without systematic **core and mini-core sets,** breeders lack access to allelic diversity for complex traits.

4.2 Long Breeding Cycles and Evaluation Costs

Each breeding cycle in plantation crops spans **10–15 years**, encompassing seedling testing, clonal propagation, and multi-location trials. High cost and time discourage continuous breeding, leading to gaps in cultivar renewal and vulnerability to pests and market shifts.

4.3 Limited Human Resources and Infrastructure

Breeding perennial crops demands long-term institutional commitment, controlled crossing facilities, molecular labs, and field stations across diverse ecologies. Many developing countries face **budgetary and manpower constraints**, hindering continuity after project phases end.

4.4 Intellectual Property and Exchange Restrictions

The **Nagoya Protocol** and national biodiversity laws, while ensuring fair benefit-sharing, can slow international germplasm exchange. Breeders must navigate complex legal frameworks to access wild relatives, delaying collabourative research.

4.5 Poor Adoption and Diffusion of New Varieties

Farmers are often reluctant to replace established clones because:

- Establishment costs are high (plantations last 20–30 years).
- Uncertainty about long-term performance.
- Lack of supply chains for certified planting material.

This results in **genetic stagnation** at farm level even when research produces new varieties.

5. Socio-Economic and Market Constraints

5.1 Market Volatility and Breeding Priorities

Unstable global commodity prices influence national breeding budgets. When prices fall, investment in long-term breeding slows, reducing innovation. Conversely, short-term demand spikes divert attention to **yield over quality**, compromising sustainability.

5.2 Farmer Fragmentation and Resource Inequality

Smallholders dominate coffee, cocoa, cashew, and coconut sectors. Limited capital and technical knowledge restrict adoption of improved varieties and inputs, leading to **genotype × management interactions** that obscure genetic potential.

5.3 Lack of Participatory Breeding Frameworks

Traditional top-down breeding seldom integrates farmers' preferences (e.g., flavour, harvesting ease). Participatory ideotype definition and decentralized trials can overcome mismatches between research priorities and field realities.

6. Breeding-Specific Technical Challenges

6.1 Low Heritability of Complex Traits

Traits like yield and quality are controlled by many minor genes and heavily influenced by environment (low $h^2 \approx 0.3$–0.5). This necessitates **replicated, multi-season trials**, further prolonging selection.

6.2 Poor Reproductive Synchronization

Flowering asynchrony among clones hinders controlled hybridization, as seen in *Hevea* and *Cashew*. Solutions include **hormonal synchronization** (GA_3, ethrel) and **pollen storage protocols**.

6.3 Somaclonal Variation in Tissue-Culture Propagation

While micropropagation accelerates multiplication, it sometimes induces off-types. SSR/SNP fingerprinting and **flow-cytometric QC** are essential to maintain clonal integrity.

6.4 Environmental Influence on Quality Traits

In crops like tea and coffee, metabolite composition varies with altitude, shade, and season, complicating selection for stable quality. **Metabolomic fingerprinting** and **G×E interaction models** (AMMI, GGE) are required for accurate trait evaluation.

7. Genomic and Molecular Constraints

1. **Incomplete genome assemblies** for several species (*Cocos*, *Cashew*) limit QTL mapping.
2. **Large genome sizes** (1–2 Gb) complicate sequencing and annotation.
3. **Polyploid complexity** introduces homeologous recombination and dosage effects.
4. **Limited functional markers** for non-model traits (flavour, latex quality).
5. **Low funding for bioinformatics capacity-building** in producer countries.

8. Environmental and Regulatory Risks for Transgenics

Public resistance and biosafety regulations slow the adoption of genetically modified plantation crops. For instance, transgenic *Hevea* and *Coffee* lines exist experimentally but remain uncommercialized due to:

- » Unclear GMO policy frameworks.
- » Concerns about gene flow to wild relatives.
- » Lack of consumer acceptance in export markets.

4 – Breeding Achievements in Major Plantation Crops

1. Tea (*Camellia sinensis*)

1.1 Historical Perspective

Tea breeding began with clonal selection from *C. assamica* and *C. sinensis* populations during the colonial era. Organized programs were initiated by the **Tocklai Experimental Station (India), UPASI (South India),** and **TRI (Sri Lanka).** Early efforts (1920–1950) focused on identifying high-yielding seedling progenies; later decades emphasized **clonal selection and hybridization** to combine yield, quality, and stress tolerance.

1.2 Breeding Objectives

1. High yield of pluckable shoots and early flushing.
2. Superior liquor quality (catechin:theaflavin balance, aroma).
3. Resistance to *Exobasidium vexans* (blister blight).
4. Tolerance to drought and frost.
5. Suitability for mechanical harvesting.

1.3 Methods and Strategies

- » **Clonal selection** from heterogeneous seedling populations.
- » **Hybridization** between Assam and China types to combine vigor × quality.
- » **Polyploid and mutation breeding** for drought tolerance.
- » **Molecular marker screening** for catechin and quality genes.

1.4 Achievements

S. No.	Clone / Variety	Origin / Institution	Key Traits
1	**TV 1**	Tocklai (India)	High yield + quality benchmark
2	**TV 23**	Tocklai	Drought-resistant, high biomass
3	**UPASI 3**	UPASI	High quality for South Indian teas
4	**TRI 2025**	Sri Lanka	Blister-blight resistance
5	**TRFK 301/5**	Kenya	Cold-tolerant, high yield

Yield gain over seedling populations: **25–40 %.** Clonal contribution to total tea area > 80 % in India and Sri Lanka.

1.5 Recent Developments

- **SSR/SNP marker panels** for clonal identity.
- **Catechin pathway gene mapping** (*CsANR*, *LAR*).
- **CRISPR pilot edits** for caffeine-synthase gene knockout.

2. Coffee (*Coffea arabica* and *C. canephora*)

2.1 Historical Background

Early introductions of *C. arabica* from Yemen and *C. canephora* (Robusta) from Congo laid the base for Indian and African germplasm. Systematic breeding started at **Central Coffee Research Institute (CCRI)** in India, **CIRAD (France)**, and **IAC (Brazil)**.

2.2 Objectives

1. High bean yield and cup quality.
2. Resistance to coffee leaf rust (*Hemileia vastatrix*).
3. Tolerance to drought and nematodes.
4. Balanced caffeine and chlorogenic-acid content.

2.3 Breeding Approaches

- **Intraspecific selection:** e.g., *Arabica* 'Kent', 'S. 795'.
- **Interspecific hybridization:** *Arabica* × *Robusta* → 'Arabusta'.
- **Backcrossing and gene pyramiding** for SH-gene combinations.
- **Recurrent selection** in *C. canephora.*
- **Marker-assisted selection** for rust resistance.

2.4 Key Achievements

Variety/Hybrid	Type	Resistance / Quality	Origin
Kent	*Arabica*	Early rust resistance	India
S. 795	*Arabica*	High yield + flavour	India
Selection 9	*Arabica*	Superior cup, rust-resistant	India
Catimor	(*C. arabica* × *C. canephora*) × *Caturra*	Strong rust tolerance	Brazil / Central America
Robusta CXR	*Robusta*	Drought-tolerant	India
Hybrid de Timor (HdT)	Natural *Arabica* × *Robusta* hybrid	Source of SH genes	Timor

2.5 Impact

» Rust losses reduced > 60 %.

» Mean yield increase 20–30 %.

» SH-gene pyramiding → durable resistance.

» **Low-caffeine mutants** via CRISPR (XMT/DXMT genes) now under trial.

3. Cocoa (*Theobroma cacao*)

3.1 History

Originating in the Amazon basin, cocoa improvement began with selection of *Criollo* (quality) × *Forastero* (vigour) crosses, producing **Trinitario types**. In India, the Kerala Agricultural University and CPCRI lead breeding.

3.2 Objectives

1. High yield and early bearing.
2. Resistance to *Phytophthora palmivora* (black pod).
3. Superior fat content and flavour precursors.
4. Compact canopy for intercropping.

3.3 Achievements

Hybrid / Clone	Feature	Institution
TSH 1188	High yield + black pod tolerance	Trinidad & Tobago
ICS 95	Excellent flavour	Trinidad
CC 28, CC 37	India: early bearing hybrids	CPCRI
NA 32, Scavina 6	Resistant sources for breeding	Brazil/Peru

Yield gain: 40 % over local seedlings; disease losses cut by half. Molecular QTLs for Phytophthora resistance on LG 1 and 3 now guide MAS.

4. Coconut (*Cocos nucifera*)

4.1 Historical Context

Traditional tall and dwarf ecotypes formed the base germplasm. The **Coconut Research Institute (Sri Lanka)** and **CPCRI (India)** pioneered inter-varietal hybridization in 1940s, creating the revolutionary **Tall × Dwarf (T×D)** hybrids.

4.2 Objectives

1. High nut yield and copra %.
2. Early flowering and compact palm.

3. Resistance to root (wilt, eriophyid mite).
4. High lauric acid and sweet tender-nut water.
5. Adaptation to drought and salinity.

4.3 Achievements

Hybrid / Variety	Parentage	Key Features	Institution
Kalpa Sagara	West Coast Tall × Chowghat Dwarf Orange	High yield, salt tolerant	CPCRI
Kalpa Dhenu	West Coast Tall × Chowghat Dwarf Green	Early bearing	CPCRI
Kalpa Pratibha	Local Tall selection	Root-wilt tolerant	CPCRI
Kera Chandra	Selection	High copra (> 190 g/nut)	Kerala Agric. Univ.
Sri Lanka Tall × Dwarf	Inter-varietal	Commercial hybrid	CRI Sri Lanka

4.4 Impact

- Hybrid palms yield 25–30 % more copra than talls.
- First bearing at 3–4 years (vs 6–8).
- Wider adoption in South India, Sri Lanka, Philippines.
- MAS now targets drought and mite tolerance.

5. Oil Palm (*Elaeis guineensis*)

5.1 Breeding Milestones

Early work in Nigeria and Malaysia produced the landmark **Dura × Pisifera → Tenera** hybrid, with a thin shell and high mesocarp oil yield. Malaysia's MPOB and India's DOPR expanded programs for climate adaptation.

5.2 Objectives

1. High oil yield and harvest index.
2. Slow height increment ($\leq$ 40 cm yr^{-1}).
3. Disease tolerance (Basal stem rot).
4. Improved fatty-acid composition (high oleic, low palmitic).

5.3 Achievements

Hybrid / Variety	Origin / Institution	Yield (t FFB ha^{-1} yr^{-1})	Trait
Tenera (D×P)	Nigeria/Malaysia standard	25–30	High oil extraction (25–28 %)
BRS Tenera	Brazil	25	Adapted to humid Amazon climate
DOPR Hybrid 1	India	20–25	Early bearing, slow height
AVROS DP 1	Indonesia	28	High oleic oil

5.4 Genomic Progress

» Single gene (*SHELL*) controls fruit shell thickness → molecular assay for seed purity.

» Genomic selection accelerates recurrent breeding.

» Genome editing of *FAD2* increases oleic acid to 70 %.

6. Rubber (*Hevea brasiliensis*)

6.1 Historical Outline

After introduction from Brazil (1876), breeding centres in Malaysia (RRIM), India (RRII), and Thailand (TRRC) developed clonal series over decades.

6.2 Objectives

1. High latex yield and early tappability.
2. Wind and disease resistance (SALB, leaf fall).
3. Long economic life and uniform growth.

6.3 Major Achievements

Clone	Institute	Key Trait	Performance
RRIM 600	Malaysia	High yield, wind tolerant	Benchmark clone (6 t ha^{-1} yr^{-1})
PB 86	Malaysia	Thick bark, high latex flow	Widely planted
GT 1	Indonesia	Wind resistance	Stable yield
RRII 105	India	High yield, cold tolerance	> 2 t DRC ha^{-1} yr^{-1}
RRIM 2025	Malaysia	SALB resistant	New generation

6.4 Impact

» Genetic gain in yield ≈ 35 %.

» Latex quality improved (viscosity, molecular weight).

» Triploid and tetraploid lines provide breeding bridges.

» Transcriptomics identified genes for cis-1,4-polyisoprene synthesis.

7. Cashew (*Anacardium occidentale*)

7.1 Breeding Evolution

Indigenous populations of Goa and Kerala served as the base. Systematic selection began at **NRCC (Puttur)** and **BPP (Andhra Pradesh)** in 1960s.

7.2 Objectives

1. High nut yield and kernel weight.
2. Pest (*Helopeltis*) and disease tolerance.
3. Compact tree for high density.
4. Improved apple quality for processing.

7.3 Achievements

Variety	Yield (kg nuts tree^{-1})	Kernel %	Traits	Origin
Vengurla 4	12–15	30	Bold nut, compact	Maharashtra
NRCC Sel 2	10–12	29	Helopeltis tolerant	Karnataka
BPP 8	14	28	Early bearing	Andhra Pradesh
Dhana	11	30	High shelling, wind tolerant	Orissa
Ullal 3	13	29	Apple sweetness improved	Karnataka

7.4 Impact

» Yield gain ≈ 40 %.

» Pest resilience improved via integrated selection.

» Compact varieties enable 800 trees ha^{-1}.

» Processing industry benefits from uniform kernel size.

8. Comparative Summary of Achievements

Crop	Main Breeding Method	Principal Gain	Key Released Varieties / Clones
Tea	Clonal selection, hybridization	+30 % yield + quality	TV series, UPASI, TRI clones
Coffee	Hybridization, gene pyramiding	Rust resistance, cup quality	S.795, Selection 9, Catimor
Cocoa	Hybridization, QTL MAS	Black-pod tolerance + flavour	TSH 1188, ICS 95

Coconut	T×D hybridization	Early bearing + yield	Kalpa Sagara, Kalpa Dhenu
Oil Palm	D×P hybridization, GS	Oil ↑ 25 %	Tenera lines
Rubber	Clonal selection + inter-specific hybrids	Latex ↑ 30 %, wind resistance	RRIM 600, RRII 105
Cashew	Clonal + hybrid selection	Nut ↑ 40 %, pest tolerance	Vengurla 4, NRCC Sel

9. Arecanut (*Areca catechu*)

9.1 Historical Perspective

Arecanut improvement programs began with **systematic germplasm surveys in India and Southeast Asia,** followed by clonal selection and controlled hybridization at institutions such as **CPCRI (India).** Early efforts emphasized **yield, regular bearing, and quality (tannin/alkaloid profile)** alongside **adaptation to different rainfall regimes.**

9.2 Breeding Objectives

1. **Higher bunch number and nut weight** with uniform ripening.
2. **Regular bearing** (reduction of bienniality).
3. **Tolerance to yellow leaf disease and fruit rot.**
4. **Suitability for mixed cropping and different soil–moisture regimes.**
5. **Quality** (tannin profile, chewing quality).

9.3 Methods

» Mass and **clonal selection** of elite palms across traditional belts.

» **Inter-varietal hybridization** among ecotypes (South Kanara, Srilankan, Mangala types).

» Progeny testing under multi-location trials; **vegetative propagation via offshoots** is limited, so **seed gardens** are maintained for true-to-type progenies.

9.4 Achievements

S. No.	Variety/Selection	Key Features	Notes
1	**Mangala**	High bunch number; regular bearing	Widely adapted
2	**Sumangala**	Bold nuts; higher chali recovery	Coastal adaptation
3	**Sreemangala**	Stable under moisture stress	Interior tracts
4	**Swarnamangala**	Early bearing; high recovery %	Farmer-preferred
5	**VTLA selections**	Field tolerance to fruit rot	Pre-breeding parents

Impact: 15–25% yield gain over local seedlings; improved uniformity aids processing.

10. Palmyrah and Date-Type Palms (Selective Notes)

While not universally grouped with the "core" plantation set above, **palmyrah (*Borassus flabellifer*)** and **date palm (*Phoenix dactylifera*)** are often managed as perennial plantation commodities in South Asia and the Gulf/North Africa, respectively. Breeding has been slower due to dioecy, long juvenility, and offshoot-based propagation, but notable **selection and biotechnology** advances merit brief inclusion.

10.1 Objectives

- **Sex identification at seedling stage** (saves 5–7 years).
- **High sugar/fruit yield, abiotic stress tolerance** (salinity/heat).
- **Compact stature** and **offshoot production** for propagation.

10.2 Approaches and Achievements

- **Molecular sexing markers** (seedling sex determination) now routine in *Phoenix dactylifera*, cutting unproductive male:female ratios in orchards.
- **Clonal offshoot selection** of elite female date palms ('Medjool', 'Barhi', 'Khalas') and **tissue culture (somatic embryogenesis)** scaling in several countries.
- **Palmyrah**: elite tree selection for **higher neera (sap) yield** and **drought resilience**; pilot micropropagation protocols are emerging.

Outcome: Better orchard design and earlier economic returns; foundational for future ideotype breeding in arid zones.

11. Cross-Cutting Adoption & Diffusion Outcomes

11.1 Seed and Clonal Systems

- **Clonal crops (tea, rubber, cocoa)**: national foundations now run **registered mother blocks**, budwood gardens, and **identity testing (SSR/SNP)** to avoid off-types.
- **Hybrid-seed crops (coconut, oil palm): isolation seed gardens**, paternity controls, and **molecular assays** (e.g., *SHELL* diagnostic in oil palm; parentage PCR in coconut) safeguard hybridity.

11.2 Area Replacement Trends

- **Tea and Rubber**: >70–80% of area in many regions under improved clones; rolling replacement cycles every 20–25 years improve resilience.
- **Oil Palm**: Tenera is the global standard; second-cycle replanting emphasizes **slow-height and high-oleic ideotypes.**
- **Coconut and Cashew**: hybrid/elite adoption rising but constrained by seedling pipeline and replant costs; **community nurseries** and **subsidized replanting** accelerate change.

12. Economic, Environmental, and Social Impact

12.1 Yield, Quality, and Profitability

S. No.	Crop	Typical Genetic Gain vs. Local Seedlings	Primary Economic Driver
1	Tea	+25–40% made-tea yield; improved quality grades	Higher auction price/ha
2	Coffee	+20–30% beans; rust losses ↓ by >60%	Stable output + quality premiums
3	Cocoa	+30–40% pods; disease losses halved	Reliable supply for processors
4	Coconut	+25–30% copra; earlier bearing (–2–3 yrs)	Oil/tender-nut markets
5	Oil Palm	+25–40% oil/ha; extraction ↑	Estate profitability
6	Rubber	+30–35% latex; earlier tappability	Higher DRC/ha/year
7	Cashew	+30–40% nut yield; kernel grade ↑	Export grade improvements
8	Arecanut	+15–25% chali recovery	Price realization

12.2 Sustainability Gains

- » **Chemical-use reduction** via disease resistance (coffee rust, cocoa Phytophthora, tea blister blight).
- » **Climate resilience**: drought/heat-tolerant clones (tea, cashew, coconut) stabilize incomes during stress years.
- » **Carbon and biodiversity**: perennial canopies act as **long-lived carbon sinks**; **shade-based systems** (coffee/cocoa) enhance on-farm biodiversity and microclimate regulation.

12.3 Social Outcomes

- » **Smallholder inclusivity** through compact ideotypes and participatory clone testing.
- » **Gender and labour benefits**: shorter palms/compact trees ease harvesting and reduce risk.
- » **Value-chain upgrading**: uniform quality supports GI/branding (e.g., specialty coffee/tea; fine-flavour cocoa).

13. Lessons from Successful Programs

1. **Ideotype clarity** up-front (compact, early, stress-ready) speeds convergence.
2. **Multi-location, multi-year testing** prevents G×E surprises at scale.
3. **Diversity stewardship**, release **clone/hybrid packages** (5–10 per region) to avoid genetic bottlenecks.
4. **Molecular QA/QC** in nurseries averts costly admixtures.
5. **Breeder–processor partnerships** (tea tasters, cocoa fermenters, coffee cuppers) keep **quality** central, not an afterthought.

6. **Extension + finance** (replant subsidies, long-tenor credit) drives adoption in long-cycle crops.

5 – Integration of Conventional and Molecular Approaches in Plantation Crop Breeding

1. Introduction

Plantation crops, due to their perennial habit, complex reproductive systems, and long life cycles, require **innovative integration of conventional breeding and modern molecular tools**. While classical breeding provides the foundation, genetic recombination, hybridization, and selection, modern molecular approaches accelerate progress by enabling **early selection, precise gene introgression, and identification of quantitative trait loci (QTLs)**. This integration transforms plantation-crop breeding from a slow, observation-based process into a **predictive and knowledge-driven enterprise**, aligning genetic improvement with the needs of **quality, resilience, and sustainability**.

2. Evolution of Integrated Breeding Paradigms

Era	Key Approach	Defining Traits	Examples
Early 20th Century	Phenotypic selection	Yield, morphology	Tea, Coffee clonal selections
Mid 20th Century	Hybridization & recurrent selection	Interspecific hybrids, heterosis	Coconut, Oil Palm
Late 20th Century	Tissue culture, mutation breeding	Clonal fidelity, somaclonal variants	Tea, Hevea
21st Century	Molecular breeding, genomics, genome editing	Marker-assisted, QTL mapping, CRISPR	Coffee, Cocoa, Oil Palm, Rubber

3. Integration Framework

Integrated breeding in plantation crops follows a **three-tiered structure**:

1. **Conventional base** – population improvement, hybridization, selection of superior progenies.
2. **Molecular overlay** – marker discovery, gene tagging, and genomic prediction to accelerate selection.
3. **Biotechnological augmentation** – tissue culture, transformation, genome editing, and omics integration for trait fine-tuning.

4. Conventional Approaches – Foundation of Genetic Improvement

4.1 Clonal Selection

Clonal selection remains the cornerstone of perennial-crop improvement. Variability within natural populations is exploited to identify elite genotypes.

Crop	Example Clone	Achievement
Camellia sinensis	TV 1, UPASI 3	High yield and flavour
Hevea brasiliensis	RRIM 600, PB 86	High latex yield
Theobroma cacao	ICS 95	Fine flavour
Anacardium occidentale	Vengurla 4	Bold nuts, pest tolerance

Clonal selection fixes **heterozygous superiority**, ensures uniformity, and provides material for vegetative propagation.

4.2 Hybridization and Recurrent Selection

Hybridization creates genetic variability by combining complementary parental traits. In plantation crops, this often includes **interspecific** and **inter-varietal** hybridization.

Crop	Type of Hybridization	Purpose
Cocos nucifera	Tall × Dwarf (T×D)	Early bearing + high yield
Coffea arabica × *C. canephora*	Inter-specific	Rust resistance + flavour
Hevea brasiliensis × *H. benthamiana*	Interspecific	Disease tolerance
Elaeis guineensis × *E. oleifera*	Interspecific	High unsaturation, reduced height

Recurrent selection cycles enable **accumulation of additive genes**, gradually enhancing complex traits like yield and stress tolerance.

4.3 Mutation and Polyploidy Breeding

- **Mutation breeding** introduces novel alleles for resistance or quality.

  Example: *Camellia sinensis* mutants with high catechins; *Coffea arabica* low-caffeine mutants.
- **Polyploidy breeding** modifies plant vigor, fertility, and product quality.
  Triploid rubber clones show improved latex yield and tree stability.

4.4 Resistance Breeding through Conventional Methods

Conventional resistance breeding involves:

1. **Screening of germplasm** under natural epiphytotic conditions.
2. **Identification of resistance donors** (e.g., *Hybrid de Timor* in coffee, *Scavina 6* in cocoa).
3. **Backcross introgression** of resistance genes into elite backgrounds.

5. Molecular Breeding Approaches – Precision Enhancement

5.1 Marker-Assisted Selection (MAS)

MAS allows indirect selection of desired traits using DNA markers tightly linked to QTLs or genes of interest.

Crop	Target Trait	Marker / Gene Identified
Coffea arabica	Rust resistance	SH1–SH9 markers
Camellia sinensis	Catechin biosynthesis	*CsANR*, *LAR*, *DFR* SSRs
Hevea brasiliensis	Latex biosynthesis genes	SSRs for *HbREF*, *HbSRPP*
Cocoa	Black pod resistance	QTLs on LG 1, 3, 9
Oil palm	Shell thickness	*SHELL* gene marker
Coconut	Drought tolerance	SNPs linked to *DREB* loci

Advantages of MAS

- Early selection before field evaluation.
- Non-destructive and high throughput.
- Facilitates gene pyramiding for durable resistance.

5.2 Quantitative Trait Loci (QTL) Mapping

QTL mapping identifies genomic regions associated with quantitative traits such as yield, quality, and stress tolerance.

Examples

- *Tea:* QTLs for catechin and caffeine content on LG 2, 4, and 9.
- *Coffee:* QTLs for bean size and cup quality.
- *Cocoa:* QTLs for flavour and fat content.
- *Hevea:* QTLs for latex yield and girth increment.

QTLs act as entry points for **genomic-assisted selection and marker validation.**

5.3 Genomic Selection (GS)

GS predicts the genetic merit of individuals using genome-wide markers and statistical models, allowing **early selection** without phenotyping every generation.

Implementation Steps

1. Develop **training population** with phenotypes and genotypes.
2. Estimate marker effects using models (GBLUP, Bayesian).
3. Predict genomic estimated breeding values (GEBVs) for new seedlings.
4. Select superior candidates based on GEBVs.

Impact

- Reduces selection cycle by 30–50%.
- Enhances gain per unit time, especially in oil palm, tea, and rubber.

5.4 Genome-Wide Association Studies (GWAS)

GWAS links phenotypic variation with genome-wide SNP data in natural populations, revealing **new candidate genes.**

Recent Examples

- *Coffea arabica:* SNP loci associated with caffeine and chlorogenic acid levels.
- *Camellia sinensis:* Genes governing aroma volatiles and drought response.
- *Hevea brasiliensis:* Genes related to rubber biosynthesis pathway.

5.5 Transcriptomics and Functional Genomics

RNA sequencing identifies gene expression profiles under different conditions.

Crop	Transcriptomic Insight
Tea	Drought-induced upregulation of *DREB, LEA, HSP* genes.
Coffee	Caffeine biosynthetic pathway (*XMT, DXMT*) regulation.
Rubber	Latex-regeneration and stress genes (*HbSRPP, HbREF*).
Coconut	Expression of fatty-acid metabolism genes under salinity.

Functional genomics validates these genes for use in **gene-based selection and transgenics.**

6. Biotechnological Tools Supporting Integrated Breeding

6.1 Tissue Culture and Somaclonal Variation

Tissue culture provides clonal propagation, embryo rescue, and pathogen-free planting materials. Somaclonal variants occasionally offer novel traits (e.g., early flushing tea, high latex Hevea).

Applications

- *Tea, Coffee:* Micropropagation of elite clones.
- *Cocoa:* Somatic embryogenesis for large-scale propagation.
- *Hevea:* Regeneration from immature anther culture.
- *Coconut:* Embryo culture for hybrid recovery.

6.2 Genetic Transformation

Genetic transformation introduces specific genes for resistance, stress tolerance, or quality.

Trait	Gene / Mechanism	Crop
Fungal resistance	Chitinase / β-1,3-glucanase	*Cocoa, Hevea*
Low caffeine	Antisense *caffeine synthase*	*Coffee*
Drought tolerance	*DREB, LEA* genes	*Tea, Coconut*
Improved oil quality	*FAD2/FAD3* overexpression	*Oil palm*

Transgenic lines remain largely confined to field trials due to biosafety regulations but offer proof of concept for **trait engineering**.

6.3 Genome Editing (CRISPR/Cas Systems)

Genome editing offers **precise, marker-free modification** of target genes.

Examples

- *Coffee:* Editing *DXMT* gene → low-caffeine lines.
- *Oil Palm:* Knockout of *FAD2* → high oleic acid.
- *Tea: CsANR* gene editing → improved catechin ratio.
- *Hevea:* Targeted modifications in *HbREF* gene to improve latex yield.

6.4 Omics Integration

The integration of **genomics, transcriptomics, proteomics, metabolomics**, and **phenomics** provides a holistic understanding of trait architecture.

- **Metabolomics** in tea and coffee identifies biochemical markers linked to quality.
- **Proteomics** in rubber clarifies latex biosynthesis enzymes.
- **Phenomics** uses drone imagery and NIR spectroscopy for non-destructive screening of oil yield and canopy vigor.

7. Digital and Data-Driven Breeding

Digital breeding platforms combine **bioinformatics, AI, and remote sensing** to manage breeding pipelines.

Tool / Platform	Function	Application
Genomic Information System	Data storage and QTL mining	Coffee, Cocoa
Breeding Management System (BMS)	Cross management and trait tracking	Coconut, Rubber
Remote Sensing & Drones	Canopy health monitoring	Tea, Oil Palm
Machine Learning Models	Predict yield and G×E interaction	Rubber, Coffee

These systems ensure traceability and enhance selection efficiency.

6 – Economic, Environmental, and Social Impact of Breeding Achievements in Plantation Crops

1. Introduction

Plantation crops such as *Camellia sinensis* (tea), *Coffea arabica* and *C. canephora* (coffee), *Theobroma cacao* (cocoa), *Cocos nucifera* (coconut), *Elaeis guineensis* (oil palm), *Hevea brasiliensis* (rubber), and *Anacardium occidentale* (cashew) form the economic backbone of tropical and subtropical agriculture. These crops not only generate export revenues and rural employment but also provide critical ecosystem and livelihood services, from carbon sequestration and soil conservation to gender-inclusive value chains.

Breeding achievements over the last century have dramatically reshaped their socio-economic and environmental roles, driving **yield increases, quality enhancement, disease resilience, and ecological sustainability**.

2. Economic Impact

2.1 Contribution to National and Global Economies

Crop	Approx. Global Area (Mha)	Annual Value (USD Billion)	Leading Producers
Tea	5.1	50	China, India, Kenya, Sri Lanka
Coffee	12.0	120	Brazil, Vietnam, Colombia, India
Cocoa	11.2	14	Côte d'Ivoire, Ghana, Indonesia
Coconut	12.0	35	Indonesia, Philippines, India
Oil Palm	22.0	65	Indonesia, Malaysia, Nigeria
Rubber	12.5	40	Thailand, Indonesia, India
Cashew	6.0	7	India, Vietnam, Benin, Nigeria

Together, these crops exceed **USD 300 billion annually** in value, underpinning millions of smallholder livelihoods and influencing foreign-exchange earnings in over 70 countries.

2.2 Yield and Productivity Gains from Breeding

Breeding programs have been directly responsible for yield improvements of **25–60 %** in major plantation crops.

Crop	Baseline Yield (traditional)	Improved Variety Yield	% Increase	Key Breeding Intervention
Tea	2.0 t made tea ha^{-1}	2.8–3.2 t	+40 %	Clonal selection (TV/UPASI)
Coffee	0.8 t clean beans ha^{-1}	1.1–1.4 t	+35 %	Rust-resistant hybrids (S. 795, Catimor)
Cocoa	0.7 t dry beans ha^{-1}	1.2–1.5 t	+50 %	Hybridization + QTL-aided selection
Coconut	6,000 nuts ha^{-1}	10,000–12,000	+60 %	T×D hybrids
Oil Palm	18 t FFB ha^{-1}	25–30 t	+40 %	Dura×Pisifera → Tenera revolution
Rubber	1.6 t DRC ha^{-1}	2.4–2.8 t	+35 %	High-latex clones (RRIM/RRII)
Cashew	800 kg nuts ha^{-1}	1,200–1,400	+40 %	Compact, high-kernel selections

2.3 Value Addition through Quality Breeding

Breeding for **biochemical quality**, polyphenols in tea, flavour volatiles in coffee/cocoa, fatty-acid profiles in oil crops, has diversified markets:

» **Tea:** High-theaflavin and catechin clones fetch 15–25 % higher auction prices.

» **Coffee:** Arabica cultivars like *Selection 9* and *Catimor* command export premiums for flavour.

» **Cocoa:** Fine-flavour *Trinitario* and *ICS 95* types enable GI-tagged chocolates.

» **Coconut & Oil Palm:** Improved lauric/oleic profiles cater to health and cosmetic industries.

Thus, genetic improvement translates directly into **price incentives, market differentiation, and foreign-exchange stability**.

2.4 Cost–Benefit and Input-Use Efficiency

Enhanced varieties reduce cost per unit output:

Parameter	Conventional Cultivar	Improved Cultivar
Fertilizer requirement	100 % baseline	80–85 % (higher nutrient-use efficiency)
Pesticide expenditure	1.0 ×	0.6 × (disease-resistant clones)
Labour productivity	1.0	1.3–1.5 (compact ideotypes, mechanizable canopies)
Net return ha^{-1}	100 % baseline	140–160 %

Economic analyses from India, Sri Lanka, and Malaysia consistently show **internal rates of return > 25 %** for breeding investments in these crops.

3. Environmental Impact

3.1 Carbon Sequestration and Climate Mitigation

Perennial plantation systems function as long-term **carbon sinks**.

Crop	Average Carbon Stock (t C ha^{-1})	Equivalent CO_2 sequestration (t ha^{-1})	Remarks
Hevea brasiliensis	140–180	500–660	Latex extraction non-destructive; deep roots
Coffea arabica (shade)	80–100	290–360	Agroforestry canopy
Cocos nucifera	100–130	360–470	Long-lived palms (60 + yrs)
Elaeis guineensis	90–110	330–400	High biomass turnover
Theobroma cacao	70–90	260–320	Shade trees intensify carbon stock

Improved ideotypes, compact yet dense-foliage clones, enhance photosynthetic efficiency and carbon sequestration per unit land area, making breeding a **climate-mitigation tool**.

3.2 Soil Conservation and Hydrological Benefits

Root-architecture improvements and canopy design in new varieties mitigate erosion:

» *Coconut* hybrids with fibrous root systems stabilize coastal soils.

» *Rubber* plantations reduce sediment yield by > 80 % compared to shifting cultivation.

» *Tea* clones with high litterfall improve soil organic matter by ~0.5 % annually.

3.3 Biodiversity and Agroforestry Synergies

Breeding for **shade tolerance and intercropping adaptability** sustains biodiversity:

» *Coffee–cocoa–banana* systems harbor > 100 bird species ha^{-1}.

» *Arecanut–pepper–coconut* multistrata models sustain pollinators and soil fauna.

» Genotypes tolerant to partial shade allow inclusion of trees, enhancing ecosystem services.

3.4 Climate-Resilient Ideotypes

Modern breeding emphasizes **abiotic stress tolerance**:

Stress Type	Crop	Resistant / Tolerant Varieties	Mechanism
Drought	Tea, Cashew, Coconut	TV 29, NRCC Sel-2, Kalpa Dhenu	Deep roots, osmolyte buildup
Salinity	Coconut, Oil Palm	Kalpa Sagara, AVROS DP 1	Ion exclusion, Na^+ sequestration
Heat	Coffee, Rubber	Catimor lines, RRII 105	HSP expression, canopy adaptation
Wind	Coconut, Rubber	Kera Chandra, GT 1	Flexible trunks, low canopy drag

Breeding therefore, underpins **climate adaptation** strategies in vulnerable tropical regions.

4. Social and Livelihood Impacts

4.1 Employment and Rural Development

Plantation crops are **labour-intensive**, generating year-round employment.

Crop	Employment (person-days ha^{-1} yr^{-1})	Role of Improved Varieties
Tea	400–500	Continuous plucking ensured by high-flush clones
Coffee	250–300	Extended harvesting due to uniform ripening
Rubber	200–250	Early tappability reduces idle years
Coconut / Cashew	150–180	Early bearing, diversified by-products

With improved clones, **job stability and labour income** increase by 15–25 %, reducing seasonal migration.

4.2 Smallholder Empowerment

Over 70 % of global tea, coffee, cocoa, and cashew output now comes from smallholders.

Breeding achievements, compact ideotypes, early maturity, manageable plant height, directly empower small farmers by:

» Lowering maintenance cost.

» Enabling women participation (e.g., shorter coconut palms).

» Facilitating cooperative processing and branding of uniform produce.

Community nurseries and participatory varietal trials have strengthened **social ownership** of innovations.

4.3 Gender and Social Inclusion

Women constitute:

» ~60 % of tea pluckers,

» ~40 % of coffee and cocoa processors, and

» 50 % of cashew shelling workers.

Varieties bred for **ease of harvest, lighter pods, and synchronized flowering** directly improve women's labour efficiency and safety. Income gains enhance household nutrition, education, and healthcare access.

4.4 Nutritional and Health Benefits

Breeding has enriched nutritional profiles:

» **Tea and cocoa:** Elevated polyphenols and antioxidants support functional-food industries.

» **Cashew and coconut:** High unsaturated fatty acids and minerals improve dietary balance.

» **Coffee:** Low-caffeine, high-chlorogenic-acid lines cater to wellness markets.

This linkage between genetics and nutrition is increasingly valued in **biofortified plantation ideotypes**.

5. Sustainability and Policy Dimensions

5.1 Reduced Chemical Footprint

Resistant clones cut pesticide use drastically:

» *Coffee rust* control costs dropped from USD 350 → 120 ha^{-1} yr^{-1}.

» *Cocoa black-pod* fungicide frequency reduced by half.

» *Tea blister blight* control improved with resistant TV-series clones.

Thus, genetic resistance provides an **eco-friendly alternative** to chemical dependence.

5.2 Contribution to SDGs and Circular Economy

UN SDG	Breeding Contribution in Plantation Crops
SDG 1: No Poverty	Yield & price gains raise farm incomes.
SDG 2: Zero Hunger	Nutrient-rich products, food security.
SDG 5: Gender Equality	Women-friendly ideotypes.
SDG 8: Decent Work & Growth	Year-round plantation employment.
SDG 12: Responsible Consumption	Sustainable raw materials for industries.
SDG 13: Climate Action	Carbon-sequestering perennial systems.

Breeding hence, functions as a **policy lever** linking productivity, equity, and sustainability.

5.3 Certification and Market Access

Quality and sustainability breeding underpin entry into:

» **Fair-Trade, Rainforest Alliance, UTZ, Organic** certification schemes.

» Traceability via **molecular fingerprinting** guarantees cultivar integrity.

» GI-tagging (e.g., Darjeeling Tea, Kodagu Coffee, Ivorian Cocoa) adds geographic value.

References

Afoakwa EO. 2016. *Cocoa Production and Processing Technology*. CRC Press.

Agarwal S, Divkarasastry EV and Sharma RK. 2001. *Seed Spices, Production, Quality and Export*. Pointer Publ. Anonymous. 1985. *Rubber and its Cultivation*. The Rubber Board of India.

Barche S. 2016. *Production technology of spices, aromatic, medicinal and plantation crops*.

New India Publishing Agency, New Delhi.

Chadha KL. 2001. *Hand Book of Horticulture*. ICAR.

Chadha KL and Gupta R. 1995. *Advances in Horticulture*. Vol.XI. Malhotra Publ. House. Chopra VL and Peter KV. 2005. *Handbook of Industrial Crops*. Panima.

Choudappa P, Anitha K, Rajesh MK and Ramesh SV. 2017. *Biotechnology of Plantation Crops*. Daya Publishing House, New Delhi.

Choudappa P, NiralV, Jerard BA and Samsudeen K. 2017. *Coconut*. Daya Publishing House, New Delhi.

E-manual on *Advances in Cashew Production Technology*. ICAR -Directorate of Cashew Research, Puttur –574 202, D.K., Karnataka.

Farooqi AA, Khan MM and Vasundhara M. 2001. *Production Technology of Medicinal and Aromatic Crops*. Natural Remedies Pvt. Ltd.

Handa SS and Kaul MK. 1982. *Cultivation and Utilization of Medicinal Plants*. NISC, CSIR. Harler CR. 1963. *The Culture and Marketing of Tea*. Oxford Univ. Press.

Jain SK. 2000. *Medicinal Plants*. National Book Trust.

Joshi P. 2018. *Text Book on Fruit and Plantation Crops*. Narendra Publishing House, New Delhi.

Julia F and Charters MC. 1997. *Major Medicinal Plants–Botany, Cultures and Uses*. Thomas Publ.

Kurian A and Peter KV. 2007. *Commercial Crops Technology*. New India Publ. Agency. Marsh AC, Moss MK and Murphy EW. 1977. *Composition of Food Spices and Herbs, Raw, Processed and Prepared*. Agric. Res. Serv. Hand Book 8-2. Washinton DC.

Nair MK, Bhaskararao EVV, Nambiar KKN and Nambiar MC. 1979. *Cashew*. CPCRI, Kasaragod.

Nybe EV, Mini Raj N and Peter KV. 2007. *Spices*. New India Publ. Agency. Panda H. 2013. *The Complete Book on Cashew*. Asia Pacific Business Press Inc.

Panda H. 2016. *The Complete Book on Cultivation and Manufacture of Tea* (2nd Revised Edition). Asia Pacific Business Press Inc.

Peter KV. 2001. *Hand Book of Herbs and Spices*. Vols. I-III. Woodhead Publ. Co., UK and CRC, USA.

Peter KV. 2002. *Plantation Crops*. National Book Trust.

Pillay PNR. 1980. *Handbook of natural rubber production in India*. Rubber Research Institute, Kottayam, 668pp.

Ponnuswami *et al.* 2018. *Botany of Horticultural crops*. Narendra Publishing House, New Delhi. Ponnuswami *et al.* 2018. *Medicinal Herbs and Herbal Cure*. Narendra Publishing House, New Delhi.

Ponnuswami V *et al.* 2018. *Blossom biology of Horticultural crops*. Narendra Publishing House, New Delhi.

Ponnuswami V *et al.* 2018. *Spices*. Narendra Publishing House, New Delhi

Pradeepkumar T, Suma B, Jyothibhaskar and Satheesan KN. 2007. *Management of Horticultural Crops*. Parts I, II. New India Publ. Agency.

Prajapati ND, Purohit SS, Sharma AK and Kumar T. 2006. *A Hand book of Medicinal Plants*. Agro Bios.

Purseglove JW. 1968. *Tropical Crops–Dicotyledons*. Longman.

Purseglove JW, Brown EG, Green CL and Robbins SRJ. 1984. *Spices*. Vols. I, II. Longman. Ramachandra *et al.* 2018. *Breeding of Spices and Plantation crops*. Narendra Publishing House, New Delhi.

Ranganathan V. 1979. *Hand Book of Tea Cultivation*. UPASI, Tea Res. Stn. Cinchona. Ravindran PN. 2003. *Cinnamon and cassia*. CRC press.

Ravindran PN. 2004. *Ginger, the genus Zingiber*. CRC press.

Ravindran PN. 2007. *Turmeric, the genus curcuma*. CRC press. Medicinal and Aromatic Plants – Industrial Profiles. Routledge, UK.

For Product Safety Concerns and Information please contact our EU representative GPSR@taylorandfrancis.com
Taylor & Francis Verlag GmbH, Kaufingerstraße 24, 80331 München, Germany

.ingramcontent.com/pod-product-compliance
Content Group UK Ltd.
ilton Keynes, MK11 3LW, UK
24190926
00007B/1182
7 8 1 0 4 1 3 7 1 4 2 7 *